Praise for ARMY OF NONE

A BILL GATES TOP FIVE BOOK OF 2018

"In this comprehensive analysis, Scharre moves beyond the clichés of 'killer robots.'" —Lawrence D. Freeman, *Foreign Affairs*

"The era of autonomous weapons is upon us. In *Army of None*, Scharre combines his experience as a warrior and his insight as a policy researcher to paint a comprehensive picture of exactly what such an era will look like. He masterfully weaves together threads tying future weapon systems, artificial intelligence, and policy imperatives to deliver a book that is simply a must-read for anyone interested in military technology and its broader implications."

—Amir Husain, founder and CEO of SparkCognition and author of *The Sentient Machine*

"Technology democratizes. What was once the exclusive purview of nation-states quickly becomes accessible by everyone. *Army of None* serves as a smart primer to what's to come in warfare, but also what we're going to have to contend with in our daily lives soon after."

—Bruce Schneier, author of *Data and Goliath: The Hidden Battles to Collect Your Data and Control Your World*

"*Army of None* delivers what will likely be the most important general-audience book on this topic for at least the next decade." —*Lawfare*

"A one-stop guide book to the debates, the challenges, and yes, the opportunities that can come from autonomous warfare." —*Techcrunch*

"A sober, policy-focused primer on what is coming . . . written in an interesting and accessible manner." —*War on the Rocks*

"A clear, well-written, and richly documented discussion of an issue that deserves deep and careful study." —*Kirkus Reviews*, starred review

"This important book illuminates what may be a fundamental change in the nature of war: the possibility of a future in which the majority of the fighting is done by autonomous weapons powered by artificial intelligence. . . . This excellent primer is of interest to the general reader and a must-read for those who have a professional interest in military topics but are not up to speed on AI and robotics."   —*Publishers Weekly*

"[A] detailed, nuanced, open-minded look at an incredibly complex and technical subject. . . . *Army of None* will spark important discussions as it empowers readers with knowledge about a subject with enormous implications."   —*Booklist*

ARMY OF NONE

Autonomous Weapons and the Future of War

PAUL SCHARRE

W. W. NORTON & COMPANY

Independent Publishers Since 1923

New York | London

For information about permission to reproduce selections from this book, write to
Permissions, W. W. Norton & Company, Inc., 500 Fifth Avenue, New York, NY 10110

For information about special discounts for bulk purchases, please contact
W. W. Norton Special Sales at specialsales@wwnorton.com or 800-233-4830

Manufacturing by LSC Communications, Harrisonburg
Book design by Chris Welch
Production manager: Beth Steidle

Library of Congress Cataloging-in-Publication Data

Names: Scharre, Paul, author.
Title: Army of none: autonomous weapons and the future of war / Paul Scharre.
Other titles: Autonomous weapons and the future of war
Description: First edition. | New York: W. W. Norton & Company, [2018] |
Includes bibliographical references and index.
Identifiers: LCCN 2017053908 | ISBN 9780393608984 (hardcover)
Subjects: LCSH: Military robots—Moral and ethical aspects. | Military weapons—Technological innovations.
| Weapons systems—Technological innovations. | Robotics—Military applications. | War—Forecasting.
Classification: LCC UG479 .S37 2018 | DDC 623.4—dc23
LC record available at https://lccn.loc.gov/2017053908

ISBN 978-0-393-35658-8 pbk.

W. W. Norton & Company, Inc., 500 Fifth Avenue, New York, N. Y. 10110
www.wwnorton.com

W. W. Norton & Company Ltd., 15 Carlisle Street, London W1D 3BS

4 5 6 7 8 9 0

For Davey, William, and Ella,
that the world might be a better place.

And for Heather.
Thanks for everything.

Contents

PART V / THE FIGHT TO BAN AUTONOMOUS WEAPONS

PART VI / AVERTING ARMAGEDDON: THE WEAPON OF POLICY

Introduction

THE POWER OVER LIFE AND DEATH

THE MAN WHO SAVED THE WORLD

On the night of September 26, 1983, the world almost ended.

It was the height of the Cold War, and each side bristled with nuclear weapons. Earlier that spring, President Reagan had announced the Strategic Defense Initiative, nicknamed "Star Wars," a planned missile defense shield that threatened to upend the Cold War's delicate balance. Just three weeks earlier on September 1, the Soviet military had shot down a commercial airliner flying from Alaska to Seoul that had strayed into Soviet air space. Two hundred and sixty-nine people had been killed, including an American congressman. Fearing retaliation, the Soviet Union was on alert.

The Soviet Union deployed a satellite early warning system called Oko to watch for U.S. missile launches. Just after midnight on September 26, the system issued a grave report: the United States had launched a nuclear missile at the Soviet Union.

Lieutenant Colonel Stanislav Petrov was on duty that night in bunker Serpukhov-15 outside Moscow, and it was his responsibility to report the missile launch up the chain of command to his superiors. In the bunker, sirens blared and a giant red backlit screen flashed "launch," warning him of the detected missile, but still Petrov was uncertain. Oko was new, and he worried that the launch might be an error, a bug in the system. He waited.

Another launch. Two missiles were inbound. Then another. And another. And another—five altogether. The screen flashing "launch" switched to "missile strike." The system reported the highest confidence level. There was no ambiguity: a nuclear strike was on its way. Soviet military command would have only minutes to decide what to do before the missiles would explode over Moscow.

Petrov had a funny feeling. Why would the United States launch only five missiles? It didn't make sense. A real surprise attack would be massive, an overwhelming strike to wipe out Soviet missiles on the ground. Petrov wasn't convinced the attack was real. But he wasn't certain it was a false alarm, either.

With one eye on the computer readouts, Petrov called the ground-based radar operators for confirmation. If the missiles were real, they would show up on Soviet ground-based radars as they arced over the horizon. Puzzlingly, the ground radars detected nothing.

Petrov put the odds of the strike being real at 50/50, no easier to predict than a coin flip. He needed more information. He needed more time. All he had to do was pick up the phone, but the possible consequences were enormous. If he told Soviet command to fire nuclear missiles, millions would die. It could be the start of World War III.

Petrov went with his gut and called his superiors to inform them the system was malfunctioning. He was right: there was no attack. Sunlight reflecting off cloud tops had triggered a false alarm in Soviet satellites. The system was wrong. Humanity was saved from potential Armageddon by a human "in the loop."

What would a machine have done in Petrov's place? The answer is clear: the machine would have done whatever it was programmed to do, without ever understanding the consequences of its actions.

THE SNIPER'S CHOICE

In the spring of 2004—two decades later, in a different country, in a different war—I stared down the scope of my sniper rifle atop a mountain in Afghanistan. My sniper team had been sent to the Afghanistan-Pakistan border to scout infiltration routes where Taliban fighters were suspected of crossing back into Afghanistan. We hiked up the mountain all night, our 120-pound packs weighing heavily on the jagged and broken terrain. As the sky in the east began to lighten, we tucked ourselves in behind a rock outcropping—the best cover we could find. We hoped our position would conceal us at daybreak.

It didn't. A farmer spied our heads bobbing above the shallow rock outcropping as the village beneath us woke to start their day. We'd been spotted.

Of course, that didn't change the mission. We kept watch, tallying the movement we could see up and down the road in the valley below. And we waited.

It wasn't long before we had company.

A young girl of maybe five or six headed out of the village and up our way, two goats in trail. Ostensibly she was just herding goats, but she walked a long slow loop around us, frequently glancing in our direction. It wasn't a very convincing ruse. She was spotting for Taliban fighters. We later realized that the chirping sound we'd heard as she circled us, which we took to be her whistling to her goats, was the chirp of a radio she was carrying. She slowly circled us, all the while reporting on our position. We watched her. She watched us.

She left, and the Taliban fighters came soon after.

We got the drop on them—we spotted them moving up a draw in the mountainside that they thought hid them from our position. The crackle of gunfire from the ensuing firefight brought the entire village out of their homes. It echoed across the valley floor and back, alerting everyone within a dozen miles to our presence. The Taliban who'd tried to sneak up on us had either run or were dead, but they would return in larger numbers. The crowd of villagers swelled below our position, and they didn't look friendly. If they decided to mob us, we wouldn't have been able to hold them all off.

"Scharre," my squad leader said. "Call for exfil."

I hopped on the radio. "This is Mike-One-Two-Romeo," I alerted our quick reaction force, "the village is massing on our position. We're going to need an exfil." Today's mission was over. We would regroup and move to a new, better position under cover of darkness that night.

Back in the shelter of the safe house, we discussed what we would do differently if faced with that situation again. Here's the thing: the laws of war don't set an age for combatants. Behavior determines whether or not a person is a combatant. If a person is participating in hostilities, as the young girl was doing by spotting for the enemy, then they are a lawful target for engagement. Killing a civilian who had stumbled across our position would have been a war crime, but it would have been legal to kill the girl.

Of course, it would have been wrong. Morally, if not legally.

In our discussion, no one needed to recite the laws of war or refer to abstract ethical principles. No one needed to appeal to empathy. The horrifying notion of shooting a child in that situation didn't even come up. We all knew it would have been wrong without needing to say it. War does force awful and difficult choices on soldiers, but this wasn't one of them.

Context is everything. What would a machine have done in our place? If it had been programmed to kill lawful enemy combatants, it would have attacked the little girl. Would a robot know when it is lawful to kill, but wrong?

THE DECISION

Life-and-death choices in war are not to be taken lightly, whether the stakes are millions of lives or the fate of a single child. Laws of war and rules of engagement frame the decisions soldiers face amid the confusion of combat, but sound judgment is often required to discern the right choice in any given situation.

Technology has brought us to a crucial threshold in humanity's relationship with war. In future wars, machines may make life-and-death engagement decisions all on their own. Militaries around the globe are racing to deploy robots at sea, on the ground, and in the air—more than ninety countries have drones patrolling the skies. These robots are increasingly autonomous and many are armed. They operate under human control for now, but what happens when a Predator drone has as much autonomy as a Google car? What authority should we give machines over the ultimate decision—life or death?

This is not science fiction. More than thirty nations already have defensive supervised autonomous weapons for situations in which the speed of engagements is too fast for humans to respond. These systems, used to defend ships and bases against saturation attacks from rockets and mis- siles, are supervised by humans who can intervene if necessary—but other weapons, like the Israeli Harpy drone, have already crossed the line to full autonomy. Unlike the Predator drone, which is controlled by a human, the Harpy can search a wide area for enemy radars and, once it finds one, destroy it without asking permission. It's been sold to a handful of coun- tries and China has reverse engineered its own variant. Wider prolifera- tion is a definite possibility, and the Harpy may only be the beginning. South Korea has deployed a robotic sentry gun to the demilitarized zone bordering North Korea. Israel has used armed ground robots to patrol its Gaza border. Russia is building a suite of armed ground robots for war on the plains of Europe. Sixteen nations already have armed drones, and another dozen or more are openly pursuing development.

These developments are part of a deeper technology trend: the rise of artificial intelligence (AI), which some have called the "next industrial revolution." Technology guru Kevin Kelly has compared AI to electricity: just as electricity brings objects all around us to life with power, so too will AI bring them to life with intelligence. AI enables more sophisticated and autonomous robots, from warehouse robots to next-generation drones, and can help process large amounts of data and make decisions to power Twitter bots, program subway repair schedules, and even make medical diagnoses. In war, AI systems can help humans make decisions—or they can be delegated authority to make decisions on their own.

The rise of artificial intelligence will transform warfare. In the early twentieth century, militaries harnessed the industrial revolution to bring tanks, aircraft, and machine guns to war, unleashing destruction on an unprecedented scale. Mechanization enabled the creation of machines that were physically stronger and faster than humans, at least for cer- tain tasks. Similarly, the AI revolution is enabling the *cognitization* of machines, creating machines that are smarter and faster than humans for narrow tasks. Many military applications of AI are uncontroversial— improved logistics, cyberdefenses, and robots for medical evacuation, resupply, or surveillance—however, the introduction of AI into weapons raises challenging questions. Automation is already used for a variety of functions in weapons today, but in most cases it is still humans choosing the targets and pulling the trigger. Whether that will continue is unclear.

Most countries have kept silent on their plans, but a few have signaled their intention to move full speed ahead on autonomy. Senior Russian military commanders envision that in the near future a "fully robotized unit will be created, capable of independently conducting military operations," while the U.S. Department of Defense officials state that the option of deploying fully autonomous weapons should be "on the table."

BETTER THAN HUMAN?

Armed robots deciding who to kill might sound like a dystopian nightmare, but some argue autonomous weapons could make war more humane. The same kind of automation that allows self-driving cars to avoid pedestrians could also be used to avoid civilian casualties in war, and unlike human soldiers, machines never get angry or seek revenge. They never fatigue or tire. Airplane autopilots have dramatically improved safety for commercial airliners, saving countless lives. Could autonomy do the same for war?

New types of AI like deep learning neural networks have shown startling advances in visual object recognition, facial recognition, and sensing human emotions. It isn't hard to imagine future weapons that could outperform humans in discriminating between a person holding a rifle and one holding a rake. Yet computers still fall far short of humans in understanding context and interpreting meaning. AI programs today can identify objects in images, but can't draw these individual threads together to understand the big picture.

Some decisions in war are straightforward. Sometimes the enemy is easily identified and the shot is clear. Some decisions, however, like the one Stanislav Petrov faced, require understanding the broader context. Some situations, like the one my sniper team encountered, require moral judgment. Sometimes doing the right thing entails breaking the rules—what's legal and what's right aren't always the same.

THE DEBATE

Humanity faces a fundamental question: should machines be allowed to make life-and-death decisions in war? Should it be legal? Is it right?

I've been inside the debate on lethal autonomy since 2008. As a civilian policy analyst in the Pentagon's Office of the Secretary of Defense, I led the group that drafted the official U.S. policy on autonomy in weapons. (Spoiler alert: it doesn't ban them.) Since 2014, I've ran the Ethical Autonomy Project at the Center for a New American Security, an independent bipartisan think tank in Washington, DC, during which I've met experts from a wide range of disciplines grappling with these questions: academics, lawyers, ethicists, psychologists, arms control activists, military professionals, and pacifists. I've peered behind the curtain of government projects and met with the engineers building the next generation of military robots.

This book will guide you on a journey through the rapidly evolving world of next-generation robotic weapons. I'll take you inside defense companies building intelligent missiles and research labs doing cutting-edge work on swarming. I'll introduce the government officials setting policy and the activists striving for a ban. This book will examine the past—including things that went wrong—and look to the future, as I meet with the researchers pushing the boundaries of artificial intelligence.

This book will explore what a future populated by autonomous weapons might look like. Automated stock trading has led to "flash crashes" on Wall Street. Could autonomous weapons lead to a "flash war"? New AI methods such as deep learning are powerful, but often lead to systems that are effectively a "black box"—even to their designers. What new challenges will advanced AI systems bring?

Over 3,000 robotics and artificial intelligence experts have called for a ban on offensive autonomous weapons, and are joined by over sixty nongovernmental organizations (NGOs) in the Campaign to Stop Killer Robots. Science and technology luminaries such as Stephen Hawking, Elon Musk, and Apple cofounder Steve Wozniak have spoken out against autonomous weapons, warning they could spark a "global AI arms race."

Can an arms race be prevented, or is one already under way? If it's already happening, can it be stopped? Humanity's track record for controlling dangerous technology is mixed; attempts to ban weapons that were seen as too dangerous or inhumane date back to antiquity. Many of these attempts have failed, including early-twentieth-century attempts to ban submarines and airplanes. Even those that have succeeded, such as the ban on chemical weapons, rarely stop rogue regimes such as Bashar al-Assad's Syria or Saddam Hussein's Iraq. If an international ban cannot stop the world's most odious regimes from building killer robot armies, we may someday face our darkest nightmares brought to life.

STUMBLING TOWARD THE ROBOPOCALYPSE

No nation has stated outright that they are building autonomous weapons, but in secret defense labs and dual-use commercial applications, AI technology is racing forward. For most applications, even armed robots, humans would remain in control of lethal decisions—but battlefield pressures could drive militaries to build autonomous weapons that take the human out of the loop. Militaries could desire greater autonomy to take advantage of computers' superior speed or so that robots can continue engagements when their communications to human controllers are jammed. Or militaries might build autonomous weapons simply because of a fear that others might do so. U.S. Deputy Secretary of Defense Bob Work has asked:

> If our competitors go to Terminators . . . and it turns out the Terminators are able to make decisions faster, even if they're bad, how would we respond?

Vice Chairman of the Joint Chiefs of Staff General Paul Selva has termed this dilemma "The Terminator Conundrum." The stakes are high: AI is emerging as a powerful technology. Used the right way, intelligent machines could save lives by making war more precise and humane. Used the wrong way, autonomous weapons could lead to more killing and even greater civilian casualties. Nations will not make these choices in a vacuum. It will depend on what other countries do, as well as on the collective choices of scientists, engineers, lawyers, human rights activists, and others participating in this debate. Artificial intelligence is coming and it *will* be used in war. *How* it is used, however, is an open question. In the words of John Connor, hero of the *Terminator* movies and leader of the human resistance against the machines, "The future's not set. There's no fate but what we make for ourselves." The fight to ban autonomous weapons cuts to the core of humanity's ages-old conflicted relationship with technology: do we control our creations or do they control us?

PART I

Robopocalypse
Now

1

THE COMING SWARM

THE MILITARY ROBOTICS REVOLUTION

On a sunny afternoon in the hills of central California, a swarm takes flight. One by one, a launcher flings the slim Styrofoam-winged drones into the air. The drones let off a high-pitched buzz, which fades as they climb into the crystal blue California sky.

The drones carve the air with sharp, precise movements. I look at the drone pilot standing next to me and realize with some surprise that his hands aren't touching the controls; the drones are flying fully autonomously. It's a silly realization—after all, autonomous drone swarms are what I've come here to see—yet somehow the experience of watching the drones fly with such agility without any human controlling them is different than I'd imagined. Their nimble movements seem purposeful, and it's hard not to imbue them with intention. It's both impressive and discomfiting, this idea of the drones operating "off leash."

I've traveled to Camp Roberts, California, to see researchers from the Naval Postgraduate School investigate something no one else in the world has ever done before: swarm warfare. Unlike Predator drones, which are individually remotely piloted by human controllers on the ground, these researchers' drones are controlled en masse. Today's experiment will feature twenty drones flying simultaneously in a ten-against-ten swarm-on-swarm mock dogfight. The shooting is simulated, but the maneuvering and flying are all real.

Each drone comes off the launcher with its autopilot already switched on. Without any human direction, they climb to their assigned altitudes and form two teams, reporting back when they are "swarm ready." The Red and Blue swarms wait in their respective corners of the aerial combat arena, circling like a flock of hungry buzzards.

The pilot commanding Red Swarm rubs his hands together, anticipating the coming battle—which is funny, because his entire role is just to click the button that tells the swarm to start. After that, he's as much of a spectator as I am.

Duane Davis, the retired Navy helicopter pilot turned computer programmer who designed the swarm algorithms, counts down to the fight:

"Initiating swarm v. swarm . . . 3, 2, 1, shoot!"

Both the Red and Blue swarm commanders put their swarms into action. The two swarms close in on each other without hesitation. "Fight's on!" Duane yells enthusiastically. Within seconds, the swarms close the gap and collide. The two swarms blend together into a furball of close air combat. The swarms maneuver and swirl as a single mass. Simulated shots are tallied up at the bottom of the computer screen:

"UAV 74 fired at UAV 33
"UAV 59 fired at UAV 25
"UAV 33 hit
"UAV 25 hit . . ."

The swarms' behavior is driven by a simple algorithm called Greedy Shooter. Each drone will maneuver to get into a kill shot position against an enemy drone. A human must only choose the swarm behavior—wait, follow, attack, or land—and tell the swarm to start. After that, all of the swarm's actions are totally autonomous.

On the Red Swarm commander's computer screen, it's hard to tell who's winning. The drone icons overlap one another in a blur while, outside, the drones circle each other in a maelstrom of air combat. The whirling gyre looks like pure chaos to me, although Davis tells me he sometimes can pick out which drones are chasing each other.

A referee software called The Arbiter tracks the score. Red Swarm gains the upper hand with four kills to Blue's two. The "killed" drones' status switches from green to red as they're taken out of the fight. Then the fight falls into a lull, with the aircraft circling each other, unable to get a kill. Davis explains that because the aircraft are perfectly matched—same airframe, same flight controls, same algorithms—they sometimes fall into a stalemate where neither side can gain the upper hand.

Davis resets the battlefield for Round 2 and the swarms return to their respective corners. When the swarm commanders click go, the swarms close on each other once again. This time the battle comes out dead even, 3–3. In Round 3, Red pulls out a decisive win, 7–4. Red Swarm commander is happy to take credit for the win. "I pushed the button," he says with a chuckle.

Just as robots are transforming industries—from self-driving cars to robot vacuum cleaners and caretakers for the elderly—they are also trans-forming war. Global spending on military robotics is estimated to reach $7.5 billion per year in 2018, with scores of countries expanding their arsenals of air, ground, and maritime robots.

Robots have many battlefield advantages over traditional human-inhabited vehicles. Unshackled from the physiological limits of humans, uninhabited ("unmanned") vehicles can be made smaller, lighter, faster, and more maneuverable. They can stay out on the battlefield far beyond the limits of human endurance, for weeks, months, or even years at a time without rest. They can take more risk, opening up tactical opportunities for dangerous or even suicidal missions without risking human lives.

However, robots have one major disadvantage. By removing the human from the vehicle, they lose the most advanced cognitive processor on the planet: the human brain. Most military robots today are remotely con-trolled, or teleoperated, by humans; they depend on fragile communica-tion links that can be jammed or disrupted by environmental conditions. Without these communications, robots can only perform simple tasks, and their capacity for autonomous operation is limited.

The solution: more autonomy.

THE ACCIDENTAL REVOLUTION

No one planned on a robotics revolution, but the U.S. military stumbled into one as it deployed thousands of air and ground robots to meet urgent needs in Iraq and Afghanistan. By 2005, the U.S. Department of Defense

(DoD) had woken up to the fact that something significant was happening. Spending on uninhabited aircraft, or drones, which had hovered around the $300 million per year mark in the 1990s, skyrocketed after 9/11, increasing sixfold to over $2 billion per year by 2005. Drones proved particularly valuable in the messy counterinsurgency wars in Iraq and Afghanistan. Larger aircraft like the MQ-1B Predator can quietly surveil terrorists around the clock, tracking their movements and unraveling their networks. Smaller hand-launched drones like the RQ-11 Raven can provide troops "over-the-hill reconnaissance" on demand while on patrol. Hundreds of drones had been deployed to Iraq and Afghanistan in short order.

Drones weren't new—they had been used in a limited fashion in Vietnam—but the overwhelming crush of demand for them was. While in later years drones would become associated with "drone strikes," it is their capacity for persistent surveillance, not dropping bombs, that makes them unique and valuable to the military. They give commanders a low-cost, low-risk way to put eyes in the sky.

Soon, the Pentagon was pouring drones into the wars at a breakneck pace. By 2011, annual spending on drones had swelled to over $6 billion per year, over twenty times pre-9/11 levels. DoD had over 7,000 drones in its fleet. The vast majority of them were smaller hand-launched models, but large aircraft like the MQ-9 Reaper and RQ-4 Global Hawk were also valuable military assets.

At the same time, DoD was discovering that robots weren't just valuable in the air. They were equally important, if not more so, on the ground. Driven in large part by the rise of improvised explosive devices (IEDs), DoD deployed over 6,000 ground robots to Iraq and Afghanistan. Small robots like the iRobot Packbot allowed troops to disable or destroy IEDs without putting themselves at risk. Bomb disposal is a great job for a robot.

THE MARCH TOWARD EVER-GREATER AUTONOMY

In 2005, after DoD started to come to grips with the robotics revolution and its implications for the future of conflict, it began publishing a series of "roadmaps" for future unmanned system investment. The first road-map was focused on aircraft, but subsequent roadmaps in 2007, 2009, 2011, and 2013 included ground and maritime vehicles as well. While the lion's share of dollars has gone toward uninhabited aircraft, ground, sea surface, and undersea vehicles have valuable roles to play as well.

These roadmaps did more than simply catalog the investments DoD was making. Each roadmap looked forward twenty-five years into the future, outlining technology needs and wants in order to help inform future investments by government and industry. They covered sensors, communications, power, weapons, propulsion, and other key enabling technologies. Across all the roadmaps, autonomy is a dominant theme.

The 2011 roadmap perhaps summarized the vision best:

> For unmanned systems to fully realize their potential, they must be able to achieve a highly autonomous state of behavior and be able to interact with their surroundings. This advancement will require an ability to understand and adapt to their environment, and an ability to collaborate with other autonomous systems.

Autonomy is the cognitive engine that power robots. Without autonomy, robots are only empty vessels, brainless husks that depend on human controllers for direction.

In Iraq and Afghanistan, the U.S. military operated in a relatively "permissive" electromagnetic environment where insurgents did not generally have the ability to jam communications with robot vehicles, but this will not always be the case in future conflicts. Major nation-state militaries will almost certainly have the ability to disrupt or deny communications networks, and the electromagnetic spectrum will be highly contested. The U.S. military has ways of communicating that are more resistant to jamming, but these methods are limited in range and bandwidth. Against a major military power, the type of drone operations the United States has conducted when going after terrorists—streaming high-definition, full-motion video back to stateside bases via satellites—will not be possible. In addition, some environments inherently make communications challenging, such as undersea, where radio wave propagation is hindered by water. In these situations, autonomy is a must if robotic systems are to be effective. As machine intelligence advances, militaries will be able to create ever more autonomous robots capable of carrying out more complex missions in more challenging environments independent from human control.

Even if communications links work perfectly, greater autonomy is also desirable because of the personnel costs of remotely controlling robots. Thousands of robots require thousands of people to control them, if each robot is remotely operated. Predator and Reaper drone operations require seven to ten pilots to staff one drone "orbit" of 24/7 continuous around-the-clock coverage over an area. Another twenty people per orbit are required to operate the sensors on the drone, and scores of intelligence analysts are needed to sift through the sensor data. In fact, because of these substantial personnel requirements, the U.S. Air Force has a strong resistance to calling these aircraft "unmanned." There may not be anyone on board the aircraft, but there are still humans controlling it and supporting it.

Because the pilot remains on the ground, uninhabited aircraft free surveillance operations from the limits of human endurance—but only the physical ones. Drones can stay aloft for days at a time, far longer than a human pilot could remain effective sitting in the cockpit, but remote operation doesn't change the *cognitive* requirements on human operators. Humans still have to perform the same tasks, they just aren't physically on board the vehicle. The Air Force prefers the term "remotely piloted aircraft" because that's what today's drones are. Pilots still fly the aircraft via stick and rudder input, just remotely from the ground, sometimes even half a world away.

It's a cumbersome way to operate. Building tens of thousands of cheap robots is not a cost-effective strategy if they require even larger numbers of highly trained (and expensive) people to operate them.

Autonomy is the answer. The 2011 DoD roadmap stated:

> Autonomy reduces the human workload required to operate systems, enables the optimization of the human role in the system, and allows human decision making to focus on points where it is most needed. These benefits can further result in manpower efficiencies and cost savings as well as greater speed in decision-making.

Many of DoD's robotic roadmaps point toward the long-term goal of full autonomy. The 2005 roadmap looked toward "fully autonomous swarms." The 2011 roadmap articulated an evolution of four levels of autonomy from (1) human operated to (2) human delegated, (3) human supervised, and eventually (4) fully autonomous. The benefits of greater autonomy was the "single greatest theme" in a 2010 report from the Air Force Office of the Chief Scientist on future technology.

Although Predator and Reaper drones are still flown manually, albeit remotely from the ground, other aircraft such as Air Force Global Hawk and Army Gray Eagle drones have much more automation: pilots direct these aircraft where to go and the aircraft flies itself. Rather than being flown via a stick and rudder, the aircraft are directed via keyboard and mouse. The Army doesn't even refer to the people controlling its aircraft as "pilots"—it called them "operators." Even with this greater automation, however, these aircraft still require one human operator per aircraft for anything but the simplest missions.

Incrementally, engineers are adding to the set of tasks that uninhabited aircraft can perform on their own, moving step by step toward increasingly autonomous drones. In 2013, the U.S. Navy successfully landed its X-47B prototype drone on a carrier at sea, autonomously. The only human input was the order to land; the actual flying was done by software. In 2014, the Navy's Autonomous Aerial Cargo/Utility System (AACUS) helicopter autonomously scouted out an improvised landing area and executed a successful landing on its own. Then in 2015, the X-47B drone again made history by conducting the first autonomous aerial refueling, taking gas from another aircraft while in flight.

These are key milestones in building more fully combat-capable uninhabited aircraft. Just as autonomous cars will allow a vehicle to drive from point A to point B without manual human control, the ability to takeoff, land, navigate, and refuel autonomously will allow robots to perform tasks under human direction and supervision, but without humans controlling each movement. This can begin to break the paradigm of humans manually controlling the robot, shifting humans into a supervisory role. Humans will command the robot what action to take, and it will execute the task on its own.

Swarming, or cooperative autonomy, is the next step in this evolution.

Davis is most excited about the nonmilitary applications of swarming, from search and rescue to agriculture. Coordinated robot behavior could be useful for a wide variety of applications and the Naval Postgraduate School's research is very basic, so the algorithms they're building could be used for many purposes. Still, the military advantages in mass, coordination, and speed are profound and hard to ignore. Swarming can allow militaries to field large numbers of assets on the battlefield with a small number of human controllers. Cooperative behavior can also allow quicker reaction times, so that the swarm can respond to changing events faster than would be possible with one person controlling each vehicle.

In conducting their swarm dogfight experiment, Davis and his colleagues are pushing the boundaries of autonomy. Their next goal is to work up to a hundred drones fighting in a fifty-on-fifty aerial swarm battle, something Davis and his colleagues are already simulating on computers, and their ultimate goal is to move beyond dogfighting to a more complex game akin to capture the flag. Two swarms would compete to score the most points by landing at the other's air base without being "shot down" first. Each swarm must balance defending its own base, shooting down enemy drones, and getting as many of its drones as possible into the enemy's base. What are the "plays" to run with a swarm? What are the best tactics? These are precisely the questions Davis and his colleagues want to explore.

"If I have fifty planes that are involved in a swarm," he said, "how much of that swarm do I want to be focused on offense—getting to the other guy's landing area? How much do I want focused on defending my landing space and doing the air-to-air problem? How do I want to do assignments of tasks between the swarms? If I've got the adversary's UAVs [unmanned aerial vehicles] coming in, how do I want my swarm deciding which UAV is going to take which adversary to try to stop them from getting to our base?"

Swarm tactics are still at a very early stage. Currently, the human operator allocates a certain number of drones to a sub-swarm then tasks that sub-swarm with a mission, such as attempting to attack an enemy's base or attacking enemy aircraft. After that, the human is in a supervisory mode. Unless there is a safety concern, the human controller won't intervene to take control of an aircraft. Even then, if an aircraft began to experience a malfunction, it wouldn't make sense to take control of it until it left the swarm's vicinity. Taking manual control of an aircraft in the middle of the swarm could actually instigate a midair collision. It would be very difficult for a human to predict and avoid a collision with all of the other drones swirling in the sky. If the drone is under the swarm's command, however, it will automatically adjust its flight to avoid a collision.

Right now, the swarm behaviors Davis is using are very basic. The human can command the swarm to fly in a formation, to land, or to attack enemy aircraft. The drones then sort themselves into position for landing or formation flying to "deconflict" their actions. For some tasks, such as landing, this is done relatively easily by altitude: lower planes land first. Other tasks, such as deconflicting air-to-air combat are trickier. It wouldn't do any good, for example, for all of the drones in the swarm to go after the same enemy aircraft. They need to coordinate their behavior.

The problem is analogous to that of outfielders calling a fly ball. It wouldn't make sense to have the manager calling who should catch the ball from the dugout. The outfielders need to coordinate among themselves. "It's one thing when you've got two humans that can talk to one another and one ball," Davis explained. "It's another thing when there's fifty humans and fifty balls." This task would be effectively impossible for humans, but a swarm can accomplish this very quickly, through a variety of methods. In centralized coordination, for example, individual swarm elements pass their data back to a single controller, which then issues commands to each robot in the swarm. Hierarchical coordination, on the other hand, decomposes the swarm into teams and squads much like a military organization, with orders flowing down the chain of command.

Consensus-based coordination is a decentralized approach where all of the swarm elements communicate with one another simultaneously and collectively decide on a course of action. They could do this by using "voting" or "auction" algorithms to coordinate behavior. For example, each swarm element could place a "bid" on an "auction" to catch the fly ball. The individual that bids highest "wins" the auction and catches the ball, while the others move out of the way.

Emergent coordination is the most decentralized approach and is how flocks of birds, colonies of insects, and mobs of people work, with coordinated action arising naturally from each individual making decisions based on those nearby. Simple rules for individual behavior can lead to

Centralized Coordination

Swarm elements communicate with a centralized planner which coordinates all tasks.

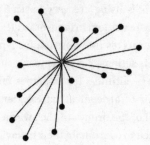

Hierarchical Coordination

Swarm elements are controlled by "squad" level agents, who are in turn controlled by higher-level controllers.

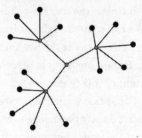

Coordination by Consensus

All swarm elements communicate to one another and use "voting" or auction-based methods to converge on a solution.

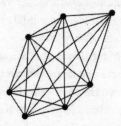

Emergent Coordination

Coordination arises naturally by individual swarm elements reacting to one another, like in animal swarms.

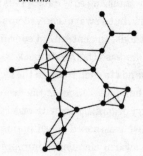

Swarm Command-and-Control Models

very complex collective action, allowing the swarm to exhibit "collective intelligence." For example, a colony of ants will converge on an optimal route to take food back to the nest over time because of simple behavior from each individual ant. As ants pick up food, they leave a pheromone trail behind them as they move back to the nest. If they come across an existing trail with stronger pheromones, they'll switch to it. More ants will arrive back at the nest sooner via the faster route, leading to a stronger pheromone trail, which will then cause more ants to use that trail. No individual ant "knows" which trail is fastest, but collectively the colony converges on the fastest route.

Communication among elements of the swarm can occur through direct signaling, akin to an outfielder yelling "I got it!"; indirect methods such as co-observation, which is how schools of fish and herds of animals stay together; or by modifying the environment in a process called *stigmergy*, like ants leaving pheromones to mark a trail.

The drones in Davis's swarm communicate through a central Wi–Fi router on the ground. They avoid collisions by staying within narrow altitude windows that are automatically assigned by the central ground controller. Their attack behavior is uncoordinated, though. The "greedy shooter" algorithm simply directs each drone to attack the nearest enemy drone, regardless of what the other drones are doing. In theory, all the drones could converge on the same enemy drone, leaving other enemies untouched. It's a terrible method for air-to-air combat, but Davis and his colleagues are still in the proof-of-concept stage. They have experimented with a more decentralized auction-based approach and found it to be very robust to disruptions, including up to a 90 percent communications loss within the swarm. As long as some communications are up, even if they're spotty, the swarm will converge on a solution.

The effect of fifty aircraft working together, rather than fighting individually or in wingman pairs as humans do today, would be tremendous. Coordinated behavior is the difference between a basketball *team* and five ball hogs all making a run at the basket themselves. It's the difference between a bunch of lone wolves and a *wolf pack*.

In 2016, the United States demonstrated 103 aerial drones flying together in a swarm that DoD officials described as "a collective organism, sharing one distributed brain for decision-making and adapting to each other like swarms in nature." (Not to be outdone, a few months later China demonstrated a 119-drone swarm.) Fighting together, a drone swarm could be far more effective than the same number of drones fighting individually. No one yet knows what the best tactics will be for swarm combat, but experiments such as these are working to tease them out. New tactics might even be evolved by the machines themselves through machine learning or evolutionary approaches.

Swarms aren't merely limited to the air. In August 2014, the U.S. Navy Office of Naval Research (ONR) demonstrated a swarm of small boats on the James River in Virginia by simulating a mock strait transit in which the boats protected a high-value Navy ship against possible threats, escorting it through a simulated high-danger area. When directed by a human controller to investigate a potential threat, a detachment of uninhabited boats moved to intercept and encircle the suspicious vessel. The human controller simply directed them to intercept the designated suspicious ship; the boats moved autonomously, coordinating their actions by sharing information. This demonstration involved five boats working together, but the concept could be scaled up to larger numbers, just as in aerial drone swarms.

Bob Brizzolara, who directed the Navy's demonstration, called the swarming boats a "game changer." It's an often-overused term, but in this case, it's not hyperbole—robotic boat swarms are highly valuable to the Navy as a potential way to guard against threats to its ships. In October 2000, the USS *Cole* was attacked by al-Qaida terrorists using a small explosive-laden boat while in port in Aden, Yemen. The blast killed seventeen sailors and cut a massive gash in the ship's hull. Similar attacks continue to be a threat to U.S. ships, not just from terrorists but also from Iran, which regularly uses small high-speed craft to harass U.S. ships near the Straits of Hormuz. Robot boats could intercept suspicious vessels further away, putting eyes (and potentially weapons) on potentially hostile boats without putting sailors at risk.

What the robot boats might do after they've intercepted a potentially hostile vessel is another matter. In a video released by the ONR, a .50 caliber machine gun is prominently displayed on the front of one of the boats. The video's narrator makes no bones about the fact that the robot boats could be used to "damage or destroy hostile vessels," but the demonstration didn't involve firing any actual bullets, and didn't include a consideration of what the rules of engagement actually would have been. Would a human be required to pull the trigger? When pressed by reporters following the demonstration, a spokesman for ONR explained that "there is always a human in the loop when it comes to the actual engagement of an enemy." But the spokesman also acknowledged that "under this swarming demonstration with multiple [unmanned surface vehicles], ONR did not study the specifics of how the human-in-the-loop works for rules of engagement."

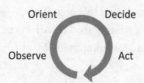

Orient   Decide

Observe   Act

In the OODA loop paradigm of combat, victory on the
battlefield goes to whichever side can complete the observe-
orient-decide-act cycle faster.

OODA Loop

The Navy's fuzzy answer to such a fundamental question reflects
a tension in the military's pursuit of more advanced robotics. Even as
researchers and engineers move to incorporate more autonomy, there is
an understanding that there are—or should be—limits on autonomy when
it comes to the use of weapons. What exactly those limits are, however, is
often unclear.

REACHING THE LIMIT

How much autonomy is too much? The U.S. Air Force laid out an ambitious vision for the future of robot aircraft in their *Unmanned Aircraft
Systems Flight Plan, 2009–2047*. The report envisioned a future where an
arms race in speed drove a desire for ever-faster automation, not unlike
real-world competition in automated stock trading.

In air combat, pilots talk about an observe, orient, decide, act (OODA)
loop, a cognitive process pilots go through when engaging enemy aircraft. Understanding the environment, deciding, and acting faster than
the enemy allows a pilot to "get inside" the enemy's OODA loop. While
the enemy is still trying to understand what's happening and decide on a
course of action, the pilot has already changed the situation, resetting the
enemy to square one and forcing him or her to come to grips with a new
situation. Air Force strategist John Boyd, originator of the OODA loop,
described the objective:

> Goal: Collapse adversary's system into confusion and disorder by
> causing him to over and under react to activity that appears simulta-
> neously menacing as well as ambiguous, chaotic, or misleading.

If victory comes from completing this cognitive process faster, then one can see the advantage in automation. The Air Force's 2009 Flight Plan saw tremendous potential for computers to exceed human decision-making speeds:

> Advances in computing speeds and capacity will change how technology affects the OODA loop. Today the role of technology is changing from supporting to fully participating with humans in each step of the process. In 2047 technology will be able to reduce the time to complete the OODA loop to micro or nanoseconds. Much like a chess master can outperform proficient chess players, [unmanned aircraft systems] will be able to react at these speeds and therefore this loop moves toward becoming a "perceive and act" vector. Increasingly humans will no longer be "in the loop" but rather "on the loop"—monitoring the execution of certain decisions. Simultaneously, advances in AI will enable systems to make combat decisions and act within legal and policy constraints without necessarily requiring human input.

This, then, is the logical culmination of the arms race in speed: autonomous weapons that complete engagements all on their own. The Air Force Flight Plan acknowledged the gravity of what it was suggesting might be possible. The next paragraph continued:

> Authorizing a machine to make lethal combat decisions is contingent upon political and military leaders resolving legal and ethical questions. These include the appropriateness of machines having this ability, under what circumstances it should be employed, where responsibility for mistakes lies and what limitations should be placed upon the autonomy of such systems.... Ethical discussions and policy decisions must take place in the near term in order to guide the development of future [unmanned aircraft system] capabilities, rather than allowing the development to take its own path apart from this critical guidance.

The Air Force wasn't recommending autonomous weapons. It wasn't even suggesting they were necessarily a good idea. What it was suggesting was that autonomous systems might have advantages over humans in speed, and that AI might advance to the point where machines could carry out lethal targeting and engagement decisions without human input. If that is true, then legal, ethical, and policy discussions should take place now to shape the development of this technology.

At the time the Air Force Flight Plan was released in 2009, I was working in the Office of the Secretary of Defense as a civilian policy analyst focusing on drone policy. Most of the issues we were grappling with at the time had to do with how to manage the overwhelming demand for more drones from Iraq and Afghanistan. Commanders on the ground had a seemingly insatiable appetite for drones. Despite the thousands that had been deployed, they wanted more, and Pentagon senior leaders—particularly in the Air Force—were concerned that spending on drones was crowding out other priorities. Secretary of Defense Robert Gates, who routinely chastised the Pentagon for its preoccupation with future wars over the ongoing ones in Iraq and Afghanistan, strongly sided with warfighters in the field. His guidance was clear: send more drones. Most of my time was spent figuring out how to force the Pentagon bureaucracy to comply with the secretary's direction and respond more effectively to warfighter needs, but when policy questions like this came up, eyes turned toward me.

I didn't have the answers they wanted. There was no policy on autonomy. Although the Air Force had asked for policy guidance in their 2009 Flight Plan, there wasn't even a conversation under way.

The 2011 DoD roadmap, which I was involved in writing, took a stab at an answer, even if it was a temporary one:

> Policy guidelines will especially be necessary for autonomous systems that involve the application of force. ... For the foreseeable future, decisions over the use of force and the choice of which individual targets to engage with lethal force will be retained under human control in unmanned systems.

It didn't say much, but it was the first official DoD policy statement on lethal autonomy. Lethal force would remain under human control for the "foreseeable future." But in a world where AI technology is racing forward at a breakneck pace, how far into the future can we really see?

2

THE TERMINATOR AND THE ROOMBA

WHAT IS AUTONOMY?

Autonomy is a slippery word. For one person, "autonomous robot" might mean a household Roomba that vacuums your home while you're away. For another, autonomous robots conjure images from science fiction. Autonomous robots could be a good thing, like the friendly—if irritating—C-3PO from *Star Wars,* or could lead to rogue homicidal agents, like those Skynet deploys against humanity in the *Terminator* movies.

Science fiction writers have long grappled with questions of autonomy in robots. Isaac Asimov created the now-iconic Three Laws of Robotics to govern robots in his stories:

1. A robot may not injure a human being or, through inaction, allow a human being to come to harm.
2. A robot must obey orders given by human beings except where such orders would conflict with the first law.
3. A robot must protect its own existence as long as such protection does not conflict with the first or second law.

In Asimov's stories, these laws embedded within the robot's "positronic brain" are inviolable. The robot must obey. Asimov's stories often explore the consequences of robots' strict obedience of these laws, and loopholes in the laws themselves. In the Asimov-inspired movie *I, Robot* (spoiler alert), the lead robot protagonist, Sonny, is given a secret secondary processor that allows him to override the Three Laws, if he desires. On the outside, Sonny looks the same as other robots, but the human characters can instantly tell there is something different about him. He dreams. He questions them. He engages in humanlike dialogue and critical thought of which the other robots are incapable. There is something unmistakably human about Sonny's behavior.

When Dr. Susan Calvin discovers the source of Sonny's apparent anomalous conduct, she finds it hidden in his chest cavity. The symbolism in the film is unmistakable: unlike other robots who are slaves to logic, Sonny has a "heart."

Fanciful as it may be, *I, Robot*'s take on autonomy resonates. Unlike machines, humans have the ability to ignore instructions and make decisions for themselves. Whether robots can ever have something akin to human free will is a common theme in science fiction. In *I, Robot*'s climactic scene, Sonny makes a choice to save Dr. Calvin, even though it means risking the success of their mission to defeat the evil AI V.I.K.I., who has taken over the city. It's a choice motivated by love, not logic. In the *Terminator* movies, when the military AI Skynet becomes self-aware, it makes a different choice. Upon determining that humans are a threat to its existence, Skynet decides to eliminate them, starting global nuclear war and initiating "Judgment Day."

THE THREE DIMENSIONS OF AUTONOMY

In the real world, machine autonomy doesn't require a magical spark of free will or a soul. Autonomy is simply the ability for a machine to perform a task or function on its own.

The DoD unmanned system roadmaps referred to "levels" or a "spectrum" of autonomy, but those classifications are overly simplistic. Autonomy encompasses three distinct concepts: the type of task the machine is performing; the relationship of the human to the machine when performing that task; and the sophistication of the machine's decision-making when performing the task. This means there are three different *dimensions* of autonomy. These dimensions are independent, and a machine can be "more autonomous" by increasing the amount of autonomy along any of these spectrums.

The first dimension of autonomy is the task being performed by the machine. Not all tasks are equal in their significance, complexity, and risk: a thermostat is an autonomous system in charge of regulating temperature, while *Terminator*'s Skynet was given control over nuclear weapons. The complexity of decisions involved and the consequences if the machine fails to perform the task appropriately are very different. Often, a single machine will perform some tasks autonomously, while humans are in control of other tasks, blending human and machine control within the system. Modern automobiles have a range of autonomous features: automatic braking and collision avoidance, antilock brakes, automatic seat belt retractors, adaptive cruise control, automatic lane keeping, and self-parking. Some autonomous functions, such as autopilots in commercial airliners, can be turned on or off by a human user. Other autonomous functions, like airbags, are always ready and decide for themselves when to activate. Some autonomous systems may be designed to override the human user in certain situations. U.S. fighter aircraft have been modified with an automatic ground collision avoidance system (Auto-GCAS). If the pilot becomes disoriented and is about to crash, Auto-GCAS will take control of the aircraft at the last minute to pull up and avoid the ground. The system has already saved at least one aircraft in combat, rescuing a U.S. F-16 in Syria.

As automobiles and aircraft demonstrate, it is meaningless to refer to a system as "autonomous" without referring to the specific task that is being automated. Cars are still driven by humans (for now), but a host of autonomous functions can assist the driver, or even take control for short periods of time. The machine becomes "more autonomous" as it takes on more tasks, but some degree of human involvement and direction always exists. "Fully autonomous" self-driving cars can navigate and drive on their own, but a human is still choosing the destination.

For any given task, there are degrees of autonomy. A machine can perform a task in a semiautonomous, supervised autonomous, or fully autonomous manner. This is the second dimension of autonomy: the human-machine relationship.

Decide

Sense   Act

The machine performs a task and
then waits for the human user to
take an action before continuing.

Semiautonomous Operation (human in the loop)

In semiautonomous systems, the machine performs a task and then waits for a human user to take an action before continuing. A human is "in the loop." Autonomous systems go through a sense, decide, act loop similar to the military OODA loop, but in semiautonomous systems the loop is broken by a human. The system can sense the environment and recommend a course of action, but cannot carry out the action without human approval.

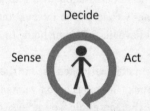

Decide

Sense   Act

The machine can sense, decide, and act
on its own. The human user supervises its
operation and can intervene, if desired.

Supervised Autonomous Operation (human on the loop)

In supervised autonomous systems, the human sits "on" the loop. Once put into operation, the machine can sense, decide, and act on its own, but a human user can observe the machine's behavior and intervene to stop it, if desired.

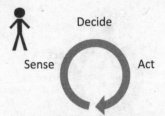

The machine can sense, decide, and
act on its own. The human cannot
intervene in a timely fashion.

Fully Autonomous Operation (human out of the loop)

Fully autonomous systems sense, decide, and act entirely without human intervention. Once the human activates the machine, it conducts the task without communication back to the human user. The human is "out of the loop."

Many machines can operate in different modes at different times. A Roomba that is vacuuming while you are home is operating in a supervised autonomous mode. If the Roomba becomes stuck—my Roomba frequently trapped itself in the bathroom—then you can intervene. If you're out of the house, then the Roomba is operating in a fully autonomous capacity. If something goes wrong, it's on its own until you come home. More often than I would have liked, I came home to a dirty house and a spotless bathroom.

It wasn't the Roomba's fault it had locked itself in the bathroom. It didn't even know that it was stuck (Roombas aren't very smart). It had simply wandered into a location where its aimless bumping would nudge the door closed, trapping it. Intelligence is the third dimension of autonomy. More sophisticated, or more intelligent, machines can be used to take on more complex tasks in more challenging environments. People often use terms like "automatic," "automated," or "autonomous" to refer to a spectrum of intelligence in machines.

Automatic systems are simple machines that don't exhibit much in the way of "decision-making." They sense the environment and act. The relationship between sensing and action is immediate and linear. It is also highly predictable to the human user. An old mechanical thermostat is an example of an automatic system. The user sets the desired temperature

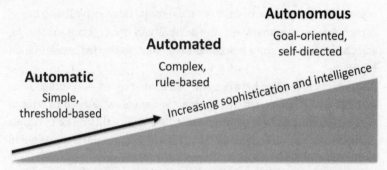

As machines become more sophisticated, they become more capable and able to accomplish more complex tasks in more open-ended environments. The downside is that their specific actions may be less predictable, even to trained users.

Spectrum of Intelligence in Machines

and when the temperature gets too high or too low, the thermostat activates the heat or air conditioning.

Automated systems are more complex, and may consider a range of inputs and weigh several variables before taking an action. Nevertheless, the internal cognitive processes of the machine are generally traceable by the human user, at least in principle. A modern digital programmable thermostat is an example of an automated system. Whether the heat or air conditioning turns on is a function of the house temperature as well as what day and time it is. Given knowledge of the inputs to the system and its programmed parameters, the system's behavior should be predictable to a trained user.

"Autonomous" is often used to refer to systems sophisticated enough that their internal cognitive processes are less intelligible to the user, who understands the task the system is supposed to perform, but not necessarily *how* the system will perform that task. Researchers often refer to autonomous systems as being "goal-oriented." That is to say, the human user specifies the goal, but the autonomous system has flexibility in how it achieves that goal.

Take a self-driving car, for example. The user specifies the destination and other goals, such as avoiding accidents, but can't possibly specify in advance every single action the autonomous car is supposed to perform. The user doesn't know where there will be traffic or obstacles in the road, when lights will change, or what other cars or pedestrians will do. The car is therefore programmed with the flexibility to decide when to stop, go, and change lanes in order to accomplish its goal: getting to the destination safely.

In practice, the line between automatic, automated, and autonomous systems is still blurry. Often, the term "autonomous" is used to refer to future systems that have not yet been built, but once they do exist, people describe those same systems as "automated." This is similar to a trend in artificial intelligence where AI is often perceived to encompass only tasks that machines cannot yet do. Once a machine conquers a task, then it is merely "software."

Autonomy doesn't mean the system is exhibiting free will or disobeying its programming. The difference is that unlike an automatic system where there is a simple, linear connection from sensing to action, autonomous systems take into account a range of variables to consider the best action in any given situation. Goal-oriented behavior is essential for autonomous systems in uncontrolled environments. If a self-driving car were on a closed track with no pedestrians or other vehicles, each movement could be programmed into the car in advance—when to go, stop, turn, etc. But such a car would not be very useful, as it could only drive in a simple environment where every action could be predicted. In more complex environments or when performing more complex tasks, it is crucial that the machine be able to make decisions based on the specific situation.

This greater complexity in autonomous systems is a double-edged sword. The downside to more sophisticated systems is that the user may not be able to predict its specific actions in advance. The feature of increased autonomy can become a flaw if the user is surprised in an unpleasant way by the machine's behavior. For simple automatic or automated systems, this is less likely. But as the complexity of the system increases, so does the difficulty of predicting how the machine will act.

It can be exciting, if a little scary, to hand over control to an autonomous system. The machine is like a black box. We specify its goal and, like magic, the machine overcomes obstacles to reach the goal. The inner workings of how it did so are often mysterious to us; the distinction between "automated" and "autonomous" is principally in the mind of the user. A new machine only feels "autonomous" because we don't yet have a good mental model for how it "thinks." As we gain experience with the machine and begin to better understand it, the layers of fog hiding the inner workings of the black box dissipate, revealing the complex logic driving its behavior. We come to decide the machine is merely "automated" after all. In understanding the machine, we have tamed it; the humans are back in control. That process of discovery, however, can be a rocky one.

A few years ago, I purchased a Nest "learning thermostat." The Nest tracks your behavior and adjusts the house's temperature as needed, "learning" your preferences over time. There were bumps along the way as I discovered various aspects of the Nest's functionality and occasionally the house was temporarily too warm or too cold, but I was sufficiently enamored of the technology that I was willing to push through these growing pains. My wife, Heather, was less tolerant of the Nest. Every time it changed the temperature on its own, disregarding an instruction she had given, she viewed it more and more suspiciously. (Unbeknownst to her, the Nest was following other guidance I had given it previously.)

The final straw for the Nest was when we came home from summer vacation to find the house a toasty 84 degrees, despite my having gone online the night before and set the Nest to a comfortable 70. With sweat dripping off our faces, we set our bags down in the foyer and I ran to the Nest to see what had happened. As it turned out, I had neglected to turn off the "auto-away feature." After the Nest's hallway sensor detected no movement and discerned we were not home, it reverted—per its programming—to the energy-saving "away" setting of 84 degrees. One look from Heather told me it was too late, though. She had lost trust in the Nest. (Or, more accurately, in my ability to use it.)

The Nest wasn't broken, though. The human-machine connection was. The same features that made the Nest "smarter" also made it harder for me to anticipate its behavior. The disconnect between my expectations of what the Nest would do and what it was actually doing meant the autonomy that was supposed to be working for me ended up, more often than not, working against my goals.

HOW MUCH SHOULD WE TRUST AUTONOMOUS SYSTEMS?

All the Nest did was control the thermostat. The Roomba merely vacuumed. Coming home to a Roomba locked in the bathroom or an overheated house might be annoying, but it wasn't a catastrophe. The tasks entrusted to these autonomous systems weren't critical ones.

What if I was dealing with an autonomous system performing a truly critical function? What if the Nest was a weapon, and my inability to understand it led to failure?

What if the task I was delegating to an autonomous system was the decision whether or not to kill?

3

MACHINES THAT KILL

WHAT IS AN AUTONOMOUS WEAPON?

The path to autonomous weapons began 150 years ago in the mid-nineteenth century. As the second industrial revolution was bringing unprecedented productivity to cities and factories, the same technology was bringing unprecedented efficiency to killing in war.

At the start of the American Civil War in 1861, inventor Richard Gatling devised a new weapon to speed up the process of firing: the Gatling gun. A forerunner of the modern machine gun, the Gatling gun employed automation for loading and firing, allowing more bullets to be fired in a shorter amount of time. The Gatling gun was a significant improvement over Civil War–era rifled muskets, which had to be loaded by hand through the muzzle in a lengthy process. Well-trained troops could fire three rounds per minute with a rifled musket. The Gatling gun fired over 300 rounds per minute.

In its time, the Gatling gun was a marvel. Mark Twain was an early enthusiast:

[T]he Gatling gun ... is a cluster of six to ten savage tubes that carry great conical pellets of lead, with unerring accuracy, a distance of two and a half miles. It feeds itself with cartridges, and you work it with a crank like a hand organ; you can fire it faster than four men can count. When fired rapidly, the reports blend together like the clattering of a

watchman's rattle. It can be discharged four hundred times a minute!
I liked it very much.

The Gatling gun was not an autonomous weapon, but it began a long
evolution of weapons automation. In the Gatling gun, the process of
loading bullets, firing, and ejecting cartridges was all automatic, provided a human kept turning the crank. The result was a tremendous
expansion in the amount of destructive power unleashed on the battlefield. Four soldiers were needed to operate the Gatling gun, but by dint
of automation, they could deliver the same lethal firepower as more than
a hundred men.

Richard Gatling's motivation was not to accelerate the process of killing, but to save lives by reducing the number of soldiers needed on the battlefield. Gatling built his device after watching waves of young men return
home wounded or dead from the unrelenting bloodshed of the American
Civil War. In a letter to a friend, he wrote:

> It occurred to me that if I could invent a machine—a gun—which could
> by its rapidity of fire, enable one man to do as much battle duty as a
> hundred, that it would, to a great extent, supersede the necessity of
> large armies, and consequently, exposure to battle and disease be
> greatly diminished.

Gatling was an accomplished inventor with multiple patents to his
name for agricultural implements. He saw the gun in a similar light—
machine technology harnessed to improve efficiency. Gatling claimed his
gun "bears the same relation to other firearms that McCormack's reaper
does to the sickle, or the sewing machine to the common needle."

Gatling was more right than he knew. The Gatling gun did indeed lay
the seeds for a revolution in warfare, a break from the old ways of killing
people one at a time with rifled muskets and shift to a new era of mechanized death. The future Gatling wrought was not one of less bloodshed,
however, but unimaginably more. The Gatling gun laid the foundations
for a new class of machine: the automatic weapon.

AUTOMATIC WEAPONS: MACHINE GUNS

Automatic weapons came about incrementally, with inventors building on and refining the work of those who came before. The next tick in the gears of progress came in 1883 with the invention of the Maxim gun. Unlike the Gatling gun, which required a human to hand-crank the gun to power it, the Maxim gun harnessed the physical energy from the recoil of the gun's firing to power the process of reloading the next round. Hand-cranking was no longer needed, and once firing was initiated, the gun could continue firing on its own. The machine gun was born.

The machine gun was a marvelous and terrible invention. Unlike semiautomatic weapons, which require the user to pull the trigger for each bullet, automatic weapons will continue firing so long as the trigger remains held down. Modern machine guns come in all shapes and sizes, from the snub-nosed Uzi that plainclothes security personnel can tuck under their suit jackets to massive chain guns that rattle off thousands of rounds per minute. Regardless of their form, their power is palpable when firing one.

As a Ranger, I carried an M249 Squad Automatic Weapon, or SAW, a single-person light machine gun carried in infantry fire teams. Weighing seventeen pounds without ammunition, the SAW is on the hefty side of what can be considered "hand held." With training, the SAW can be fired from the shoulder standing up in short controlled bursts, but is best used lying on the ground. The SAW comes equipped with two metal bipod legs that can be flipped down to allow the gun to stand elevated off the dirt. One does not simply lay on the ground and fire the SAW, however. The SAW has to be managed; it has to be controlled. When fired, the weapon bucks and moves like a wild animal from the rapid-fire recoil. At a cyclic rate of fire, with the trigger held down, the SAW will fire 800 rounds per minute. That's thirteen bullets streaming out of the barrel per second. At that rate of fire, a gunner will rip through his entire stash of ammunition in under two minutes. The barrel will overheat and begin to melt.

Using the SAW effectively requires discipline. The gunner must lean into the weapon to control it, putting his weight behind it and digging the bipod legs into the dirt to pin the weapon down as it is fired. The gunner fires in short bursts of five to seven rounds at a time to conserve ammunition, keep the weapon on target, and prevent the barrel from overheating. Under heavy firing, the SAW's barrel will glow red hot—the barrel may need to be removed and replaced with a spare before it begins to melt. The gun can't handle its own power.

On the other end of the spectrum of infantry machine guns is the M2 .50 caliber heavy machine gun, the "ma deuce." Mounted on military trucks, the .50 cal is the gun that turns a simple off-road truck into a piece of lethal machinery, the "gun truck." At eighty pounds—plus a fifty-pound tripod—the gun is a behemoth. To fire it, the gunner leans back in the turret to brace him or herself and thumbs down the trigger with both hands. The gun unleashes a powerful THUK THUK THUK as the rounds exit. The half inch–wide bullets can sail over a mile.

Machine guns changed warfare forever. In the late 1800s, the British Army used the Maxim gun to aid in their colonial conquest of Africa, allowing them to take on and defeat much larger forces. For a time, to the British at least, machine guns might have seemed like a weapon that lessened the cost of war. In World War I, however, both sides had machine guns and the result was bloodshed on an unprecedented scale. At the Battle of the Somme, Britain lost 20,000 men in a single day, mowed down by automatic weapons. Millions died in the trenches of World War I, an entire generation of young men.

Machine guns accelerated the process of killing by harnessing industrial age efficiency in the service of war. Men weren't merely killed by machine guns; they were mowed down, like McCormack's mechanical reaper cutting down stalks of grain. Machine guns are dumb weapons, however. They still have to be aimed by the user. Once initiated, they can continue firing on their own, but the guns have no ability to sense targets. In the twentieth century, weapons designers would take the next step to add rudimentary sensing technologies into weapons—the initial stages of intelligence.

THE FIRST "SMART" WEAPONS

From the first time a human threw a rock in anger until the twentieth century, warfare was fought with unguided weapons. Projectiles—whether shot from a sling, a bow, or a cannon—follow the laws of gravity once released. Projectiles are often inaccurate, and the degree of inaccuracy increases with range. With unguided weapons, destroying the enemy hinged on getting close enough to deliver overwhelming barrages of fire to blanket an area.

In World War II, as rockets, missiles, and bombs increased the range at which combatants could target one another—but not their accuracy—militaries sought to develop methods for precision guidance that would allow weapons to accurately strike targets from long distances. Some attempts to insert intelligence into weapons were seemingly comical, such as behaviorist B. F. Skinner's efforts to control a bomb by the pecking of a pigeon on a target image. Skinner's pigeon-guided bomb might have worked, but it never saw combat. Other attempts to implement onboard guidance measures did, giving birth to the first "smart" weapons: precision-guided munitions (PGMs).

The first successful PGM was the German G7e/T4 *Falke* ("Falcon") torpedo, introduced in 1943. The Falcon torpedo incorporated a new technological innovation: an acoustic homing seeker. Unlike regular torpedoes that traveled in a straight line and could very well miss a passing ship, the Falcon used its homing seeker to account for aiming errors. After traveling 400 meters from the German U-boat (submarine) that launched it, the Falcon would activate its passive acoustic sensors, listening for any nearby merchant ships. It would then steer toward any ships, detonating once it reached them.

The Falcon was used by only three U-boats in combat before being replaced by the upgraded G7es/T5 *Zaunkönig* ("Wren"), which had a faster motor and therefore could hit faster moving Allied navy ships in addition to merchant vessels. Using a torpedo that could home in on targets rather than travel in a straight line had clear military advantages, but it also immediately created complications. Two U-boats were sunk in December 1943 (*U-972*) and January 1944 (*U-377*) when their torpedoes circled back on them, homing in on the sound of their own propeller. In response to this problem, Germany instituted a 400-meter safety limit before activating the homing mechanism. To more fully mitigate against the dangers of a homing torpedo turning back on oneself, German U-boats also began incorporating a tactic of diving immediately after launch and then going completely silent.

The Allies quickly developed a countermeasure to the Wren torpedo. The Foxer, an acoustic decoy towed behind Allied ships, was intended to lure away the Wren so that it detonated harmlessly against the decoy, not the ship itself. The Foxer introduced other problems; it loudly broadcast the Allied convoy's position to other nearby U-boats, and it wasn't long before the Germans introduced the Wren II with an improved acoustic seeker. Thus began the arms race in smart weapons and countermeasures against them.

PRECISION-GUIDED MUNITIONS

The latter half of the twentieth century saw the expansion of PGMs like the Wren into sea, air, and ground combat. Today, they are widely used by militaries around the world in a variety of forms. Sometimes called "smart missiles" or "smart bombs," PGMs use automation to correct for aiming errors and help guide the munition (missile, bomb, or torpedo) onto the intended target. Depending on their guidance mechanism, PGMs can have varying degrees of autonomy.

Some guided munitions have very little autonomy at all, with the human controlling the aimpoint of the weapon throughout its flight. Command-guided weapons are manually controlled by a human remotely via a wire or radio link. For other weapons, a human operator "paints" the target with a laser or radar and the missile or bomb homes in on the laser or radar reflection. In these cases, the human doesn't directly control the movements of the munition, but does control the weapon's aimpoint in real time. This allows the human controller to redirect the munition in flight or potentially abort the attack.

Other PGMs are "autonomous" in the sense that they cannot be recalled once launched, but the munition's flight path and target are predetermined. These munitions can use a variety of guidance mechanisms. Nuclear-tipped ballistic missiles use inertial navigation systems consisting of gyroscopes and accelerometers to guide the missile to its preselected target point. Submarine-launched nuclear ballistic missiles use star-tracking celestial navigation systems to orient the missile, since the undersea launching point varies. Many cruise missiles look down to earth rather than up to the stars for navigation, using radar or digital scene mapping to follow the contours of the Earth to their preselected target. GPS-guided weapons rely on signals from the constellation of U.S. global positioning system satellites to determine their position and guidance to their target. While many of these munitions cannot be recalled or redirected after launch, the munitions do not have any freedom to select their own targets or even their own navigational route. In terms of the task they are performing, they have very little autonomy, even if they are beyond human control once launched. Their movements are entirely predetermined. The guidance systems, whether internal such as inertial navigation or external such as GPS, are only designed to ensure the munition stays on path to its preprogrammed target. The limitation of these guidance systems, however, is that they are only useful against fixed targets.

Homing weapons are a type of PGM used to track onto moving targets. By necessity since the target is moving, homing munitions have the ability to sense the target and adapt to its movements. Some homing munitions use passive sensors to detect their targets, as the Wren did. Passive sensors listen to or observe the environment and wait for the target to indicate its position by making noise or emitting in the electromagnetic spectrum. Active seekers send out signals, such as radar, to sense a target. An early U.S. active homing munition was the Bat anti-ship glide bomb, which had an active radar seeker to target enemy ships.

Some homing munitions "lock" onto a target, their seeker sensing the target before launch. Other munitions "lock on" after launch; they are launched with the seeker turned off, then it activates to begin looking for the moving target.

An attack dog is a good metaphor for a fire-and-forget homing munition. U.S. pilots refer to the tactic of launching the AIM-120 AMRAAM air-to-air missile in "lock on after launch" mode as going "maddog." After the weapon is released, it turns on its active radar seeker and begins looking for targets. Like a mad dog in a meat locker, it will go after the first target it sees. Similar to the problem German U-boats faced with the Wren, pilots need to take care to ensure that the missile doesn't track onto friendly targets. Militaries around the world often use tactics, techniques, and procedures ("TTPs" in military parlance) to avoid homing munitions turning back on themselves or other friendlies, such as the U-boat tactic of diving immediately after firing.

HOMING MUNITIONS HAVE LIMITED AUTONOMY

Homing munitions have some autonomy, but they are not "autonomous weapons"—a human still decides which specific target to attack. It's true that many homing munitions are "fire and forget." Once launched, they cannot be recalled. But this is hardly a new development in war. Projectiles have always been "fire and forget" since the sling and stone. Rocks, arrows, and bullets can't be recalled after being released either. What makes homing munitions different is their rudimentary onboard intelligence to guide their behavior. They can sense the environment (the target), determine the right course of action (which way to turn), and then act (maneuvering to hit the target). They are, in essence, a simple robot.

The autonomy given to a homing munition is tightly constrained, however. Homing munitions aren't designed to search for and hunt potential targets on their own. The munition simply uses automation to ensure it hits the specific target the human intended. They are like an attack dog sent by police to run down a suspect, not like a wild dog roaming the streets deciding on its own whom to attack.

In some cases, automation is used to ensure the munition does not hit unintended targets. The Harpoon anti-ship missile has a mode where the seeker stays off while the missile uses inertial navigation to fly a zigzag pattern toward the target. Then, at the designated location, the seeker activates to search for the intended target. This allows the missile to fly past other ships in the environment without engaging them. Because the autonomy of homing munitions is tightly constrained, the human operator needs to be aware of a specific target in advance. There must be some kind of intelligence informing the human of that *particular target* at that *specific time and place*. This intelligence could come from radars based on ships or aircraft, a ping on a submarine's sonar, information from satellites, or some other indicator. Homing munitions have a very limited ability in time and space to search for targets, and to launch one without knowledge of a specific target would be a waste. This means homing munitions must operate as part of a broader *weapon system* to be useful.

THE WEAPON SYSTEM

A weapon system consists of a sensor to search for and detect enemy targets, a decision-making element that decides whether to engage the target, and a munition (or other effector, such as a laser) that engages the target. Sometimes the weapon system is contained on a single platform, such as an aircraft. In the case of an Advanced Medium-Range Air-to-Air Missile (or AMRAAM), for example, the weapon system consists of the aircraft, radar, pilot, and missile. The radar searches for and senses the target, the human decides whether to engage, and the missile carries out the engagement. All of these elements are necessary for the engagement to work.

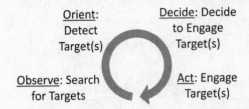

A *weapon system* consists of the components necessary to complete an entire combat OODA loop: searching for and detecting enemy targets, deciding whether to engage them, and engaging the targets.

Weapon System OODA Loop

In other cases, components of the weapon system may be distributed across multiple physical platforms. For example, a maritime patrol aircraft might detect an enemy ship and pass the location data to a nearby friendly ship, which launches a missile. Defense strategists refer to this larger, distributed system with multiple components as a *battle network*. Defense analyst Barry Watts described the essential role battle networks play in making precision-guided weapons effective:

Because "precision munitions" require detailed data on their intended targets or aim-points to be militarily useful—as opposed to wasteful— they require "precision information." Indeed, the tight linkage between guided munitions and "battle networks," whose primary reason for existence is to provide the necessary targeting information, was one of the major lessons that emerged from careful study of the US-led air campaign during Operation Desert Storm in 1991. ... [It] is *guided munitions together with the targeting networks that make these munitions "smart."* [emphasis in the original]

Automation is used for many engagement-related tasks in weapon systems and battle networks: finding, identifying, tracking, and prioritizing potential targets; timing when to fire; and maneuvering munitions to the target. For most weapon systems in use today, a human makes the decision whether to engage the target. If there is a human in the loop deciding which target(s) to engage, it is a *semiautonomous weapon system*.

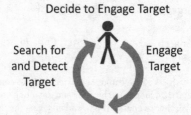

Decide to Engage Target

Search for and Detect Target

Engage Target

In semiautonomous weapons, automation may be used to search for and detect targets and carry out the engagement, but the human makes the decision to engage specific targets.

Semiautonomous Weapon System (human in the loop)

In *autonomous weapon systems*, the entire engagement loop—searching, detecting, deciding to engage, and engaging—is automated. (For ease of use, I'll often shorten "autonomous weapon system" to "autonomous weapon." The terms should be treated as synonymous, with the understanding that "weapon" refers to the entire system: sensor, decision-making element, and munition.) Most weapon systems in use today are semiautonomous, but a few cross the line to autonomous weapons.

SUPERVISED AUTONOMOUS WEAPON SYSTEMS

Because homing munitions can precisely target ships, bases, and vehicles, they can overwhelm defenders through saturation attacks with waves, or "salvos" of missiles. In an era of unguided ("dumb") munitions, defenders could simply ride out an enemy barrage, trusting that most of the incoming rounds would miss. With precision-guided ("smart") weapons, however, the defender must find a way to actively intercept and defeat incoming munitions before they impact. More automation—this time for defensive purposes—is the logical response.

At least thirty nations currently employ supervised autonomous weapon systems of various types to defend ships, vehicles, and bases from attack. Once placed in automatic mode and activated, these systems will engage incoming rockets, missiles, or mortars all on their own without further human intervention. Humans are on the loop, however, supervising their operation in real time.

Decide to Engage Target

Search for and Detect Target

Engage Target

Once activated, supervised autonomous weapons can search for, detect, decide to engage, and engage targets all on their own, but the human can intervene, if necessary.

Supervised Autonomous Weapon System (human on the loop)

These supervised autonomous weapons are necessary for circumstances in which the speed of engagements could overwhelm human operators. Like in the Atari game *Missile Command*, saturation attacks from salvos of simultaneous incoming threats could overwhelm human operators. Automated defenses are a vital part of surviving attacks from precision-guided weapons. They include ship-based defenses, such as the U.S. Aegis combat system and Phalanx Close-In Weapon System (CIWS); land-based air and missile defense systems, such as the U.S. Patriot; counter-rocket, artillery, and mortar systems such as the German MANTIS; and active protection systems for ground vehicles, such as the Israeli Trophy or Russian Arena system.

While these weapon systems are used for a variety of different situations—to defend ships, land bases, and ground vehicles—they operate in similar ways. Humans set the parameters of the weapon, establishing which threats the system should target and which it should ignore. Depending on the system, different rules may be used for threats coming from different directions, angles, and speeds. Some systems may have multiple modes of operation, allowing human in-the-loop (semiautonomous) or on-the-loop (supervised autonomous) control.

These automated defensive systems are autonomous weapons, but they have been used to date in very narrow ways—for immediate defense of human-occupied vehicles and bases, and generally targeting objects (like missiles, rockets, or aircraft), not people. Humans supervise their operation in real time and can intervene, if necessary. And the humans supervising the system are physically colocated with it, which means in principle they could physically disable it if the system stopped responding to their commands.

FULLY AUTONOMOUS WEAPON SYSTEMS

Do any nations have fully autonomous weapons that operate with no human supervision? Generally speaking, fully autonomous weapons are not in wide use, but there are a few select systems that cross the line. These weapons can search for, decide to engage, and engage targets on their own and no human can intervene. Loitering munitions are one example.

Loitering munitions can circle overhead for extended periods of time, searching for potential targets over a wide area and, once they find one, destroy it. Unlike homing munitions, loitering munitions do not require precise intelligence on enemy targets before launch. Thus, a loitering munition is a complete "weapon system" all on its own. A human can launch a loitering munition into a "box" to search for enemy targets without knowledge of any specific targets beforehand. Some loitering muni-

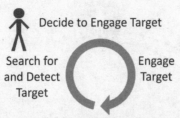

Once activated, fully autonomous weapons can search for, detect, decide to engage, and engage targets all on their own and the human cannot intervene.

Fully Autonomous Weapon System (human out of the loop)

tions keep humans in the loop via a radio connection to approve targets before engagement, making them semiautonomous weapon systems. Some, however, are fully autonomous.

The Israeli Harpy is one such weapon. No human approves the specific target before engagement. The Harpy has been sold to several countries—Chile, China, India, South Korea, and Turkey—and the Chinese are reported to have reverse engineered their own variant.

HARM vs. Harpy

	Type of weapon	Target	Time to search	Distance	Degree of autonomy
HARM	Homing missile	Radars	Approx. 4.5 minutes	90+ km	Semiautonomous weapon
Harpy	Loitering munition	Radars	2.5 hours	500 km	Fully autonomous weapon

The difference between a fully autonomous loitering munition and a semiautonomous homing munition can be illustrated by comparing the Harpy with the High-speed Anti-Radiation Missile (HARM). Both go after the same type of target (enemy radars), but their freedom to search for targets is massively different. The semiautonomous HARM has a range of 90-plus kilometers and a top speed of over 1,200 kilometers per hour, so it is only airborne for approximately four and a half minutes. Because it

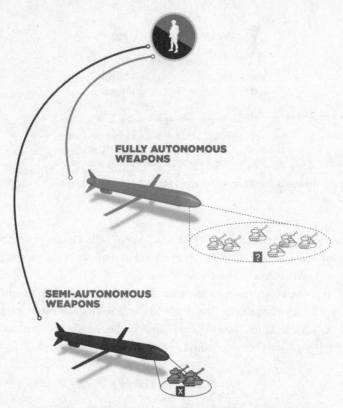

Semiautonomous vs. Fully Autonomous Weapons *For semiautonomous weapons, the human operator launches the weapon at a specific known target or group of targets. The human chooses the target and the weapon carries out the attack. Fully autonomous weapons can search for and find targets over a wide area, allowing human operators to launch them without knowledge of specific targets in advance. The human decides to launch the fully autonomous weapon, but the weapon itself chooses the specific target to attack.*

cannot loiter, the HARM has to be launched at a specific enemy radar in order to be useful. The Harpy can stay aloft for over two and a half hours covering up to 500 kilometers of ground. This allows the Harpy to operate independently of a broader battle network that gives the human targeting information before launch. The human launching the Harpy decides to destroy *any* enemy radars within a general area in space and time, but the Harpy itself chooses the specific radar it destroys.

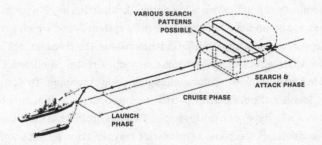

Tomahawk Anti-Ship Missile Mission Profile *A typical mission for a Tomahawk Anti-Ship Missile (TASM). After being launched from a ship or submarine, the TASM would cruise to the target area. Once over the target area, it would fly a search pattern to look for targets and, if it found one, attack the target on its own.*

Despite conventional thinking that fully autonomous weapons are yet to come, isolated cases of fully autonomous loitering munitions go back decades. In the 1980s, the U.S. Navy deployed a loitering anti-ship missile that could hunt for, detect, and engage Soviet ships on its own. The Tomahawk Anti-Ship Missile (TASM) was intended to be launched over the horizon at possible locations of Soviet ships, then fly a search pattern over a wide area looking for their radar signatures. If it found a Soviet ship, TASM would attack it. (Despite the name, the TASM was quite different from the Tomahawk Land Attack Missile [TLAM], which uses digital scene mapping to follow a preprogrammed route to its target.) The TASM was taken out of Navy service in the early 1990s. While it was never fired in anger, it has the distinction being the first operational fully autonomous weapon, a significance that was not recognized at the time.

In the 1990s, the United States began development on two experimental loitering munitions: Tacit Rainbow and the Low Cost Autonomous Attack System (LOCAAS). Tacit Rainbow was intended to be a persistent anti-radiation weapon to target land-based radars, like the Harpy. LOCAAS had an even more ambitious goal: to search for and destroy enemy tanks, which are harder targets than radars because they are not emitting in the electromagnetic spectrum. Neither Tacit Rainbow nor LOCAAS were ever deployed; both were cancelled while still in development.

These examples shine a light on a common misperception about autonomous weapons, which is the notion that intelligence is what makes a weapon "autonomous." How intelligent a system is and which tasks it performs autonomously are different dimensions. It is freedom, not intelligence, that defines an autonomous weapon. Greater intelligence can be added into weapons without changing their autonomy. To date, the target identification algorithms used in autonomous and semiautonomous weapons have been fairly simple. This has limited the usefulness of fully autonomous weapons, as militaries may not trust giving a weapon very much freedom if it isn't very intelligent. As machine intelligence advances, however, autonomous targeting will become technically possible in a wider range of situations.

UNUSUAL CASES— MINES, ENCAPSULATED TORPEDO MINES, AND SENSOR FUZED WEAPON

There are a few unusual cases of weapons that blur the lines between semiautonomous and fully autonomous weapons: mines and the Sensor Fuzed Weapon deserve special mention.

Placed on land or at sea, mines wait for their target to approach, at which point the mine explodes. While mines are automatic devices that will detonate on their own once triggered, they have no freedom to maneuver and search for targets. They simply sit in place. (For the most part—some naval mines can drift with the current.) They also generally have very limited methods for "deciding" whether or not to fire. Mines typically have a simple method for sensing a target and, when the threshold for the sensor is reached, the mine explodes. (Some naval mines and antitank mines employ a counter so that they will let the first few targets pass unharmed before detonating against a ship or vehicle later in the convoy.) Mines deserve special mention because their freedom in time is virtually unbounded, however. Unless specifically designed to self-deactivate after a certain period of time, mines can lay in wait for years, sometimes remaining active long after a war has ended.

The fact that mines are often unbounded in time has had devastating humanitarian consequences. By the mid-1990s, an estimated more than

110 million land mines lay hidden in sixty-eight countries around the globe, accumulated from scores of conflicts. Land mines have killed thousands of civilians, many of them children, and maimed tens of thousands more, sparking the global movement to ban land mines that culminated in the Ottawa Treaty in 1997. Adopted by 162 nations, the Ottawa Treaty prohibits the production, stockpiling, transfer, or use of antipersonnel land mines. Antitank land mines and naval mines are still permitted.

Mines can sense and act on their own, but do not search for targets. Encapsulated torpedo mines are a special type of naval mine that acts more like an autonomous weapon, however. Rather than simply exploding once activated, encapsulated torpedo mines release a torpedo that homes in on the target. This gives encapsulated torpedo mines the freedom to engage targets over a much wider area than a traditional mine, much like a loitering munition. The U.S. Mk 60 CAPTOR encapsulated torpedo mine had a published range of 8,000 yards. By contrast, a ship would have to pass over a regular mine for it to detonate. Even though encapsulated torpedo mines are moored in place to the seabed, their ability to launch a torpedo to chase down targets gives them a much greater degree of autonomy in space than a traditional naval mine. As with loitering munitions, examples of encapsulated torpedo mines are rare. The U.S. CAPTOR mine was in service for throughout the 1980s and 1990s but has been retired. The only encapsulated torpedo mine still in service is the Russian PMK-2, used by Russia and China.

The Sensor Fuzed Weapon (SFW) is an air-delivered antitank weapon that defies categorization. Released from an aircraft, an SFW can destroy an entire column of enemy tanks within seconds. The SFW functions through a series of Rube Goldberg machine–like steps: First, the aircraft releases a bomb-shaped canister than glides toward the target area. As the canister approaches the target area, the outer casing releases, exposing ten submunitions which are ejected from the canister. Each submunition releases a drogue parachute slowing its descent. At a certain height above the ground, the submunition springs into action. It opens its outer case, exposing four internally held "skeets" which are then rotated out of the inner casing and exposed. The parachute releases and the submunition fires retrojets that cause it to climb in altitude while spinning furiously. The hockey-puck-shaped skeets are then released, flung outward violently from the force of the spinning. Each skeet carries onboard laser and infra-red sensors that it uses to search for targets beneath it. Upon detecting a vehicle beneath it, the skeet fires an explosively formed penetrator—a metal slug—downward into the vehicle. The metal slug strikes the vehi-cle on top, where armored vehicles have the thinnest armor, destroying the vehicle. In this manner, a single SFW can take out a group of tanks or other armored vehicles simultaneously, with the skeets targeting each vehicle precisely.

Similar to the distinction between Harpy and HARM, the critical vari-able in the evaluating SFW's autonomy is its freedom in time and space. While the weapon distributes forty skeets over several acres, the time the weapon can search for targets is minuscule. Each skeet can hover with its sensor active for only a few seconds before firing. Unlike the Harpy, the SFW cannot loiter for an extended period over hundreds of kilometers. The human launching the SFW must know that there is a group of tanks at a particular point in space and time. Like a homing munition, the SFW must be part of a wider weapon system that provides targeting data in order to be useful. The SFW is different than a traditional homing muni-tion, because the SFW can hit multiple objects. This makes the SFW like a salvo of forty homing munitions launched at a tightly geographically clus-tered set of targets.

PUSHING "START"

Autonomous weapons are defined by the ability to complete the engage-ment cycle—searching for, deciding to engage, and engaging targets—on their own. Autonomous weapons, whether supervised or fully autono-mous, are still built and put into operation by humans, though. Humans are involved in the broader process of designing, building, testing, and deploying weapons.

The fact that there are humans involved at some stage does not change the significance of a weapon that could complete engagements entirely on its own. Even the most highly autonomous system would still have been borne out of a process initiated by humans at some point. In the climactic scene of *Terminator 3: Rise of the Machines*, an Air Force general pushes the button to start Skynet. (Absurdly, this is done with an old "EXECUTE

Y/N?" prompt like the kind used in MS-DOS in the 1980s.) From that point forward, Skynet embarks on its path to exterminate humanity, but at least at the beginning a human was in the loop. The question is not whether there was ever a human involved, but rather how much freedom the system has once it is activated.

WHY AREN'T THERE MORE AUTONOMOUS WEAPONS?

Automation has been used extensively in weapons around the world for decades, but the amount of freedom given to weapons has been, up to now, fairly limited. Homing munitions have seekers, but their ability to search for targets is narrowly constrained in time and space. Supervised autonomous weapons have only been used for limited defensive purposes. The technology to build simple fully autonomous loitering munitions like TASM and Harpy has existed for decades, yet there is only one example in use today.

Why aren't there more fully autonomous weapons? Homing munitions and even semiautonomous loitering munitions are widely used, but militaries have not aggressively pursued fully autonomous loitering munitions. The U.S. experience with TASM may shed some light on why. TASM was in service in the U.S. Navy from 1982 to 1994, when it was retired. To understand better why TASM was taken out of service, I spoke with naval strategist Bryan McGrath.

McGrath, a retired Navy officer, is well known in Washington defense circles. He is a keen strategist and unabashed advocate of sea power who thinks deeply about the past, present, and future of naval warfare. McGrath is familiar with TASM and other anti-ship missiles such as the Harpoon, and was trained on TASM in the 1980s when it was in the fleet.

McGrath explained to me that TASM could outrange the ship's own sensors. That meant that initial targeting had to come from another sensor, such as a helicopter or maritime patrol aircraft that detected an enemy ship. The problem, as McGrath described it, was a "lack of confidence in how the targeting picture would change from the time you fired the missile until you got it downrange." Because the target could move, unless there was an "active sensor" on the target, such as a helicopter with eyes on the target the whole time, the area of uncertainty of where the target was would grow over time.

The ability of the TASM to search for targets over a wide area mitigated, to some extent, this large area of uncertainty. If the target had moved, the TASM could simply fly a search pattern looking for it. But TASM didn't have the ability to accurately discriminate between enemy ships and merchant vessels that just happened to be in its path. As the search area widened, the risk increased that the TASM might run across a merchant ship and strike it instead. In an all-out war with the Soviet Navy, that risk might be acceptable, but in any situations short of that, getting approval to shoot the TASM was unlikely. TASM was, according to McGrath, "a weapon we just didn't want to fire."

Another factor was that if a TASM was launched and there wasn't a valid target within the search area of the weapon, the weapon would be wasted. McGrath would be loath to launch a weapon on scant evidence that there was a valid target in the search area. "I would want to know that there's something there, even if there was some kind of end-game autonomy in place." Why? "Because the weapons cost money," he said, "and I don't have a lot of them. And I may have to fight tomorrow."

Modern missiles can cost upwards of a million dollars apiece. As a practical matter, militaries will want to know that there is, in fact, a valid enemy target in the area before using an expensive weapon. One of the reasons militaries have not used fully autonomous loitering munitions more may be the fact that the advantage they bring—the ability to launch a weapon without precise targeting data in advance—may not be of much value if the weapon is not reusable, since the weapon could be wasted.

FUTURE WEAPONS

The trend of creeping automation that began with Gatling's gun will continue. Advances in artificial intelligence will enable smarter weapons, which will be capable of more autonomous operation. At the same time, another facet of the information revolution is greater networking. German U-boats couldn't control the Wren torpedo once it was launched, not because they didn't want to; they simply had no means to do so.

Modern munitions are increasingly networked to allow them to be

controlled or retargeted after they've been launched. Wire-guided munitions have existed for decades, but are only feasible for short distances.
Long-range weapons are now incorporating datalinks to allow them to
be controlled via radio communication, even over satellites. The Block
IV Tomahawk Land Attack Missile (TLAM-E, or Tactical Tomahawk)
includes a two-way satellite communications link that allows the weapon
to be retargeted in flight. The Harpy 2, or Harop, has a communications
link that allows it to be operated in a human-in-the-loop mode so that the
human operator can directly target the weapon.

When I asked McGrath what feature he would most desire in a future
weapon, it wasn't autonomy—it was a datalink. "You've got to talk to the
missile," he explained. "The missiles have to be part of a network." Connecting the weapons to the network would allow you to send updates
on the target while in flight. As a result, "confidence in employing that
weapon would dramatically increase."

A networked weapon is a far more valuable weapon than one that is on
its own. By connecting a weapon to the network, the munition becomes
part of a broader system and can harness sensor data from other ships,
aircraft, or even satellites to assist its targeting. Additionally, the commander can keep control of the weapon while in flight, making it less
likely to be wasted. One advantage to the networked Tactical Tomahawk,
for example, is the ability for humans to use sensors on the missile to
do battle damage assessment (BDA) of potential targets before striking.
Without the ability to conduct BDA of the target, commanders might have
to launch several Tomahawks at a target to ensure its destruction, since
the first missile might not completely destroy the target. Onboard BDA
allows the commander to look at the target after the first missile hits. If
more strikes are needed, more missiles can be used. If not, then subsequent missiles can be diverted in flight to secondary targets.

Everything has a countermeasure, though, and increased networking
runs counter to another trend in warfare, the rise of electronic attack.
The more that militaries rely on the electromagnetic spectrum for communications and sensing targets, the more vital it will be to win the invisible electronic war of jamming, spoofing, and deception fought through
the electromagnetic spectrum. In future wars between advanced militaries, communications in contested environments is by no means assured.

Advanced militaries have ways of communicating that are resistant to jamming, but they are limited in range and bandwidth. When communications are denied, missiles or drones will be on their own, reliant on their onboard autonomy.

Due to their expensive cost, even highly advanced loitering munitions are likely to fall into the same trap as TASM, with commanders hesitant to fire them unless targets are clearly known. But drones change this equation. Drones can be launched, sent on patrol, and can return with their weapons unused if they do not find any targets. This simple feature—reusability—dramatically changes how a weapon could be used. Drones could be sent to search over a wide area in space and time to hunt for enemy targets. If none were found, the drone could return to base to hunt again another day.

More than ninety nations and non-state groups already have drones, and while most are unarmed surveillance drones, an increasing number are armed. At least sixteen countries already possess armed drones and another dozen or more nations are working on arming their drones. A handful of countries are even pursuing stealth combat drones specifically designed to operate in contested areas. For now, drones are used as part of traditional battle networks, with decision-making residing in the human controller. If communications links are intact, then countries can keep a human in the loop to authorize targets. If communications links are jammed, however, what will the drones be programmed to do? Will they return home? Will they carry out surveillance missions, taking pictures and reporting back to their human operators? Will the drones be authorized to strike fixed targets that have been preauthorized by humans, much like cruise missiles today? What if the drones run across emerging targets of opportunity that have not been authorized in advance by a human—will they be authorized to fire? What if the drones are fired upon? Will they be allowed to fire back? Will they be authorized to shoot first?

These are not hypothetical questions for the future. Engineers around the globe are programming the software for these drones today. In their hands, the future of autonomous weapons is being written.

PART II

Building the
Terminator

4

THE FUTURE BEING BUILT TODAY

AUTONOMOUS MISSILES, DRONES, AND ROBOT SWARMS

ew actors loom larger in the robotics revolution than the U.S. Department of Defense. The United States spends 600 billion dollars annually on defense, more than the next seven countries combined. Despite this, U.S. defense leaders are concerned about the United States falling behind. In 2014, the United States launched a "Third Offset Strategy" to reinvigorate America's military technological advantage. The name harkens back to the first and second "offset strategies" in the Cold War, where the U.S. military invested in nuclear weapons in the 1950s and later precision-guided weapons in the 1970s to offset the Soviet Union's numerical advantages in Europe. The centerpiece of DoD's Third Offset Strategy is robotics, autonomy, and human-machine teaming.

Many applications of military robotics and autonomy are noncontroversial, such as uninhabited logistics convoys, tanker aircraft, or reconnaissance drones. Autonomy is also increasing in weapon systems, though, with next-generation missiles and combat aircraft pushing the boundaries of autonomy. A handful of experimental programs show how the U.S. military is thinking about the role of autonomy in weapons. Collectively, they are laying the foundations for the military of the future.

SALTY DOGS: THE X-47B DRONE

The X-47B experimental drone is one of the world's most advanced aircraft. Only two have been ever built, named Salty Dog 501 and Salty Dog 502. With a sleek bat-winged shape that looks like something out of the 1980s sci-fi flick *Flight of the Navigator*, the X-47B practically screams "the future is here." In their short life-span as experimental aircraft from 2011 to 2015, Salty Dog 501 and 502 repeatedly made aviation history. The X-47B was the first uninhabited (unmanned) aircraft to autonomously take off and land on an aircraft carrier and the first uninhabited aircraft to autonomously refuel from another plane while in flight. These are key milestones to enabling future carrier-based combat drones. However, the X-47B was not a combat aircraft. It was an experimental "X-plane," a demonstration program designed to mature technologies for a follow-on aircraft. The focus of technology development was automating the physical movement of the aircraft—takeoff, landing, flight, and aerial refueling. The X-47B did not carry weapons or sensors that would permit it to make engagements.

The Navy has stated their first operational carrier-based drone will be the MQ-25 Stingray, a future aircraft that is still on the drawing board. While the specific design has yet to be determined, the MQ-25 is envisioned primarily as a tanker, ferrying fuel for manned combat aircraft such as the F-35 Joint Strike Fighter, with possibly a secondary role in reconnaissance. It is not envisioned as a combat aircraft. In fact, over the past decade the Navy has moved steadily away from any notion of uninhabited aircraft in combat roles.

The origin of the X-47 was in the Joint Unmanned Combat Air Systems (J-UCAS) program, a joint program between DARPA, the Navy, and the Air Force in the early 2000s to design an uninhabited combat aircraft. J-UCAS led to the development of two experimental X-45A aircraft, which in 2004 demonstrated the first drone designed for combat missions. Most drones today are intended for surveillance missions, which means they are designed for soaring and staying aloft for long periods of time. The X-45A, however, sported the same sharply angled wings and smooth top surfaces that define stealth aircraft like the F-117, B-2 bomber, and F-22 fighter. Designed to penetrate enemy air defenses, the intent was for the X-45A to perform close in jamming and strike missions in support of manned aircraft. The program was never completed, though. In the Pentagon's 2006 Quadrennial Defense Review, a major strategy and budget review conducted every four years, the J-UCAS program was scrapped and restructured.

J-UCAS's cancellation was curious because it came at the height of the post-9/11 defense budget boom and at a time when the Defense Department was waking up to the potential of robotic systems more broadly. Even while the military was deploying thousands of drones to Iraq and Afghanistan, the Air Force was highly resistant to the idea of uninhabited aircraft taking on combat roles in future wars. In the ensuing decade since J-UCAS's cancellation, despite repeated opportunities, the Air Force has not restarted a program to build a combat drone. Drones play important roles in reconnaissance and counterterrorism, but when it comes to dogfighting against other enemy aircraft or taking down another country's air defense network, those missions are currently reserved for traditional manned aircraft.

The reality is that what may look from the outside like an unmitigated rush toward robotic weapons is, in actuality, a much more muddled picture inside the Pentagon. There is intense cultural resistance within the U.S. military to handing over combat jobs to uninhabited systems. Robotic systems are frequently embraced for support roles such as surveillance or logistics, but rarely for combat applications. The Army is investing in logistics robots, but not frontline armed combat robots. The Air Force uses drones heavily for surveillance, but is not pursing air-to-air combat drones. Pentagon vision documents such as the Unmanned Systems Roadmaps or the Air Force's 2013 *Remotely Piloted Aircraft Vector* often articulate ambitious dreams for robots in a variety of roles, but these documents are often disconnected from budgetary realities. Without funding, these visions are more hallucinations than reality. They articulate goals and aspirations, but do not necessarily represent the most likely future path.

The downscoping of the ambitious J-UCAS combat aircraft to the plodding MQ-25 tanker is a great case in point. In 2006 when the Air Force abandoned the J-UCAS experimental drone program, the Navy continued a program to develop a combat aircraft. The X-47B was supposed to mature the technology for a successor stealth drone, but in a series of internal Pentagon memoranda issued in 2011 and 2012, Navy took a sharp turn away from a combat aircraft. Designs were scaled back in favor of a less ambitious nonstealthy surveillance drone. Concept sketches shifted from looking like the futuristic sleek and stealthy X-45A and X-47B to the more pedestrian Predator and Reaper drones, already over a decade old at that point. The Navy, it appears, wasn't immune to the same cultural resistance to combat drones found in the Air Force.

The Navy's resistance to developing an uninhabited combat aerial vehicle (UCAV) is particularly notable because it comes in the face of pressure from Congress and a compelling operational need. China has developed anti-ship ballistic and cruise missiles that can outrange carrier-based F-18 and F-35 aircraft. Only uninhabited aircraft, which can stay aloft far longer than would be possible with a human in the airplane, have sufficient range to keep the carrier relevant in the face of advanced Chinese missiles. Sea power advocates outside the Navy in Congress and think tanks have argued that without a UCAV on board, the aircraft carrier itself would be of limited utility against a high-technology opponent. Yet the Navy's current plan is for its carrier-based drone, the MQ-25, to ferry gas for human-inhabited jets. For now, the Navy is deferring any plans for a future UCAV.

The X-47B is an impressive machine and, to an outside observer, it may seem to portend a future of robot combat aircraft. Its appearance belies the reality that within the halls of the Pentagon, however, there is little enthusiasm for combat drones, much less fully autonomous ones that would target on their own. Neither the Air Force nor the Navy have programs under way to develop an operational UCAV. The X-47B is a bridge to a future that, at least for now, doesn't exist.

THE LONG-RANGE ANTI-SHIP MISSILE

The Long-Range Anti-Ship Missile (LRASM) is a state-of-the-art missile pushing the boundaries of autonomy. It is a joint DARPA-Navy-Air Force project intended to fill a gap in the U.S. military's ability to strike enemy ships at long ranges. Since the retirement of the TASM, the Navy has relied on the shorter-range Harpoon anti-ship missile, which has a range of only 67 nautical miles. The LRASM, on the other hand, can fly up to 500 nautical miles. LRASM also sports a number of advanced survivability features, including the ability to autonomously detect and evade threats while en route to its target.

LRASM uses autonomy in several novel ways, which has alarmed some opponents of autonomous weapons. The LRASM has been featured in no less than three *New York Times* articles, with some critics claiming it exhibits "artificial intelligence outside human control." In one of the articles, Steve Omohundro, a physicist and leading thinker on advanced artificial intelligence, stated "an autonomous weapons arms race is already taking place." It is a leap, though, to assume that these advances in autonomy mean states intend to pursue autonomous weapons that would hunt for target on their own.

The actual technology behind LRASM, while cutting edge, hardly warrants these breathless treatments. LRASM has many advanced features, but the critical question is who chooses LRASM's targets—a human or the missile itself? On its website, Lockheed Martin, the developer of LRASM, states:

> LRASM employs precision routing and guidance. ... The missile employs a multi-modal sensor suite, weapon data link, and enhanced digital anti-jam Global Positioning System to detect and destroy specific targets within a group of numerous ships at sea.... This advanced guidance operation means the weapon can use gross target cueing data to find and destroy its pre-defined target in denied environments.

While the description speaks of advanced precision guidance, it doesn't say much that would imply artificial intelligence that would hunt for targets on its own. What was the genesis of the criticism? Well ... Lockheed used to describe LRASM differently.

Before the first *New York Times* article in November 2014, Lockheed's description of LRASM boasted much more strongly of its autonomous features. It used the word "autonomous" three times in the description, describing it as an "autonomous, precision-guided anti-ship" missile that "cruises autonomously" and has an "autonomous capability." What exactly the weapon was doing autonomously was somewhat ambiguous, though.

After the first *New York Times* article, the description changed, substituting "semi-autonomous" for "autonomous" in multiple places. The new description also clarified the nature of the autonomous features, stating "The semi-autonomous guidance capability gets LRASM safely to the enemy area." Eventually, even the words "semi-autonomous" were removed, leading to the description online today which only speaks of "precision routing and guidance" and "advanced guidance." Autonomy isn't mentioned at all.

What should we make of this shifting story line? Presumably the weapon's functionality hasn't changed, merely the language used to describe it. So how autonomous is LRASM?

Lockheed has described LRASM as using "gross target cueing data to find and destroy its predefined target in denied environments." If "predefined" target means that the specific target has been chosen in advance by a human operator, LRASM would be a semiautonomous weapon. On the other hand, if "predefined" means that the human has chosen only a general class of targets, such as "enemy ships," and given the missile the freedom to hunt for these targets over a wide area and engage them on its own, then it would be an autonomous weapon.

Helpfully, Lockheed posted a video online that explains LRASM's functionality. In a detailed combat simulation, the video shows precisely which engagement-related functions would be done autonomously and which by a human. In the video, a satellite identifies a hostile surface action group (SAG)—a group of enemy ships—and relays their location to a U.S. destroyer. The video shows a U.S. sailor looking at the enemy ships on his console. He presses a button and two LRASMs leap from their launching tubes in a blast of flame into the air. The text on the video explains the LRASMs have been launched against the enemy cruiser, part of the hostile SAG. Once airborne, the LRASMs establish a line-of-sight datalink with the ship. As they continue to fly out toward the enemy SAG, they transition to satellite communications. A U.S. F/A-18E fighter aircraft then fires a third LRASM (this one air-launched) against an enemy destroyer, another ship in the SAG. The LRASMs enter a "communications and GPS-denied environment." They are now on their own.

The LRASMs maneuver via planned navigational routing, moving from one predesignated way point to another. Then, unexpectedly, the LRASMs encounter a "pop-up threat." In the video, a large red bubble appears in the sky, a no-go zone for the missiles. The missiles now execute "autonomous routing," detouring around the red bubble on their own. A second pop-up threat appears and the LRASMs modify their route again, moving around the threat to continue on their mission.

As the LRASMs approach their target destination, the video shifts to a new perspective focusing on a single missile, simulating what the missile's sensors see. Five dots appear on the screen representing objects detected by the missile's sensors, labeled "ID:71, ID:56, ID:44, ID:24, ID:19." The missile begins a process the video calls "organic [area of uncertainty] reduction." That's military jargon for a bubble of uncertainty. When the missile was launched, the human launching it knew where the enemy ship was located, but ships move. By the time the missile arrives at the ship, the ship could be somewhere else. The "area of uncertainty" is the bubble within which the enemy ship could be, a bubble that gets larger over time.

Since there could be multiple ships in this bubble, the LRASM begins to narrow down its options to determine which ship was the one it was sent to destroy. How this occurs is not specified, but on the video a large "area of uncertainty" appears around all the dots, then quickly shrinks to surround only three of them: ID:44, ID:24, and ID:19. The missile then moves to the next phase of its targeting process: "target classification." The missile scans each object, finally settling on ID:24. "Criteria match," the video states, "target classified." ID:24, the missile has determined, is the ship it was sent to destroy.

Having zeroed in on the right target, the missiles begin their final maneuvers. Three LRASMs descend below the enemy ships' radars to skim just above the water's surface. On their final approach, the missiles scan the ships one last time to confirm their targets. The enemy ships fire their defenses to try to hit the incoming missiles, but it's too late. Two enemy ships are hit.

The video conveys the LRASM's impressive autonomous features, but is it an autonomous weapon? The autonomous/semiautonomous/advanced guidance described on the website is clearly on display. In the video, midway through the flight the missiles enter a "communications and GPS denied environment." Within this bubble, the missiles are on their own; they cannot call back to human controllers. Any actions they

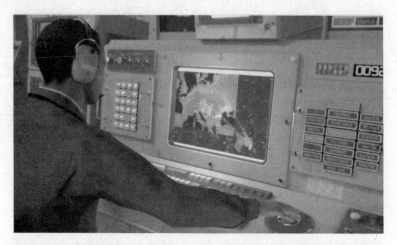

Screenshots from LRASM Video *In a video simulation depicting how the LRASM functions, a satellite transmits the location of enemy ships to a human, who authorizes the attack on those specific enemy ships.*

The LRASMs are launched against specific enemy ships, in this case a "SAG Cruiser."

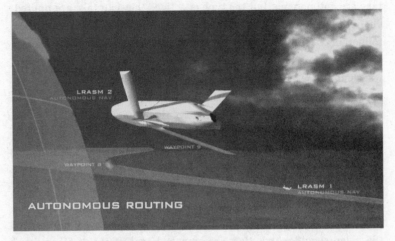

While en route to their human-designated targets, the LRASMs employ autonomous routing around pop-up threats (shown as a bubble).

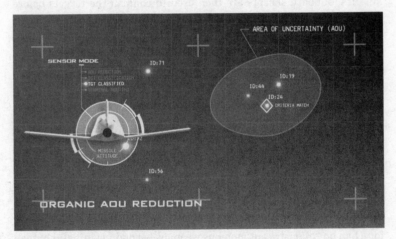

Because the human-designated target is a moving ship, by the time the LRASM arrives at the target area there is an "area of uncertainty" that defines the ship's possible location. Multiple objects are identified within this area of uncertainty. LRASM uses its onboard ("organic") sensors to reduce the area of uncertainty and identify the human-designated target. LRASM confirms "ID:24" is the target it was sent to destroy. While the missile has many advanced features, it does not choose its own target. The missile uses its sensors to confirm the human-selected target.

take are autonomous, but the type of actions they can take are limited. Just because the weapon is operating without a communications link to human controllers doesn't mean it has the freedom to do anything it wishes. The missile isn't a teenager whose parents have left town for the weekend. It has only been programmed to perform certain tasks autonomously. The missile can identify pop-up threats and autonomously reroute around them, but it doesn't have the freedom to choose its own targets. It can identify and classify objects to confirm which object was the one it was sent to destroy, but that isn't the same as being able to *choose* which target to destroy.

It is the human who decides which enemy ship to destroy. The critical point in the video isn't at the end of the missile's flight as it zeroes in on the ship—it's at the beginning. When the LRASMs are launched, the video specifies that they are launched against the "SAG cruiser" and "SAG destroyer." The humans are launching the missiles at specific ships, which the humans have tracked and identified via satellites. The missiles' onboard sensors are then used to confirm the targets before completing the attack. LRASM is only one piece of a weapon *system* that consists of the satellite, ship/aircraft, human, and missile. The human is "in the loop," deciding which specific targets to engage in the broader decision cycle of the weapon system. The LRASM merely carries out the engagement.

BREAKING THE SPEED LIMIT: FAST LIGHTWEIGHT AUTONOMY

Dr. Stuart Russell is a pioneering researcher in artificial intelligence. He literally wrote the textbook that is used to teach AI researchers around the world. Russell is also one of the leaders in the AI community calling for a ban on "offensive autonomous weapons beyond meaningful human control." One research program Russell has repeatedly raised concerns about is DARPA's Fast Lightweight Autonomy (FLA).

FLA is a research project to enable high-speed autonomous navigation in congested environments. Researchers outfit commercial off-the-shelf quadcopters with custom sensors, processors, and algorithms with the goal of making them autonomously navigate through the interior of a cluttered warehouse at speeds up to forty-five miles per hour. In a press release, DARPA compared the zooming quadcopters to the Millennium Falcon zipping through the hull of a crashed Star Destroyer in *Star Wars: The Force Awakens*. (I would have gone with the Falcon maneuvering through the asteroid field in *The Empire Strikes Back* . . . or the Falcon zipping through the interior of Death Star II in *The Return of the Jedi*. But you get the idea: fast = awesome.) In a video accompanying the press release, shots of the flying quadcopters are set to peppy instrumental music. It's incongruous because in the videos released so far the drones aren't actually moving through obstacles at 45 mph . . . yet. For now, they are creeping their way around obstacles, but they are doing so fully autonomously. FLA's quadcopters use a combination of high-definition cameras, sonar, and laser light detection and ranging (LIDAR) to sense obstacles and avoid them all on their own.

Autonomous navigation around obstacles, even at slow speeds, is no mean feat. The quadcopter's sensors need to detect potential obstacles and track them as the quadcopter moves, a processor-hungry task. Because the quadcopter can only carry so much computing power, it is limited in how quickly it can process the obstacles it sees. The program aims in the coming months to speed it up. As DARPA program manager Mark Micire explained in a press release, "The challenge for the teams now is to advance the algorithms and onboard computational efficiency to extend the UAVs' perception range and compensate for the vehicles' mass to make extremely tight turns and abrupt maneuvers at high speeds." In other words, to pick up the pace.

FLA's quadcopters don't look menacing, but it isn't because of the up-tempo music or the cutesy *Star Wars* references. It's because there's nothing in FLA that has anything to do with weapons engagements. Not only are the quadcopters unarmed, they aren't performing any tasks associated with searching for and identifying targets. DARPA explains FLA's intended use as indoor reconnaissance:

> FLA technologies could be especially useful to address a pressing surveillance shortfall: Military teams patrolling dangerous overseas urban environments and rescue teams responding to disasters such as earthquakes or floods currently can use remotely piloted unmanned

aerial vehicles (UAVs) to provide a bird's-eye view of the situation, but to know what's going on inside an unstable building or a threatening indoor space often requires physical entry, which can put troops or civilian response teams in danger. The FLA program is developing a new class of algorithms aimed at enabling small UAVs to quickly navigate a labyrinth of rooms, stairways and corridors or other obstacle-filled environments without a remote pilot.

To better understand what FLA was doing, I caught up with one of the project's research teams from the University of Pennsylvania's General Robotics Automation Sensing and Perception (GRASP) lab. Videos of GRASP's nimble quadcopters have repeatedly gone viral online, showing swarms of drones artfully zipping through windows, seemingly dancing in midair, or playing the James Bond theme song on musical instruments. I asked Dr. Daniel Lee and Dr. Vijay Kumar, the principal investigators of GRASP's work on FLA, what they thought about the criticism that the program was paving the way toward autonomous weapons. Lee explained that GRASP's research was "very basic" and focused on "fundamental capabilities that are generally applicable across all of robotics, including industrial and consumer uses." The technology GRASP was focused on "localization, mapping, obstacle detection and high-speed dynamic navigation." Kumar added that their motivations for this research were "applications to search and rescue and first response where time-critical response and navigation at high speeds are critical."

Kumar and Lee aren't weapons designers, so it may not be at the forefront of their minds, but it's worth pointing out that the technologies FLA is building aren't even the critical ones for autonomous weapons. Certainly, fast-moving quadcopters could have a variety of applications. Putting a gun or bomb on an FLA-empowered quadcopter isn't enough to make it an autonomous weapon, however. It would still need the ability to find targets on its own. Depending on the intended target, that may not be particularly complicated, but at any rate that's a separate technology. All FLA is doing is making quadcopters maneuver faster indoors. Depending on one's perspective, that could be cool or could be menacing, but either way FLA doesn't have anything more to do with autonomous weapons than self-driving cars do.

DARPA's description of FLA didn't seem to stack up against Stuart Russell's criticism. He has written that FLA and another DARPA program "foreshadow planned uses of [lethal autonomous weapon systems]." I first met Russell on the sidelines of a panel we both spoke on at the United Nations meetings on autonomous weapons in 2015. We've had many discussions on autonomous weapons since then and I've always found him to be thoughtful, unsurprising given his prominence in his field. So I reached out to Russell to better understand his concerns. He acknowledged that FLA wasn't "cleanly directed only at autonomous weapon capability," but he saw it as a stepping stone toward something truly terrifying.

FLA is different from projects like the X-47B, J-UCAS, or LRASM, which are designed to engage highly sophisticated adversaries. Russell has a very different kind of autonomous weapon in mind, a swarm of millions of small, fast-moving antipersonnel drones that could wipe out an entire urban population. Russell described these lethal drones used en masse as a kind of "weapon of mass destruction." He explained, "You can make small, lethal quadcopters an inch in diameter and pack several million of them into a truck and launch them with relatively simple software and they don't have to be particularly effective. If 25 percent of them reach a target, that's plenty." Used in this way, even small autonomous weapons could devastate a population.

There's nothing to indicate that FLA is aimed at developing the kind of people-hunting weapon Russell describes, something he acknowledges. Nevertheless, he sees indoor navigation as laying the building blocks toward antipersonnel autonomous weapons. "It's certainly one of the things you'd like to do if you were wanting to develop autonomous weapons," he said.

It's worth nothing that Russell isn't opposed to the military as a whole or even military investments in AI or autonomy in general. He said that some of his own AI research is funded by the Department of Defense, but he only takes money for basic research, not weapons. Even a program like FLA that isn't specifically aimed at weapons still gives Russell pause, however. As a researcher, he said, it's something that he would "certainly think twice" about working on.

WEAPONS THAT HUNT IN PACKS: COLLABORATIVE OPERATIONS IN DENIED ENVIRONMENTS

Russell also raised concerns about another DARPA program: Collaborative Operations in Denied Environments (CODE). According to DARPA's official description, CODE's purpose is to develop "collaborative autonomy—the capability of groups of [unmanned aircraft systems] to work together under a single person's supervisory control." In a press release, CODE's program manager, Jean-Charles Ledé, described the project more colorfully as enabling drones to work together "just as wolves hunt in coordinated packs with minimal communication."

The image of drones hunting in packs like wolves might be a little unsettling to some. Ledé clarified that the drones would remain under the supervision of a human: "multiple CODE-enabled unmanned aircraft would collaborate to find, track, identify and engage targets, all under the command of a single human mission supervisor." Graphics on DARPA's website depicting how CODE might work show communications relay drones linking the drone pack back to a manned aircraft removed from the edge of the battlespace. So, in theory, a human would be in the loop.

CODE is designed for "contested electromagnetic environments," however, where "bandwidth limitations and communications disruptions" are likely to occur. The means that the communications link to the human-inhabited aircraft might be limited or might not work at all. CODE aims to overcome these challenges by giving drones greater intelligence and autonomy so that they can operate with minimal supervision. Cooperative behavior is central to this concept. With cooperative behavior, one person can tell a group of drones to achieve a goal, and the drones can divvy up tasks on their own.

In CODE, the drone team finds and engages "mobile or rapidly relocatable targets," that is, targets whose locations cannot be specified in advance by a human operator. If there is a communications link to a human, then the human could authorize targets for engagement once CODE air vehicles find them. Communications are challenging in contested electromagnetic environments, but not impossible. U.S. fifth-generation fighter aircraft use low probability of intercept / low probability of detection (LPI/LPD) methods of communicating stealthily inside enemy air space. While these communications links are limited in range and bandwidth, they do exist. According to CODE's technical specifications, developers should count on no more than 50 kilobits per second of communications back to the human commander, essentially the same as a 56K dial-up modem circa 1997.

Keeping a human in the loop via a connection on par with a dial-up modem would be a significant change from today, where drones stream back high-definition full-motion video. How much bandwidth is required for a human to authorize targets? Not much, in fact. The human brain is extremely good at object recognition and can recognize objects even in relatively low resolution images. Snapshots of military objects and the surrounding area on the order of 10 to 20 kilobytes in size may be fuzzy to the human eye, but are still of sufficiently high resolution that an untrained person can discern trucks or military vehicles. A 50 kilobit per second connection could transmit one image of this size every two to three seconds (1 kilobyte = 8 kilobits). This would allow the CODE air vehicles to identify potential targets and send them back to a human supervisor who would approve (or disapprove) each specific target before attack.

But is this what CODE intends? CODE's public description explains that the aircraft will operate "under a single person's supervisory control," but does not specify that the human would need to approve each target before engagement. As is the case with all of the systems encountered so far, from thermostats to next-generation weapons, the key is which tasks are being performed by the human and which by the machine. Publicly available information on CODE presents a mixed picture.

A May 2016 video released online of the human-machine interface for CODE shows a human authorizing each specific individual target. The human doesn't directly control the air vehicles. The human operator commands four groups of air vehicles, labeled Aces, Badger, Cobra, and Disco groups. The groups, each composed of two to four air vehicles, are given high-level commands such as "orbit here" or "follow this route." Then the vehicles coordinate among themselves to accomplish the task.

Disco Group is sent on a search and destroy mission: "Disco Group search and destroy all [anti-aircraft artillery] in this area." The human operator sketches a box with his cursor and the vehicles in Disco Group move into the box. "Disco Group conducting search and destroy at Area One," the computer confirms.

As the air vehicles in Disco Group find suspected enemy targets, they cue up their recommended classification to the human for confirmation. The human clicks "Confirm SCUD" and "Confirm AAA" [antiaircraft artillery] on the interface. But confirmation does not mean approval to fire. A few seconds later, a beeping tone indicates that Disco Group has drawn up a strike plan on a target and is seeking approval. Disco Group has 90 percent confidence it has found an SA-12 surface-to-air missile system and includes a photo for confirmation. The human clicks on the strike plan for more details. Beneath the picture of the SA-12 is a small diagram showing estimated collateral damage. A brown splotch surrounds the target, showing potential damage to anything in the vicinity. Just outside of the splotch is a hospital, but it is outside of the anticipated area of collateral damage. The human clicks "Yes" to approve the engagement. In this video, a human is clearly in the loop. Many tasks are automated, but a human approves each specific engagement.

In other public information, however, CODE seems to leave the door open to removing the human from the loop. A different video shows two teams of air vehicles, Team A and Team B, sent to engage a surface-to-air missile. As in the LRASM video, the specific target is identified by a human ahead of time, who then launches the missiles to take it out. Similar to LRASM, the air vehicles maneuver around pop-up threats, although this time the air vehicles work cooperatively, sharing navigation and sensor data while in flight. As they maneuver to their target, something unexpected happens: a "critical pop-up target" emerges. It isn't their primary target, but destroying it is a high priority. Team A reprioritizes to engage the pop-up target while Team B continues to the primary target. The video makes clear this occurs under the supervision of the human commander. This implies a different type of human-machine relationship, though, than the earlier CODE video. In this one, instead of the human being *in* the loop, the human is *on* the loop, at least for pop-up threats. For their primary target, they operate in a semiautonomous fashion. The human chose the primary target. But when a pop-up threat emerges, the missiles have the authority to operate as supervised autonomous weapons. They don't need to ask additional permission to take out the target. Like a quarterback calling an audible at the scrimmage line to adapt to the defense, they have the freedom to adapt to unexpected situations that arise. The human operator is like the coach standing on the sidelines—able to call a time-out to intervene, but otherwise merely supervising the action.

DARPA's description of CODE online seems to show a similar flexibility for whether the human or air vehicles themselves approve targets. The CODE website says: "Using collaborative autonomy, CODE-enabled unmanned aircraft would find targets and engage them as appropriate under established rules of engagement . . . and adapt to dynamic situations such as . . . the emergence of unanticipated threats." This appears to leave the door open to autonomous weapons that would find and engage targets on their own.

The detailed technical description issued to developers provides additional information, but little clarity. DARPA explains that developers should:

> Provide a concise but comprehensive targeting chipset so the mission commander can exercise appropriate levels of human judgment over the use of force or evaluate other options.

The specific wording used, "appropriate levels of human judgment," may sound vague and squishy, but it isn't accidental. This guidance directly quotes the official DoD policy on autonomy in weapons, DoD Directive 3000.09, which states:

> Autonomous and semi-autonomous weapon systems shall be designed to allow commanders and operators to exercise appropriate levels of human judgment over the use of force.

Notably, that policy does not prohibit autonomous weapons. "Appropriate levels of human judgment" could include autonomous weapons. In fact, the DoD policy includes a path through which developers could seek approval to build and deploy autonomous weapons, with appropriate safeguards and testing, should they be desired.

At a minimum, then, CODE would seem to allow for the possibility of autonomous weapons. The aim of the project is not to build autonomous weapons necessarily. The aim is to enable collaborative autonomy. But in a contested electromagnetic environment where communications links to the human supervisor might be jammed, the program appears to allow for the possibility that the drones could be delegated the authority to engage pop-up threats on their own.

In fact, CODE even hints at one way that collaborative autonomy might aid in target identification. Program documents list one of the advantages of collaboration as "providing multi-modal sensors and diverse observation angles to improve target identification." Historically, automatic target recognition (ATR) algorithms have not been good enough to trust with autonomous engagements. This poor quality of ATR algorithms could be compensated for by bringing together multiple different sensors to improve the confidence in target identification or by viewing a target from multiple angles, building a more complete picture. One of the CODE videos actually shows this, with air vehicles viewing the target from multiple directions and sharing data. Whether target identification could be improved enough to allow for autonomous engagements is unclear, but if CODE is successful, DoD will have to confront the question of whether to authorize autonomous weapons.

THE DEPARTMENT OF MAD SCIENTISTS

At the heart of many of these projects is the Defense Advanced Research Projects Agency (DARPA), or what writer Michael Belfiore called "the Department of Mad Scientists." DARPA, originally called ARPA, the Advanced Research Projects Agency, was founded in 1958 by President Eisenhower in response to Sputnik. DARPA's mission is to prevent "strategic surprise." The United States was surprised and shaken by the Soviet Union's launch of Sputnik. The small metal ball hurdling through space overhead was a wake-up call to the reality that the Soviet Union could now launch intercontinental ballistic missiles that could hit anywhere in the United States. In response, President Eisenhower created two organizations to develop breakthrough technologies, the National Aeronautics and Space Administration (NASA) and ARPA. While NASA had the mission of winning the space race, ARPA had a more fundamental mission of investing in high-risk, high-reward technologies so the United States would never again be surprised by a competitor.

To achieve its mission, DARPA has a unique culture and organization distinct from the rest of the military-industrial complex. DARPA only invests in projects that are "DARPA hard," challenging technology problems that others might deem impossible. Sometimes, these bets don't pan out. DARPA has a mantra of "fail fast" so that if projects fail, they do so before investing massive resources. Sometimes, however, these investments in game-changing technologies pay huge dividends. Over the past five decades, DARPA has time and again laid the seeds for disruptive technologies that have given the United States decisive advantages. Out of ARPA came ARPANET, an early computer network that later developed into the internet. DARPA helped develop basic technologies that underpin the global positioning system (GPS). DARPA funded the first-ever stealth combat aircraft, HAVE Blue, which led to the F-117 stealth fighter. And DARPA has consistently advanced the horizons of artificial intelligence and robotics.

DARPA rarely builds completed weapon systems. Its projects are small, focused efforts to solve extremely hard problems, such as CODE's efforts to get air vehicles to collaborate autonomously. Stuart Russell said that he found these projects concerning because, from his perspective, they seemed to indicate that the United States was expecting to be in a position to deploy autonomous weapons at a future date. Was that, in fact, their intention, or was that simply an inevitability of the technology? If projects like CODE were successful, did DARPA intend to turn the key to full auto or was the intention to always keep a human in the loop?

It was clear that if I was going to understand the future of autonomous weapons, I would need to talk to DARPA.

5

INSIDE THE PUZZLE PALACE

IS THE PENTAGON BUILDING AUTONOMOUS WEAPONS?

DARPA sits in a nondescript office building in Ballston, Virginia, just a few miles from the Pentagon. From the outside, it doesn't look like a "Department of Mad Scientists." It looks like just another glass office building, with no hint of the wild-eyed ideas bubbling inside.

Once you're inside DARPA's spacious lobby, the organization's gravitas takes hold. Above the visitors' desk on the marble wall, raised metal letters that are both simple and futuristic announce: DEFENSE ADVANCED RESEARCH PROJECTS AGENCY. Nothing else. No motto or logo or shield. The organization's confidence is apparent. The words seem to say, "the future is being made here."

As I wait in the lobby, I watch a wall of video monitors announce DARPA's latest project to go public: the awkwardly named Anti-Submarine Warfare (ASW) Continuous Trail Unmanned Vessel (ACTUV). The ship's christened name, Sea Hunter, is catchier. The project is classic DARPA—not only game-changing, but paradigm-bending: the Sea Hunter is an entirely unmanned ship. Sleek and angular, it looks like something time-warped in from the future. With a long, narrow hull and two outriggers, the Sea Hunter carves the oceans like a three-pointed dagger, tracking enemy submarines. At the ship's christening, Deputy Secretary of Defense Bob Work compared it to a Klingon Bird of Prey from *Star Trek*.

There are no weapons on board the Sea Hunter, for now. There should be no mistake, however: the Sea Hunter is a warship. Work called it a "fighting ship," part of the Navy's future "human machine collaborative battle fleet." At $21 million apiece, the Sea Hunter is a fraction of the cost of a new $1.6-billion Arleigh Burke destroyer. The low price allows the Navy to purchase scores of the sub-hunting ships on the cheap. Work laid out his vision for flotillas of Sea Hunters roaming the seas:

> You can imagine anti-submarine warfare pickets, you can imagine anti-submarine warfare wolfpacks, you can imagine mine warfare flotillas, you can imagine distributive anti-surface warfare surface action groups . . . We might be able to put a six pack or a four pack of missiles on them. Now imagine 50 of these distributed and operating together under the hands of a flotilla commander, and this is really something.

Like many other robotic systems, the Sea Hunter can navigate autonomously and might someday be armed. There is no indication that DoD has any intention of authorizing autonomous weapons engagements. Nevertheless, the video on DARPA's lobby wall is a reminder that the robotics revolution continues at a breakneck pace.

BEHIND THE CURTAIN: INSIDE DARPA'S TACTICAL TECHNOLOGY OFFICE

DARPA is organized into six departments focusing on different technology areas: biology, information science, microelectronics, basic sciences, strategic technologies, and tactical technologies. CODE, FLA, LRASM, and the Sea Hunter fall into DARPA's Tactical Technology Office (TTO), the division that builds experimental vehicles, ships, airplanes, and spacecraft. Other TTO projects include the XS-1 Experimental Spaceplane, designed to fly to the edge of space and back; the Blue Wolf undersea robotic vehicle; an R2-D2-like robotic copilot for aircraft called ALIAS; the Mach 20 Falcon Hypersonic Technology Vehicle, which flies fast enough to zip from New York to Los Angeles in 12 minutes; and the Vulture program to build an ultra-long endurance drone that can stay in the air for up to five years without refueling. Mad science, indeed.

TTO's offices look like a child's dream toy room. Littered around the offices are models and even some actual prototype pieces of hardware from past TTO projects—missiles, robots, and stealth aircraft. I can't help but wonder what TTO is building today that will be the stealth of tomorrow.

Bradford Tousley, TTO's director, graciously agreed to meet with me to discuss CODE and other projects. Tousley began his government career as an Army armor officer during the Cold War. His first tour was in an armored cavalry unit on the German border, being ready for a Soviet invasion that might kick off World War III. Later in his career, when the Army sent him back for a secondary education, Tousley earned a doctorate in electrical engineering. His career shifted from frontline combat units to research and development in lasers and optics, working to ensure the U.S. military had the best possible technology. Tousley's career has covered multiple stints at DARPA as well as time in the intelligence community on classified satellite payloads, so he has a breadth of understanding in technology beyond merely robotics.

Tousley pointed out that DARPA was founded in response to the strategic surprise of Sputnik: "DARPA's fundamental mission is unchanged: Enabling pivotal early investments for breakthrough capabilities for national security to achieve or prevent strategic surprise." Inside DARPA, they weigh these questions heavily. "Within the agency, we talk about every single program we begin and we have spirited discussions. We talk about the pros and cons. Why? Why not? ... How far are we willing to go?" Tousley made clear, however, that answering those questions isn't DARPA's job. "Those are fundamental policy and concept and military employment considerations" for others to decide. "Our fundamental job is to take that technical question off the table. It's our job to make the investments to show the capabilities can exist" to give the warfighter options. In other words, to prevent another Sputnik.

If machines improved enough to reliably take out targets on their own, what the role was for humans in warfare? Despite his willingness to push the boundaries of technology, Tousley still saw humans in command of the mission: "That final decision is with humans, period." That might not mean requiring human authorization for every single target, but autonomous weapons would still operate under human direction, hunting and attacking targets at the direction of a human commander. At least for the foreseeable future, Tousley explained, humans were better than machines at identifying anomalies and reacting to unforeseen events. This meant that keeping humans involved at the mission level was critical to understand the broader context and make decisions. "Until the machine processors equal or surpass humans at making abstract decisions, there's always going to be mission command. There's always going to be humans in the loop, on the loop—whatever you want to call it."

Tousley painted a picture for me of what this might look like in a future conflict: "Groups of platforms that are unmanned that you are willing to attrit [accept some losses] may do extremely well in an anti-access air defense environment ... How do I take those platforms and a bunch of others and knit them together in architectures that have manned and unmanned systems striking targets in a congested and contested environment? You need that knitted system because you're going to be GPS-jammed; communications are going to be going in and out; you're going to have air defenses shooting down assets, manned and unmanned. In order to get in and strike critical targets, to control that [anti-access] environment, you're going to have to have a system-of-systems architecture that takes advantage of manned and unmanned systems at different ranges with some amount of fidelity in the ability of the munition by itself to identify the target—could be electronically, could be optically, could be infrared, could be [signals intelligence], could be different ways to identify the target. So that system-of-systems architecture is going to be necessary to knit it all together."

Militaries especially need autonomy in electronic warfare. "We're using physical machines and electronics, and the electronics themselves are becoming machines that operate at machine speed. ... I need the cognitive electronic warfare to adapt in microseconds. ... If I have radars trying to jam other radars but they're frequency hopping [rapidly changing radio frequencies] back and forth, I've got to track with it. So [DARPA's Microsystems Technology Office] is thinking about, how do I operate at machine speed to allow these machines to conduct their functions?"

Tousley compared the challenge of cognitive electronic warfare to Google's *go*-playing AlphaGo program. What happens when that program plays another version of AlphaGo at "machine speed?" He explained, "As humans ascend to the higher-level mission command and I've got machines doing more of that targeting function, those machines are going to be challenged by machines on the adversary's side and a human can't respond to that. It's got to be machines responding to machines.... That's one of the trends of the Third Offset, that machine on machine." Humans, therefore, shift into a "monitoring" role, watching these systems and intervening, if necessary. In fact, Tousley argues that a difficult question will be whether humans *should* intervene in these machine-on-machine contests, particularly in cyberspace and electronic warfare where the pace of interactions will far exceed human reaction times.

I pointed out that having a human involved in a monitoring role still implies some degree of connectivity, which might be difficult in a contested environment with jamming. Tousley was unconcerned. "We expect that there will be jamming and communications denial going on, but it won't be necessarily everywhere, all the time," he said. "It's one thing to jam my communication link over 1,000 miles, it's another thing to jam two missiles that are talking in flight that may be three hundred meters apart flying in formation." Reliable communications in contested areas, even short range, would still permit a human being to be involved, at least in some capacity.

So, what role would that person play? Would this person need to authorize every target before engagement, or would human control sit at a higher level? "I think that will be a rule of engagement-dependent decision," Tousley said. "In an extremely hot peer-on-peer conflict, the rules of engagement may be more relaxed.... If things are really hot and heavy, you're going to rely on the fact that you built some of that autonomous capability in there." Still, even in this intense battlefield environment, he attested, the human plays the important role of overseeing the combat action. "But you still want some low data rate" to keep a person involved.

It took me a while to realize that Tousley wasn't shrugging off my questions about whether the human would be required to authorize each target because he was being evasive or trying to conceal a secret program, it was because he genuinely didn't see the issue the same way. Automation had been increasing in weapons for decades—from Tousley's perspective, programs like CODE were merely the next step. Humans would remain involved in lethal decision-making, albeit at a higher level overseeing and directing the combat action. The precise details of how much freedom an autonomous system might be granted to choose its own targets and in which situations wasn't his primary concern. Those were questions for military commanders to address. His job as a researcher was to, as he put it, "take that technical question off the table." His job was to build the options. That meant building swarms of autonomous systems that could go into a contested area and conduct a mission with as minimal human supervision as possible. It also meant building in resilient communications so that humans could have as much bandwidth and connectivity to oversee and direct the autonomous systems as possible. How exactly those technologies were implemented—which specific decisions were retained for the human and which were delegated to the machine—wasn't his call to make.

Tousley acknowledged that delegating lethal decision-making came with risks. "If [CODE] enables software that can enable a swarm to execute a mission, would that same swarm be able to execute a mission against the wrong target? Yeah, that is a possibility. We don't want that to happen. We want to build in all the fail-safe systems possible." For this reason, his number-one concern with autonomous systems was actually test and evaluation: "What I worry about the most is our ability to effectively test these systems to the point that we can quantify that we trust them." Trust is essential to commanders being willing to employ autonomous systems. "Unless the combatant commander feels that that autonomous system is going to execute the mission with the trust that he or she expects, they'll never deploy it in the first place." Establishing that trust was all about test and evaluation, which could mean putting an autonomous system through millions of computer simulations to test its behavior. Even still, testing all of the possible situations an autonomous system might encounter and its potential behaviors in response could be very difficult. "One of the concerns I have," he said, "is that the technology for autonomy and the technology for human-machine integration and understanding is going too far surpass our ability to test it. . . . That worries me."

TARGET RECOGNITION AND ADAPTION IN
CONTESTED ENVIRONMENTS (TRACE)

Tousley declined to comment on another DARPA program, Target Recognition and Adaption in Contested Environments (TRACE), because it
fell under a different department he wasn't responsible for. And although
DARPA was incredibly open and helpful throughout the research for
this book, the agency declined to comment on TRACE beyond publicly
available information. If there's one program that seems to be a linchpin
for enabling autonomous weapons, it's TRACE. The CODE project aims
to compensate for poor automatic target recognition (ATR) algorithms
by leveraging cooperative autonomy. TRACE aims to improve ATR algorithms directly.

TRACE's project description explains the problem:

> In a target-dense environment, the adversary has the advantage of
> using sophisticated decoys and background traffic to degrade the
> effectiveness of existing automatic target recognition (ATR) solu-
> tions. . . . the false-alarm rate of both human and machine-based radar
> image recognition is unacceptably high. Existing ATR algorithms
> also require impractically large computing resources for airborne
> applications.

TRACE's aim is to overcome these problems and "develop algorithms and
techniques that rapidly and accurately identify military targets using
radar sensors on manned and unmanned tactical platforms." In short,
TRACE's goal is to solve the ATR problem.

To understand just how difficult ATR is—and how game-changing
TRACE would be if successful—a brief survey of sensing technologies is
in order. Broadly speaking, military targets can be grouped into two categories: "cooperative" and "non-cooperative" targets. Cooperative targets
are those that are actively emitting a signal, which makes them easier to
detect. For example, radars, when turned on, emit energy in the electromagnetic spectrum. Radars "see" by observing the reflected energy from
their signal. This also means the radar is broadcasting its own position,

however. Enemies looking to target and destroy the radar can simply home in on the source of the electromagnetic energy. This is how simple autonomous weapons like the Harpy find radars. They can use passive sensors to simply wait and listen for the cooperative target (the enemy radar) to broadcast its position, and then home in on the signal to destroy the radar.

Non-cooperative targets are those that aren't broadcasting their location. Examples of non-cooperative targets could be ships, radars, or aircraft operating with their radars turned off; submarines running silently; or ground vehicles such as tanks, artillery, or mobile missile launchers. To find non-cooperative targets, active sensors are needed to send signals out into the environment to find targets. Radar and sonar are examples of active sensors; radar sends out electromagnetic energy and sonar sends out sound waves. Active sensors then observe the reflected energy and attempt to discern potential targets from the random noise of background clutter in the environment. Radar "sees" reflected electromagnetic energy and sonar "hears" reflected sound waves.

Militaries are therefore like two adversaries stumbling around in the dark, each listening and peering fervently into the darkness to hear and see the other while remaining hidden themselves. Our eyes are passive sensors; they simply receive light. In the darkness, however, an external source of light like a flashlight is needed. Using a flashlight gives away one's own position, though, making one a "cooperative target" for the enemy. In this contest of hiding and finding, zeroing in on the enemy's cooperative targets is like finding a person waving a flashlight around in the darkness. It isn't hard; the person waving the flashlight is going to stand out. Finding the non-cooperative targets who keep their flashlights turned off can be very, very tricky.

When there is little background clutter, objects can be found relatively easily through active sensing. Ships and aircraft stand out easily against their background—a flat ocean and an empty sky. They stand out like a person standing in an open field. A quick scan with even a dim light will pick out a person standing in the open, although discerning friend from foe can be difficult. In cluttered environments, however, even finding targets in the first place can be hard. Moving targets can be discerned via

Doppler shifting—essentially the same method that police use to detect speeding vehicles. Moving objects shift the frequency of the return radar signal, making them stand out against a stationary background. Stationary targets in cluttered environments can be as hard to see as a deer hiding in the woods, though. Even with a light shined directly on them, they might not be noticed.

Humans have challenges seeing stationary, camouflaged objects and human visual cognitive processing is incredibly complex. We take for granted how computationally difficult it is to see objects that blend into the background. While radars and sonars can "see" and "hear" in frequencies that humans are incapable of, military ATR is nowhere near as good as humans at identifying objects amid clutter.

Militaries currently sense many non-cooperative targets using a technique called synthetic aperture radar, or SAR. A vehicle, typically an aircraft, flies in a line past a target and sends out a burst of radar pulses as the aircraft moves. This allows the aircraft to create the same effect as having an array of sensors, a powerful technique that enhances image resolution. The result is sometimes grainy images composed of small dots, like a black-and-white pointillist painting. While SAR images are generally not as sharp as images from electro-optical or infrared cameras, SAR is a powerful tool because radar can penetrate through clouds, allowing all-weather surveillance. Building algorithms that can automatically identify SAR images is extremely difficult, however. Grainy SAR images of tanks, artillery, or airplanes parked on a runway often push the limits of human abilities to recognize objects, and historically ATR algorithms have fallen far short of human abilities.

The poor performance of military ATR stands in stark contrast to recent advances in computer vision. Artificial intelligence has historically struggled with object recognition and perception, but the field has seen rapid gains recently due to deep learning. Deep learning uses neural networks, a type of AI approach that is analogous to biological neurons in animal brains. Artificial neural networks don't directly mimic biology, but are inspired by it. Rather than follow a script of *if-then* steps for how to perform a task, neural networks work based on the strength of connections within a network. Thousands or even millions of data samples are fed into the network and the weights of various connections between nodes in

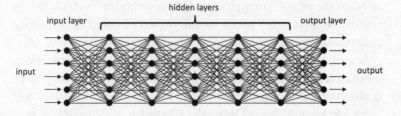

A deep neural network has hidden layers between the input and output layers. Some deep neural networks can have as many as 150 or more hidden layers.

Deep Neural Network

the network are constantly adjusted to "train" the network on the data. In this way, neural networks "learn." Network settings are refined until the correct output, such as the correct image category (for example, cat, lamp, car) is achieved.

Deep neural networks are those that have multiple "hidden" layers between the input and output, and have proven to be a very powerful tool for machine learning. Adding more layers in the network between the input data and output allows for a much greater complexity of the network, enabling the network to handle more complex tasks. Some deep neural nets have over a hundred layers.

This complexity is, it turns out, essential for image recognition, and deep neural nets have made tremendous progress. In 2015, a team of researchers from Microsoft announced that they had created a deep neural network that for the first time surpassed human performance in visual object identification. Using a standard test dataset of 150,000 images, Microsoft's network achieved an error rate of only 4.94 percent, narrowly edging out humans, who have an estimated 5.1 percent error rate. A few months later, they improved on their own performance with a 3.57 percent rate by a 152-layer neural net.

TRACE intends to harness these advances and others in machine learning to build better ATR algorithms. ATR algorithms that performed on par with or better than humans in identifying non-cooperative targets such as tanks, mobile missile launchers, or artillery would be a game changer in terms of finding and destroying enemy targets. If the resulting target recognition system was of sufficiently low power to be located on board the missile or drone itself, human authorization would not be required, at least from a purely technical point of view. The technology would enable weapons to hunt and destroy targets all on their own.

Regardless of whether DARPA was intending to build autonomous weapons, it was clear that programs like CODE and TRACE were putting in place the building blocks that would enable them in the future. Tousley's view was that it wasn't DARPA's call whether to authorize that next fateful step across the line to weapons that would choose their own targets. But if it wasn't DARPA's call whether to build autonomous weapons, then whose call was it?

6

CROSSING THE THRESHOLD

APPROVING AUTONOMOUS WEAPONS

The Department of Defense has an official policy on the role of autonomy in weapons, DoD Directive 3000.09, "Autonomy in Weapon Systems." (Disclosure: While at DoD, I led the working group that drafted the policy.) Signed in November 2012, the directive is published online so anyone can read it.

The directive includes some general language on principles for design of semiautonomous and autonomous systems, such as realistic test and evaluation and understandable human-machine interfaces. The meat of the policy, however, is the delineation of three classes of systems that get the "green light" for approval in the policy. These are: (1) semiautonomous weapons, such as homing munitions; (2) defensive supervised autonomous weapons, such as the ship-based Aegis weapon system; and (3) non-lethal, nonkinetic autonomous weapons, such as electronic warfare to jam enemy radars. These three types of autonomous systems are in wide use today. The policy essentially says to developers, "If you want to build a weapon that uses autonomy in ways consistent with existing practices, you're free to do so." Normal acquisition rules apply, but those types of systems do not require any additional approval.

Any future weapon system that would use autonomy in a novel way outside of those three categories gets a "yellow light." Those systems need to be reviewed before beginning formal development (essentially the point at which large sums of money would be spent) and again before fielding.

The policy outlines who participates in the review process—the senior defense civilian officials for policy and acquisitions and the chairman of the Joint Chiefs of Staff—as well as the criteria for review. The criteria are lengthy, but predominantly focus on test and evaluation for autonomous systems to ensure they behave as intended—the same concern Tousley expressed. The stated purpose of the policy is to "minimize the probability and consequences of failures in autonomous and semiautonomous weapon systems that could lead to unintended engagements." In other words, to minimize the chances of armed robots running amok.

Lethal autonomous weapons are not prohibited by the policy directive. Instead, the policy provides a process by which new uses of autonomy could be reviewed by relevant officials before deployment. The policy helps ensure that if DoD were to build autonomous weapons that they weren't developed and deployed without sufficient oversight, but it doesn't help answer the question of whether DoD might actually approve such systems. On that question, the policy is silent. All the policy says is that if an autonomous weapon met all of the criteria, such as reliability under realistic conditions, then in principle it could be authorized.

GIVING THE GREEN LIGHT TO AUTONOMOUS WEAPONS

But *would* it be authorized? DARPA programs are intended to explore the art of the possible, but that doesn't mean that DoD would necessarily turn those experimental projects into operational weapon systems. To better understand whether the Pentagon might actually approve autonomous weapons, I sat down with then-Pentagon acquisition chief, Under Secretary of Defense Frank Kendall. As the under secretary of defense for acquisition, technology and logistics, Kendall was the Pentagon's chief technologist and weapons buyer under the Obama Administration. When it came to major weapons systems like the X-47B or LRASM, the decision whether or not to move forward was in Kendall's hands. In the process laid out under the DoD Directive, Kendall was one of three senior officials, along with the under secretary for policy and the chairman of the Joint Chiefs, who all had to agree in order to authorize developing an autonomous weapon.

Kendall has a unique background among defense technologists. In addition to a distinguished career across the defense technology enterprise, serving in a variety of roles from vice president of a major defense firm to several mid-level bureaucratic jobs within DoD, Kendall also has worked pro bono as a human rights lawyer. He has worked with Amnesty International, Human Rights First, and other human rights groups, including as an observer at the U.S. prison at Guantánamo Bay. Given his background, I was hopeful that Kendall might be able to bridge the gap between technology and policy.

Kendall made clear, for starters, that there had never been a weapon autonomous enough even to trigger the policy review. "We haven't had anything that was even remotely close to autonomously lethal." If he were put in that position, Kendall said his chief concerns would be ensuring that it complied with the laws of war and that the weapon allowed for "appropriate human judgment," a phrase that appears in the policy directive. Kendall admitted those terms weren't defined, but conversation with him began to elucidate his thinking.

Kendall started his career as an Army air defender during the Cold War, where he learned the value of automation first hand. "We had an automatic mode for the Hawk system that we never used, but I could see in an extreme situation where you'd turn it on, because you just couldn't do things fast enough otherwise," he said. When you have "fractions of a second" to decide—that's a role for machines.

Kendall said that automatic target recognition and machine learning were improving rapidly. As they improve, it should become possible for the machine to select its own targets for engagement. In some settings, such as taking out an enemy radar, he thought it could be done "relatively soon."

This raises tricky questions. "Where do you want the human intervention to be?" he asked. "Do you want it to be the actual act of employing the lethality? Do you want it to be the acceptance of the rules that you set for identifying something as hostile?" Kendall didn't have the answers. "I think we're going to have to sort through all that."

One important factor was the context. "Are you just driving down the street or are you actually in a war, or you're in an insurgency? The context matters." In some settings, using autonomy to select and engage targets might be appropriate. In others, it might not.

Kendall saw using an autonomous weapon to target enemy radars as fairly straightforward and something he didn't see many people objecting to. There were other examples that pushed the boundaries. Kendall said on a trip to Israel, his hosts from the Israel Defense Forces had him sit in a Merkava tank that was outfitted with the Trophy active protection system. The Israelis fired a rocket propelled grenade near the tank ("offset a few meters," he said) and the Trophy system intercepted it automatically. "But suppose I also wanted to shoot back at . . . wherever the bullet had come from?" he asked. "You can automate that, right? That's protecting me, but it's the use of that weapon in a way which could be lethal to whoever, you know, was in the line of fire when I fire." He pointed out that automating a return-fire response might prevent a second shot, saving lives. Kendall acknowledged that had risks, but there were risks in not doing it as well. "How much do we want to put our own people at risk by not allowing them to use this technology? That's the other side of the equation."

Things become especially difficult if the machine is better than the person, which, at some point, will happen. "I think at that point, we'll have a tough decision to make as to how we want to go with that." Kendall saw value in keeping a human in the loop as a backup, but, "What if it's a situation where there isn't that time? Then aren't you better off to let the machine do it? You know, I think that's a reasonable question to ask."

I asked him for his answer to the question—after all, he was the person who would decide in DoD. But he didn't know.

"I don't think we've decided that yet," he said. "I think that's a question we'll have to confront when we get to where technology supports it."

Kendall wasn't worried, though. "I think we're a long way away from the Terminator idea, the killer robots let loose on the battlefield idea. I don't think we're anywhere near that and I don't worry too much about that." Kendall expressed confidence in how the United States would address this technology. "I'm in my job because I find my job compatible with being a human rights lawyer. I think the United States is a country which has high values and it operates consistent with those values. . . . I'm confident that whatever we do, we're going to start from the premise that we're going to follow the laws of war and obey them and we're going to follow humanitarian principles and obey them."

Kendall was worried about other countries, but he was most concerned about what terrorists might do with commercially available technology. "Automation and artificial intelligence are one of the areas where the commercial developments I think dwarf the military investments in R&D. They're creating capabilities that can easily be picked up and applied for military purposes." As one example, he asked, "When [ISIS] doesn't have to put a person in that car and can just send it out on its own, that's a problem for us, right?"

THE REVOLUTIONARY

Kendall's boss was Deputy Secretary of Defense Bob Work, the Pentagon's number-two bureaucrat—and DoD's number-one robot evangelist. As deputy secretary from 2014–17, Work was the driving force behind the Pentagon's Third Offset Strategy and its focus on human-machine teaming. In his vision of future conflicts, AI will work in concert with humans in human-machine teams. This blended human-plus-machine approach could take many forms. Humans could be enhanced through exoskeleton suits and augmented reality, enabled by machine intelligence. AI systems could help humans make decisions, much like in "centaur chess," where humans are assisted by chess programs that analyze possible moves. In some cases, AI systems may perform tasks on their own with human oversight, particularly when speed is an advantage, similar to automated stock trading. Future weapons will be more intelligent and cooperative, swarming adversaries.

Collectively, Work argues these advances may lead to a "revolution" in warfare. Revolutions in warfare, Work explained in a 2014 monograph, are "periods of sharp, discontinuous change [in which] . . . existing military regimes are often upended by new more dominant ones, leaving old ways of warfare behind."

In defense circles, this is a bold claim. The U.S. defense community of the late 1990s and early 2000s became enamored with the potential of information technology to lead to a revolution in warfare. Visions of "information dominance" and "network-centric warfare" foundered in the mountains of Afghanistan and the dusty streets of Iraq as the United States became mired in messy counterinsurgency wars. High-tech investments in next-generation weapon systems such as F-22 fighter jets were overpriced or simply irrelevant for finding and tracking insurgents or winning the hearts and minds of civilian populations. And yet . . .

The information revolution continued, leading to more advanced computer processors and ever more sophisticated machine intelligence. And even while warfare in the information age might not have unfolded the way Pentagon futurists might have envisioned, the reality is information technology dramatically shaped how the United States fought its counterinsurgency wars. Information became the dominant driver of counter-network operations as the United States sought to find insurgents hiding among civilians, like finding a needle in a stack of needles.

Sweeping technological changes like the industrial revolution or the information revolution unfold in stages over time, over the course of decades or generations. As they do, they inevitably have profound effects on warfare. Technologies like the internal-combustion engine that powered civilian automobiles and airplanes in the industrial revolution led to tanks and military aircraft. Tanks and airplanes, along with other industrial-age weaponry such as machine guns, profoundly changed World War I and World War II.

Work is steeped in military history and a student of Pentagon futurist Andy Marshall, who for decades ran DoD's Office of Net Assessment and championed the idea that another revolution in warfare was unfolding today. Work understands the consequences of falling behind during periods of revolutionary change. Militaries can lose battles and even wars. Empires can fall, never to recover. In 1588, the mighty Spanish Armada was defeated by the British, who had more expertly exploited the revolutionary technology of the day: cannons. In the interwar period between World War I and World War II, Germany was more successful in capitalizing on innovations in aircraft, tanks, and radio technology and the result was the blitzkrieg—and the fall of France. The battlefield is an unforgiving environment. When new technologies upend old ways of fighting, militaries and nations don't often get second chances to get it right.

If Work is right, and a revolution in warfare is under way driven in part by machine intelligence, then there is an imperative to invest heavily in AI, robotics, and automation. The consequences of falling behind could be disastrous for the United States. The industrial revolution led to machines that were stronger than humans, and the victors were those who best capitalized on that technology. Today's information revolution is leading to machines that are smarter and faster than humans. Tomorrow's victors will be those who best exploit AI.

Right now, AI systems can outperform humans in narrow tasks but still fall short of humans in general intelligence, which is why Work advocates human-machine teaming. Such teaming allows the best of both human and machine intelligence. AI systems can be used for specific, tailored tasks and for their advantages in speed while humans can understand the broader context and adapt to novel situations. There are limitations to this approach. In situations where the advantages in speed are overwhelming, delegating authority entirely to the machine is preferable.

When it comes to lethal force, in a March 2016 interview, Work stated, "We will not delegate lethal authority for a machine to make a decision." He quickly caveated that statement a moment later, however, adding, "The only time we will . . . delegate a machine authority is in things that go faster than human reaction time, like cyber or electronic warfare."

In other words, we won't delegate lethal authority to a machine . . . unless we have to. In the same interview, Work said, "We might be going up against a competitor that is more willing to delegate authority to machines than we are and as that competition unfolds, we'll have to make decisions about how to compete." How long before the tightening spiral of an ever-faster OODA loop forces that decision? Perhaps not long. A few weeks later in another interview, Work stated it was his belief that "within the next decade or decade and a half it's going to become clear when and where we delegate authority to machines." A principal concern of his was the fact that while in the United States we debate the "moral, political, legal, ethical" issues surrounding lethal autonomous weapons, "our potential competitors may not."

There was no question that if I was going to understand where the robotics revolution was heading, I needed to speak to Work. No single individual had more sway over the course of the U.S. military's investments in autonomy than he did, both by virtue of his official position in the bureaucracy as well as his unofficial position as the chief thought-leader on autonomy. Work may not be an engineer writing the code for the next generation of robotic systems, but his influence was even broader and deeper. Through his public statements and internal policies, Work was shaping the course of DoD's investments, big and small. He had championed the concept of human-machine teaming. How he framed the technology would influence what engineers across the defense enterprise chose to build. Work immediately agreed to an interview.

THE FUTURE OF LETHAL AUTONOMY

The Pentagon is an imposing structure. At 6.5 million square feet, it is one of the largest buildings in the world. Over 20,000 people enter the Pentagon every day to go to work. As I moved through the sea of visitors clearing security, I was reminded of the ubiquity of the robotics revolution. I heard the man in line behind me explain to Pentagon security that the mysterious item in his briefcase raising alarms in their x-ray scanners was a drone. "It's a UAV," he said. "A drone. I have clearance to bring it in," he added hastily.

The drones are literally everywhere, it would seem.

Work's office was in the famed E-ring where the Pentagon's top executives reside, and he was kind enough to take time out of his busy schedule to talk with me. I started with a simple question, one I had been searching to answer in vain in my research: Is the Department of Defense building autonomous weapons?

Underscoring the definitional problem, Work wanted to clarify what I meant by "autonomous weapon" before answering. I explained I was defining an autonomous weapon as one that could search for, select, and engage targets on its own. Work replied, "We, the United States, have had a lethal autonomous weapon, using your definition, since 1945: the Bat [radar-guided anti-ship bomb]." He said, "I would define it as a narrow lethal autonomous weapon in that the original targeting of the Japanese destroyer that we fired at was done by a Navy PBY maritime patrol aircraft ... they knew [the Japanese destroyer] was hostile—and then they launched the weapon. But the weapon itself made all of the decisions on the final engagement using an S-band radar seeker." Despite his use of the term "autonomous weapon" to describe a radar-guided homing munition, Work clarified he was comfortable with that use of autonomy. "I see absolutely no problem in those types of weapons. It was targeted on a specific capability by a man in the loop and all the autonomy was designed to do was do the terminal endgame engagement." He was also comfortable with how autonomy was used in a variety of modern weapons, from torpedoes to the Aegis ship combat system.

Painting a picture of the future, Work said, "We are moving to a world in which the autonomous weapons will have smart decision trees that will be completely preprogrammed by humans and completely targeted by humans. So let's say we fire a weapon at 150 nautical miles because our off-board sensors say a Russian battalion tactical group is operating in this area. We don't know exactly what of the battalion tactical group this weapon will kill, but we know that we're engaging an area where there are hostiles." Work explained that the missile itself, following its programming logic, might prioritize which targets to strike—tanks, artillery, or infantry fighting vehicles. "We're going to get to that level. And I see no problem in that," he said. "There's a whole variety of autonomous weapons that do end-game engagement decisions after they have been targeted and launched at a specific target or target area." (Here Work is using "autonomous weapon" to refer to fire-and-forget homing munitions.)

Loitering weapons, Work acknowledged, were qualitatively different. "The thing that people worry about is a weapon we fire at range and it loiters in the area and it decides when, where, how, and what to kill without anything other than the human launching it in the general direction." Work acknowledged that, regardless of the label used, these loitering munitions were qualitatively different than homing munitions that had to be launched at a specific target. But Work didn't see any problem with loitering munitions either. "People start to get nervous about that, but again, I don't worry about that at all." He said he didn't believe the United States would ever fire such a weapon into an area unless it had done the appropriate estimates for potential collateral damage. If, on the other hand, "we are relatively certain that there are no friendlies in the area: weapons free. Let the weapon decide."

These search-and-destroy weapons didn't bother Work, even if they were choosing their own targets, because they were still "narrow AI systems." These weapons would be "programmed for a certain effect against a certain type of target. We can tell them the priorities. We can even delegate authority to the weapon to determine how it executes end game attack." With these weapons, there may be "a lot of prescribed decision trees, but the human is always firing it into a general area and we will do [collateral damage estimation] and we will say, 'Can we accept the risk that in this general area the weapon might go after a friendly?' And we will do the exact same determination that we have right now."

Work said the key question is, "What is your comfort level on target location error?" He explained, "If you are comfortable firing a weapon into an area in which the target location error is pretty big, you are starting to take more risks that it might go against an asset that might be a friendly asset or an allied asset or something like that.... So, really what's happening is because you can put so much more processing power onto the weapon itself, the [acceptable degree of] target location error is growing. And we will allow the weapon to search that area and figure out the endgame." An important factor is what else is in the environment and the acceptable level of collateral damage. "If you have real low collateral damage [requirements]," he said, "you're not going to fire a weapon into an area where the target location is so large that the chances of collateral damage go up."

In situations where that risk was acceptable, Work saw no problems with such weapons. "I hear people say, 'This is some terrible thing. We've got killer robots.' No we don't. Robots ... will only hit the targets that you program in. ... The human is still launching the weapon and specifying the type of targets to be engaged, even if the weapon is choosing the specific targets to attack within that wide area. There's always going to be a man or woman in the loop who's going to make the targeting decision," he said, even if that targeting decision was now at a higher level.

Work contrasted these narrow AI systems with artificial general intelligence (AGI), "where the AI is actually making these decisions on its own." This is where Work would draw the line. "The danger is if you get a general AI system and it can rewrite its own code. That's the danger. We don't see ever putting that much AI power into any given weapon. But that would be the danger I think that people are worried about. What happens if Skynet rewrites its own code and says, 'humans are the enemy now'? But that I think is very, very, very far in the future because general AI hasn't advanced to that." Even if technology did get there, Work was not so keen on using it. "We will be extremely careful in trying to put general

AI into an autonomous weapon," he said. "As of this point I can't get to a place where we would ever launch a general AI weapon ... [that] makes all the decisions on its own. That's just not the way that I would ever foresee the United States pursuing this technology. [Our approach] is all about empowering the human and making sure that the humans inside the battle network has tactical and operational overmatch against their enemies."

Work recognized that other countries may use AI technology differently. "People are going to use AI and autonomy in ways that surprise us," he said. Other countries might deploy weapons that "decide who to attack, when to attack, how to attack" all on their own. If they did, then that could change the U.S. calculus. "The only way that we would go down that path, I think, is if it turns out our adversaries do and it turns out that we are at an operational disadvantage because they're operating at machine speed and we're operating at human speeds. And then we might have to rethink our theory of the case." Work said that challenge is something he worries about. "The nature of the competition about how people use AI and autonomy is really going to be something that we cannot control and we cannot totally foresee at this point."

THE PAST AS A GUIDE TO THE FUTURE

Work forthrightly answered every question I put to him, but I still found myself leaving the interview unsatisfied. He had made clear that he was comfortable using narrow AI systems to perform the kinds of tasks we're doing today: endgame autonomy to confirm a target chosen by a human or defensive human-supervised autonomy like at Aegis. He was comfortable with loitering weapons that might operate over a wider area or smarter munitions that could prioritize targets, but he continued to see humans playing a role in launching and directing those weapons. There were some technologies Work wasn't comfortable with—artificial general intelligence or "boot-strapping" systems that could modify their own code. But there was a wide swath of systems in between. What about a uninhabited combat aircraft that made its own targeting decisions? How much target error was acceptable? He simply didn't know. Those are questions future defense leaders would have to address.

To help shed light on how future leaders might answer those questions, I turned to Dr. Larry Schuette, director of research at the Office of Naval Research. Schuette is a career scientist with the Navy and has a doctorate in electrical engineering, so he understands the technology intimately. ONR has repeatedly been at the forefront of advancements in autonomy and robotics, and Schuette directs much of this research. He is also an avid student of history, so I hoped he could help me understand what the past might tell us about the shape of things to come.

As a researcher, Schuette made it clear to me that autonomous weapons are not an area of focus for ONR. There are a lot of areas where uninhabited and autonomous systems could have value, but his perspective was to focus on the mundane tasks. "I'm always looking for: what's the easiest thing with the highest return on investment that we could actually go do where people would thank us for doing it. . . . Don't go after the hard missions. . . . Let's do the easy stuff first." Schuette pointed to thankless jobs like tanking aircraft or cleaning up oil spills. "Be the trash barge. . . . The people would love you." His view was that even tackling these simple, unobjectionable missions was a big enough challenge. "I know that what is simple to imagine in science and technology isn't as simple to do."

Schuette also emphasized that he didn't see a compelling operational need for autonomous weapons. Today's model of "The man pushes a button and the weapon goes autonomous from there but the man makes the decision" was a "workable framework for some large fraction of what you would want to do with unmanned air, unmanned surface, unmanned underwater, unmanned ground vehicles. . . . I don't see much need in future warfare to get around that model," he said.

As a student of history, however, Schuette had a somewhat different perspective. His office looked like a naval museum, with old ship's logs scattered on the bookshelves and black-and-white photos of naval aviators on the walls. While speaking, Schuette would frequently leap out of his chair to grab a book about unrestricted submarine warfare or the Battle of Guadalcanal to punctuate his point. The historical examples weren't about autonomy, rather they were about a broader pattern in warfare. "History is full of innovations and asymmetric responses," he said. In World War II, the Japanese were "amazed" at U.S. skill at naval surface gunfire. In response, they decided to fight at night, resulting in devastating nighttime naval surface action at the Battle of Guadalcanal. The lesson is that "the threat gets a vote." Citing Japanese innovations in long-range torpedoes, Schuette said, "We had not planned on fighting a torpedo war. . . . The Japanese had a different idea."

This dynamic of innovation and counter-innovation inevitably leads to surprises in warfare and can often change what militaries see as ethical or appropriate. "We've had these debates before about ethical use of X or Y," Schuette pointed out. He compared today's debates about autonomous weapons to debates in the U.S. Navy in the interwar period between World War I and World War II about unrestricted submarine warfare. "We went all of the twenties, all the thirties, talking about how unrestricted submarine warfare was a bad idea we would never do it. And when the shit hit the fan the first thing we did was begin executing unrestricted submarine warfare." Schuette grabbed a book off his shelf and quoted the order issued to all U.S. Navy ship and submarine commanders on December 7, 1941, just four and a half hours after the attack at Pearl Harbor:

EXECUTE AGAINST JAPAN UNRESTRICTED AIR
AND SUBMARINE WARFARE

The lesson from history, Schuette said, was that "we are going to be violently opposed to autonomous robotic hunter-killer systems until we decide we can't live without them." When I asked him what he thought would be the decisive factor, he had a simple response: "Is it December eighth or December sixth?"

7

WORLD WAR R

ROBOTIC WEAPONS AROUND THE WORLD

The robotics revolution isn't American-made. It isn't even American-led. Countries around the world are pushing the envelope in autonomy, many further and faster than the United States. Conversations in U.S. research labs and the Pentagon's E-ring are only one factor influencing the future of autonomous weapons. Other nations get a vote too. What they do will influence how the technology develops, proliferates, and how other nations—including the United States—react.

The rapid proliferation of drones portends what is to come for increasingly autonomous systems. Drones have spread to nearly a hundred countries around the globe, as well as non-state groups such as Hamas, Hezbollah, ISIS, and Yemeni Houthi rebels. Armed drones are next. A growing number of countries have armed drones, including nations that are not major military powers such as South Africa, Nigeria, and Iraq.

Armed robots are also proliferating on the ground and at sea. South Korea has deployed a robot sentry gun to its border with North Korea. Israel has sent an armed robotic ground vehicle, the Guardium, on patrol near the Gaza border. Russia is building an array of ground combat robots and has plans for a robot tank. Even Shiite militias in Iraq have gotten in on the game, fielding an armed ground robot in 2015.

Armed robots are heading to sea as well. Israel has also developed an

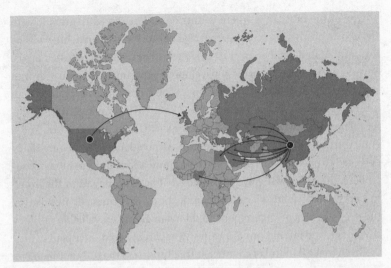

Armed Drone Proliferation *As of June 2017, sixteen countries possessed armed drones: China, Egypt, Iran, Iraq, Israel, Jordan, Kazakhstan, Myanmar, Nigeria, Pakistan, Saudi Arabia, Turkey, Turkmenistan, United Arab Emirates, the United Kingdom, and the United States. Some nations developed armed drones indigenously, while others acquired the technology from abroad. Over 90 percent of international armed drone transfers (shown on the map via arrows) have been from China.*

armed uninhabited boat, the Protector, to patrol its coast. Singapore has purchased the Protector and deployed it for counterpiracy missions in the Straits of Malacca. Even Ecuador has an armed robot boat, the ESGRUM, produced entirely indigenously. Armed with a rifle and rocket launcher, the ESGRUM will patrol Ecuadorian waterways to counter pirates.

As in the United States, the key question will be whether these nations plan to cross the line to full autonomy. No nation has stated they plan to build autonomous weapons. Few have ruled them out either. Only twenty-five nations have said they support a ban on lethal autonomous weapons: Pakistan, Ecuador, Egypt, the Holy See, Cuba, Ghana, Bolivia, Palestine, Zimbabwe, Algeria, Costa Rica, Mexico, Chile, Nicaragua, Panama, Peru, Argentina, Venezuela, Guatemala, Brazil, Iraq, Uganda, Austria, Colombia, and Djibouti (as of October 2018). None of these states are major military powers and some, such as Costa Rica or the Holy See, lack a military entirely.

One of the first areas where countries will be forced to grapple with the choice of whether to delegate lethal authority to the machine will be for uninhabited combat aircraft designed to operate in contested areas. Several nations are reportedly developing experimental combat drones similar to the X-47B, although for operation from land bases rather than aircraft carriers. These include the United Kingdom's Taranis, China's Sharp Sword, Russia's Skat, France's nEUROn, India's Aura, and a rumored unnamed Israeli stealth drone. Although these drones are likely designed to operate with protected communications links to human controllers, militaries will have to decide what actions they want the drone to carry out if (and when) communications are jammed. Restricting the drone's rules of engagement could mean giving up valuable military advantage, and few nations are being transparent about their plans.

Given that a handful of countries already possess the fully autonomous Harpy, it isn't a stretch to imagine them and others authorizing a similar level of autonomy with a recoverable drone. Whether countries are actually building those weapons today is more difficult to discern. If understanding what's happening inside the U.S. defense industry is difficult, peering behind the curtain of secret military projects around the globe is even harder. Are countries like Russia, China, the United Kingdom, and Israel building autonomous weapons? Or are they still keeping humans in the loop, walking right up to the line of autonomous weapons but not crossing it? Four high-profile international programs, a South Korean robot gun, a British missile, a British drone, and a Russian fleet of armed ground robots, show the difficulty in uncovering what nations around the globe are doing.

THE CURIOUS CASE OF THE AUTONOMOUS SENTRY BOT

South Korea's Samsung SGR-A1 robot is a powerful example of the challenge in discerning how much autonomy weapon systems have. The SGR-A1 is a stationary armed sentry robot designed to defend South Korean's border against North Korea. In 2007, when the robot was revealed, the electrical engineering magazine *IEEE Spectrum* reported it had a fully autonomous mode for engaging targets on its own. In an interview with the magazine, Samsung principal research engineer Myung Ho Yoo said, "the ultimate decision about shooting should be made by a human, not the robot." But the article made clear that Yoo's "should" was not a requirement, and that the robot did have a fully automatic option.

The story was picked up widely, with the SGR-A1 cited as an example of a real-world autonomous weapon by *The Atlantic*, the BBC, NBC, *Popular Science*, and *The Verge*. The SGR-A1 made *Popular Science*'s list of "Scariest Ideas in Science" with PopSci asking, "WHY, GOD? WHY?" Several academic researchers conducting in-depth reports on military robotics similarly cited the SGR-A1 as fully autonomous.

In the face of this negative publicity, Samsung backpedaled, saying that in fact a human *was* required to be in the loop. In 2010, a spokesperson for Samsung clarified that "the robots, while having the capability of automatic surveillance, cannot automatically fire at detected foreign objects or figures." Samsung and the South Korean government have been tight-lipped about details, though, and one can understand why. The SGR-A1 is designed to defend South Korea's demilitarized zone along its border with North Korea, with whom South Korea is technically still at war. Few countries on earth face as immediate and intense a security threat. One million North Korean soldiers and the threat of nuclear weapons loom over South Korea like a menacing shadow. In the same interview in which he asserted a human will always remain in the loop, the Samsung spokesperson asserted, "the SGR-1 can and will prevent wars."

What are the actual specifications and design parameters for the SGR-A1? It's essentially impossible to know without directly inspecting the robot. If Samsung says a human is in the loop, all we can do is take their word for it. If South Korea is willing to delegate more autonomy to their robots than other nations, however, it wouldn't be surprising. Defending the DMZ against North Korea is a matter of survival for South Korea. Accepting the risks of a fully autonomous sentry gun may be more than worth it for South Korea if it enhances deterrence against North Korea.

THE BRIMSTONE MISSILE

Similar to the U.S. LRASM, the United Kingdom's Brimstone missile has come under fire from critics who have questioned whether it has too much autonomy. The Brimstone is an aircraft-launched fire-and-forget missile designed to destroy ground vehicles or small boats. It can accomplish this mission in a variety of ways.

Brimstone has two primary modes of operation: Single Mode and Dual Mode. In Single Mode, a human "paints" the target with a laser and the missile homes in on the laser reflection. The missile will go wherever the human points the laser, allowing the human to provide "guidance all the way to the target." Dual Mode combines the laser guidance with a millimeter-wave (MMW) radar seeker for "fast moving and maneuvering targets and under narrow Rules of Engagement." The human designates the target with a laser, then there is a "handoff" from the laser to the MMW seeker at the final stage so the weapon can home in on fast moving targets. In both modes of operation, the missile is clearly engaging targets that have been designated by a human, making it a semiautonomous weapon.

However, the developer also advertises another mode of operation, "a previously-developed fire-and-forget, MMW-only mode" that can be enabled "via a software role change." The developer explains:

> This mode provides through-weather targeting, kill box-based discrimination and salvo launch. It is highly effective against multitarget armor formations. Salvo-launched Brimstones self-sort based on firing order, reducing the probability of overkill for increased one-pass lethality.

This targeting mode would allow a human to launch a salvo of Brimstones against a group of enemy tanks, letting the missiles sort out which missiles hit which tank. According to a 2015 *Popular Mechanics* article, in this mode the Brimstone is fairly autonomous:

> It can identify, track, and lock on to vehicles autonomously. A jet can fly over a formation of enemy vehicles and release several Brimstones to find targets in a single pass. The operator sets a "kill box" for Brimstone, so it will only attack within a given area. In one demonstration, three missiles hit three target vehicles while ignoring nearby neutral vehicles.

On the Brimstone's spec sheet, the developer also describes a similar functionality against fast-moving small boats, also called fast inshore attack craft (FIAC):

> In May 2013, multiple Brimstone missiles operating in an autonomous [millimeter] wave (MMW) mode completed the world's first single button, salvo engagement of multiple FIAC, destroying three vessels (one moving) inside a kill box, while causing no damage to nearby neutral vessels.

When operating in MMW-only mode, is the Brimstone an autonomous weapon? While the missile has a reported range in excess of 20 kilometers, it cannot loiter to search for targets. This means that the human operator must know there are valid targets—ground vehicles or small boats—within the kill box before launch in order for the missile to be effective.

The Brimstone can engage these targets using some innovative features. A pilot can launch a salvo of multiple Brimstones against a group of targets within a kill box and the missiles themselves "self-sort based on firing order" to hit different targets. This makes the Brimstone especially useful for defending against enemy swarm attacks. For example, Iran has harassed U.S. ships with swarming small boats that could overwhelm ship defenses, causing a USS *Cole*-type suicide attack. Navy helicopters armed with Brimstones would be an extremely effective defense against boat swarms, allowing pilots to take out an entire group of enemy ships at once without having to individually target each ship.

Even with all of the Brimstone's features, the human user still needs to launch it at a known group of targets. Because it cannot loiter, if there weren't targets in the kill box when the missile activated its seeker, the missile would be wasted. Unlike a drone, the missile couldn't return to base. The salvo launch capability allows the pilot to launch multiple missiles against a swarm of targets, rather than select each one individually. This makes a salvo of Brimstones similar to the Sensor Fuzed Weapon that is used to take out a column of tanks. Even though the missiles themselves might self-sort which missile hits which target, the human is still deciding to attack that specific cluster of targets. Even in MMW-only mode, the Brimstone is a semiautonomous weapon.

The line between the semiautonomous Brimstone and a fully autonomous weapon that would choose its own targets is a thin one. It isn't based on the seeker or the algorithms. The same seeker and algorithms could be used on a future weapon that *could* loiter over the battlespace—a missile with an upgraded engine or a drone that could patrol an area. A future weapon that patrolled a kill box, rather than entered one at a snapshot in time, would be an autonomous weapon, because the human could send the weapon to monitor the kill box without knowledge of any specific targets. It would allow the human to fire the weapon "blind" and let the weapon decide if and when to strike targets.

Even if the Brimstone doesn't quite cross the line to an autonomous weapon, it takes one more half step toward it, to the point where all that is needed is a light shove to cross the line. A MMW-only Brimstone could be converted into a fully autonomous weapon simply by upgrading the missile's engine so that it could loiter for longer. Or the MMW-only mode algorithms and seeker could be placed on a drone. Notably, the MMW-only mode is enabled in the missile by a software change. As autonomous technology continues to advance, more missiles around the globe will step right up to—or cross—that line.

Would the United Kingdom be willing to cross that line? The debate surrounding another British program, the Taranis drone, shows the difficulty in ascertaining how far the British might be willing to push the technology.

THE TARANIS DRONE

The Taranis is a next-generation experimental combat drone similar to those being developed by the United States, India, Russia, China, France, and Israel. BAE Systems, developer of the Taranis, has given one of the most extensive descriptions of how a combat drone's autonomy might work for weapons engagements. Similar to the X-47B, the Taranis is a demonstrator airplane, but the British military intends to carry the demonstration further than the United States and conduct simulated weapons engagements with the Taranis.

Information released by BAE shows how Taranis might be employed. It explains a simulated weapons test that "will demonstrate the ability of [an unmanned combat aircraft system] to: fend off hostile attack; deploy weapons deep in enemy territory and relay intelligence information." In the test:

1. Taranis would reach the search area via a preprogrammed flight path in the form of a three-dimensional corridor in the sky. Intelligence would be relayed to mission command.
2. When Taranis identifies a target it would be verified by mission command.
3. On the authority of mission command, Taranis would carry out a simulated firing and then return to base via the programmed flight path.

At all times, Taranis will be under the control of a highly-trained ground crew. The Mission Commander will both verify targets and authorise simulated weapons release.

This protocol keeps the human in the loop to approve each target, which is consistent with other statements by BAE leadership. In a 2016 panel at the World Economic Forum in Davos, BAE Chairman Sir Roger Carr described autonomous weapons as "very dangerous" and "fundamentally wrong." Carr made clear that BAE only envisioned developing weapons that kept a connection to a human who could authorize and remain responsible for lethal decision-making.

In a 2016 interview, Taranis program manager Clive Marrison made a similar statement that "decisions to release a lethal mechanism will always require a human element given the Rules of Engagement used by the UK in the past." Marrison then hedged, saying, "but the Rules of Engagement could change."

The British government reacted swiftly. Following multiple media articles alleging BAE was building in the option for Taranis to "attack targets of its own accord," the UK government released a statement the next day stating:

The UK does not possess fully autonomous weapon systems and has no intention of developing or acquiring them. The operation of our weapons will always be under human control as an absolute guarantee of human oversight, authority and accountability for their use.

The British government's full-throated denial of autonomous weapons would appear to be as clear a policy statement as there could be, but an important asterisk is needed regarding how the United Kingdom defines an "autonomous weapon system." In its official policy expressed in the UK Joint Doctrine Note 2/11, "The UK Approach to Unmanned Aircraft Systems," the British military describes an autonomous system as one that "must be capable of achieving the same level of situational understanding as a human." Short of that, a system is defined as "automated." This definition of autonomy, which hinges on the complexity of the system rather than its function, is a different way of using the term "autonomy" than many others in discussions on autonomous weapons, including the U.S. government. The United Kingdom's stance is not a product of sloppy language; it's a deliberate choice. The UK doctrine note continues:

> As computing and sensor capability increases, it is likely that many systems, using very complex sets of control rules, will appear and be described as autonomous systems, but as long as it can be shown that the system logically follows a set of rules or instructions and is not capable of human levels of situational understanding, then they should only be considered to be automated.

This definition shifts the lexicon on autonomous weapons dramatically. When the UK government uses the term "autonomous system," they are describing systems with human-level intelligence that are more analogous to the "general AI" described by U.S. Deputy Defense Secretary Work. The effect of this definition is to shift the debate on autonomous weapons to far-off future systems and away from potential near-term weapon systems that may search for, select, and engage targets on their own—what others might call "autonomous weapons." Indeed, in its 2016 statement to the United Nations meetings on autonomous weapons, the United Kingdom stated: "The UK believes that [lethal autonomous weapon systems] do not, and may never, exist." That is to say, Britain may develop weapons that would search for, select, and engage targets on their own; it simply would call them "automated weapons," not "autonomous weapons." In fact, the UK doctrine note refers to systems such as the Phalanx gun (a supervised autonomous weapon) as "fully automated weapon systems." The doctrine note leaves open the possibility of their development, provided they pass a legal weapons review showing they can be used in a manner compliant with the laws of war.

In practice, the British government's stance on autonomous weapons is not dissimilar from that expressed by U.S. defense officials. Humans will remain involved in lethal decision-making . . . at some level. That might mean a human operator launching an autonomous/automated weapon into an area and delegating to it the authority to search for and engage targets on its own. Whether the public would react differently to such a weapon if it were rebranded an "automated weapon" is unclear.

Even if the United Kingdom's stance retains some flexibility, there is still a tremendous amount of transparency into how the U.S. and UK governments are approaching the question of autonomous weapons. Weapons developers like BAE, MBDA, and Lockheed Martin have detailed descriptions of their weapon systems on their websites, which is not uncommon for defense companies in democratic nations. DARPA describes its research programs publicly and in detail. Defense officials in both countries openly engage in a dialogue about the boundaries of autonomy and the appropriate role of humans and machines in lethal force. This transparency stands in stark contrast to authoritarian regimes.

RUSSIA'S WAR BOTS

While the United States has been very reluctant to arm ground robots, with only one short-lived effort during the Iraq war and no developmental programs for armed ground robots, Russia has shown no such hesitation. Russia is developing a fleet of ground combat robots for a variety of missions, from protecting critical installations to urban combat. Many of Russia's ground robots are armed, ranging from small robots to augment infantry troops to robotic tanks. How much autonomy Russia is willing to place into its ground robots will have a profound impact on the future of land warfare.

The Platform-M, a tracked vehicle roughly the size of a four-wheeler armed with a grenade launcher and an assault rifle, is on the smaller scale of Russian war bots. In 2014, the Platform-M took part in an urban combat exercise alongside Russian troops. According to an official statement from the Russian military, "the military robots were assigned to eliminate provisional illegal armed formations in urban conditions and striking stationary and mobile targets." The Russian military did not describe the degree of the Platform-M's autonomy, although according to the developer:

> Platform-M . . . is used for gathering intelligence, for discovering and
> eliminating stationary and mobile targets, for firepower support, for
> patrolling and for guarding important sites. The unit's weapons can
> be guided, it can carry out supportive tasks and it can destroy targets
> in automatic or semiautomatic control systems; it is supplied with
> optical-electronic and radio reconnaissance locators.

The phrase "can destroy targets in automatic . . . control" makes it sound like an autonomous weapon. This claim should be viewed with some skepticism. For one, videos of Russian robots show soldiers selecting targets on a computer screen. More importantly, the reality is that detecting targets autonomously in a ground combat environment is far more technically challenging than targeting enemy radars as the Harpy does or enemy ships on the high seas like TASM. The weapons Platform-M carries—a grenade launcher and assault rifle—would be effective against people, not armored vehicles like tanks or armored personnel carriers. People don't emit in the electromagnetic spectrum like radars. They aren't "cooperative targets." At the time this claim was made in 2014, autonomously finding a person in a cluttered ground combat environment would have been difficult. Advances in neural nets have changed this in the past few years, making it easier to identify people. But discerning friend from foe would still be a challenge.

The autonomous target identification problem Russian war bots face is far more challenging than the South Korean sentry gun on the DMZ. In a demilitarized zone such as that separating North and South Korea, a country might decide to place stationary sentry guns along the border and authorize them to shoot anything with an infrared (heat) signature coming across. Such a decision would not be without its potential problems. Sentry guns that lack any ability to discriminate valid military targets from civilians could senselessly murder innocent refugees attempting to flee an authoritarian regime. In general, though, a DMZ is a more controlled environment than offensive urban combat operations. Authorizing static, defensive autonomous weapons that are fixed in place would be far different than roving autonomous weapons that would be intended to maneuver in urban areas where combatants are mixed in among civilians.

Technologies exist today that could be used for automatic responses against military targets, if the Russians wanted to give such a capability to the Platform-M. The technology is fairly crude, though. For example, the Boomerang shot detection system is a U.S. system that uses an array of microphones to detect incoming bullets and calculate their origin. According to the developer, "Boomerang uses passive acoustic detection and computer-based signal processing to locate a shooter in less than a second." By comparing the relative time of arrival of a bullet's shock wave at the various microphones, Boomerang and other shot detection systems can pinpoint a shooter's direction. It can then call out the location of a shot, for example, "Shot. Two o'clock. 400 meters." Alternatively, acoustic shot detection systems can be directly connected to a camera or remote weapon station and automatically aim them at the shooter. Going the next step to allow the gun to automatically fire back at the shooter would not be technically challenging. Once the shot has been detected and the gun aimed, all that it would take would be to pull the trigger.

It's possible this is what Russia means when it says the Platform-M "can destroy targets in automatic ... control." From an operational perspective, however, authorizing automatic return-fire would be quite hazardous. It would require an extreme confidence in the ability of the shot detection system to weed out false positives and to not be fooled by acoustic reflections and echoes, especially in urban areas. Additionally, the gun would have no ability to account for collateral damage—say, to hold fire because the shooter is using human shields. Finally, such a system would be a recipe for fratricide, with robot systems potentially automatically shooting friendly troops or other friendly robots. Two robots on the same side could become trapped in a never-ending loop of automatic fire and response, mindlessly exchanging gunfire until they exhausted their ammunition or destroyed each other. It is unclear whether this is what Russia intends, but from a technical standpoint it would possible.

Russia's other ground combat robots scale up in size and sophistication from the Platform-M. The MRK-002-BG-57 "Wolf-2" is the size of a small car and outfitted with a 12.7 mm heavy machine gun. According to David Hambling of *Popular Mechanics*, "In the tank's automated mode, the operator can remotely select up to 10 targets, which the robot then bombards. Wolf-2 can act on its own to some degree (the makers are vague about what degree), but the decision to use lethal force is ultimately under human control." The Wolf-2 sits among a family of similar size robot vehicles. The amphibious Argo is roughly the size of a Mini Cooper, sports a machine gun and rocket-propelled grenade launcher, and can swim at speeds up to 2.5 knots. The A800 Mobile Autonomous Robotic System (MARS) is an (unarmed) infantry support vehicle the size of a compact car that can carry four infantry soldiers and their gear. Pictures online show Russian soldiers riding on the back, looking surprisingly relaxed as the tracked robot cruises through an off-road course.

Compact car–sized war bots aren't necessarily unique to Russia, although the Russian military seems to have a casual attitude toward arming them not seen in Western nations. The Russian military isn't stopping at midsize ground robots, though. Several Russian programs are pushing the boundaries of what is possible with robotic combat vehicles, building systems that could prove decisive in highly lethal tank-on-tank warfare.

The Uran-9 looks like something straight out of a *MechWarrior* video game, where players pilot a giant robot warrior armed with rockets and cannons. The Uran-9 is fully uninhabited, although it is controlled by soldiers remotely from a nearby command vehicle. It is the size of a small armored personnel carrier, sports a 30 mm cannon, and has an elevated platform to launch antitank guided missiles. The elevated missile platform that gives the Uran-9 a distinctive sci-fi appearance. The missiles rest on two platforms on either side of the vehicle that, when raised, look like arms reaching into the sky. The elevated platform allows the robot to fire missiles while safely sitting behind cover, for example behind the protective slope of a hillside. In an online promotional video from the developer, Rosoboronexport, slo-mo shots of the Uran-9 firing antitank missiles are set to music reminiscent of a Tchaikovsky techno remix.

The Uran-9 is a major step beyond smaller robotic platforms like the Platform-M and Wolf-2 not just because it's larger, but because its larger size allows it to carry heavier weapons capable of taking on antitank missions. Whereas the assault rifle and grenade launcher on a Platform-M would do very little to a tank, the Uran-9's antitank missiles would be potentially highly lethal. This makes the Uran-9 potentially a useful weapon in high-intensity combat against NATO forces on the plains of Europe. Uran-9s could hide behind hillsides or other protective cover and launch missiles against NATO tanks. The Uran-9 doesn't have the armor or guns to stand toe-to-toe against a modern tank, but because it's uninhabited, it doesn't have to. The Uran-9 could be a successful ambush predator. Even if firing its missiles exposed its position and led it to be taken out by NATO forces, the exchange might still be a win if it took out a Western tank. Because there's no one inside it and the Uran-9 is significantly smaller than a tank, and therefore presumably less expensive, Russia could field many of them on the battlefield. Just like many stings from a hornet can bring down a much larger animal, the Uran-9 could make the modern battlefield a deadly place for Western forces.

Russia's Vikhr "robot tank" has a similar capability. At 14 tons and lacking a main gun, it is significantly smaller and less lethal than a 50- to 70-ton main battle tank. Like the Uran-9, though, its 30 mm cannon and six antitank missiles show it is designed as a tank-killing ambush predator, not a tank-on-tank street fighter. The Vikhr is remote controlled, but news reports indicate it has the ability to "lock onto a target" and keep firing until the target is destroyed. While not the same as choosing its own target, tracking a moving target is doable today. In fact, tracking moving objects is as a standard feature on DJI's base model Spark hobby drone, which retails for under $500.

Taking the next step and allowing the Uran-9 or Vikhr to autonomously target tanks would take some additional work, but it would be more feasible than trying to accurately discriminate among human targets. With large cannons and treads, tanks are distinctive military vehicles not easily confused with civilian objects. Moreover, militaries may be more willing to risk civilian casualties or fratricide in the no-holds-

barred arena of tank warfare, where armored divisions vie for dominance and the fate of nations is at stake. In videos of the Uran-9, human
operators can be clearly seen controlling the vehicle, but the technology
is available for Russia to authorize fully autonomous antitank engagements, if it chose to do so.

Russia isn't stopping at development of the Vikhr and Uran-9, however. It envisions even more advanced robotic systems that could not
only ambush Western tanks, but stand with them toe-to-toe and win.
Russia reportedly has plans to develop a fully robotic version of its next-
generation T-14 Armata tank. The T-14 Armata, which reportedly entered
production as of 2016, sports a bevy of new defensive features, including advanced armor, an active protection system to intercept incoming
antitank missiles, and a robotic turret. The T-14 will be the first main
battle tank to sport an uninhabited turret, which will afford the crew
greater protection by sheltering them within the body of the vehicle.
Making the entire tank uninhabited would be the next logical step in
protection, enabling a crew to control the vehicle remotely. While current T-14s are human-inhabited, Russia has long-term plans to develop
a fully robotic version. Vyacheslav Khalitov, deputy director general of
UralVagonZavod, manufacturer of the T-14 Armata, has stated, "Quite
possibly, future wars will be waged without human involvement. That is
why we have made provisions for possible robotization of Armata." He
acknowledged that achieving the goal of full robotization would require
more advanced AI that could "calculate the situation on the battlefield
and, on this basis, to take the right decision."

In addition to pushing the boundaries on robots' physical characteristics, the Russian military has signaled it intends to use cutting-edge AI to
boost its robots' decision-making. In July 2017, Russian arms manufacturer Kalashnikov stated that they would soon release "a fully automated
combat module" based on neural networks. News reports indicate the
neural networks would allow the combat module "to identify targets and
make decisions." As in other cases, it is difficult to independently evaluate
these claims, but they signal a willingness to use artificial intelligence for
autonomous targeting. Russian companies' boasting of autonomous features has none of the hesitation or hedging that is often seen from American or British defense firms.

Senior Russian military commanders have stated they intend to move toward fully robotic weapons. In a 2013 article on the future of warfare, Russian military chief of staff General Valery Gerasimov wrote:

> Another factor influencing the essence of modern means of armed conflict is the use of modern automated complexes of military equipment and research in the area of artificial intelligence. While today we have flying drones, tomorrow's battlefields will be filled with walking, crawling, jumping, and flying robots. In the near future it is possible a fully robotized unit will be created, capable of independently conducting military operations.
>
> How shall we fight under such conditions? What forms and means should be used against a robotized enemy? What sort of robots do we need and how can they be developed? Already today our military minds must be thinking about these questions.

This Russian interest in pursuing fully robotic units has not escaped notice in the West. In December 2015, Deputy Secretary of Defense Bob Work mentioned Gerasimov's comments in a speech on the future of warfare. As Work has repeatedly noted, U.S. decisions may be shaped by those of Russia and other nations. This is the danger of an arms race in autonomy: that nations feel compelled to race forward and build autonomous weapons out of the fear that others are doing so, without pausing to weigh the risks of their actions.

AN ARMS RACE IN AUTONOMOUS WEAPONS?

If it is true, as some have suggested, that a dangerous arms race in autonomous weapons is under way, then it is a strange kind of race. Nations are pursuing autonomy in many aspects of weaponry but, with the exception of the Harpy, are still keeping humans in the loop for now. Some weapons like Brimstone use autonomy in novel ways, pushing the boundaries of what could be considered a semiautonomous weapon. DARPA's CODE program appears to countenance moving to human-*on*-the-loop supervisory control for some types of targets, but there is no indication of full autonomy. Developers of the SGR-A1 gun and Taranis drone have suggested full autonomy could be a future option, although higher authorities immediately disputed the claim, saying that was not their intent.

Rather than a full-on sprint to build autonomous weapons, it seems that many nations do not yet know whether they might want them in the future and are hedging their bets. One challenge in understanding the global landscape of lethal autonomy is that the degree of transparency among nations differs greatly. While the official policies of the U.S. and UK governments leave room to develop autonomous weapons (although they express this differently with the United Kingdom calling them "automated weapons") countries such as Russia don't even have a public policy. Policy discussions may be happening in private in authoritarian regimes, but we don't know what they are. Pressure from civil society for greater transparency differs greatly across countries. In 2016, the UK-based NGO Article 36, which has been a leading voice in shaping international discussions on autonomous weapons, wrote a policy brief critiquing the UK government's stance on autonomous weapons. In the United States, Stuart Russell and a number of well-respected colleagues from the AI community have met with mid-level officials from across the U.S. government to discuss autonomous weapons. In authoritarian Russia, there are no equivalent civil society groups to pressure the government to be more transparent about its plans. As a result, scrutiny focuses on the most transparent countries— democratic nations who are responsive to elements of civil society and are generally more open about their weapons development. What goes on in authoritarian regimes is far murkier, but no less relevant to the future path of lethal autonomy.

Looking across the global landscape of robotic systems, it's clear that many nations are pursuing armed robots, including combat drones that would operate in contested air space. How much autonomy some weapon systems have is unclear, but there is nothing preventing countries from crossing the line to lethal autonomy in their next-generation missiles, combat drones, or ground robots. Next-generation robotic systems such as the Taranis may give countries that option, forcing uncomfortable conversations. Even if many countries would rather not move forward with autonomous weapons, it may only take one to start a cascade of others.

With no autonomous smoking gun, it seems unnecessarily alarmist to declare that an autonomous weapons arms race is already under way, but we could very well be at the starting blocks. The technology to build autonomous weapons is widely available. Even non-state groups have armed robots. The only missing ingredient to turn a remotely controlled armed robot into an autonomous weapon is software. That software, it turns out, is pretty easy to come by.

8

GARAGE BOTS

DIY KILLER ROBOTS

A gunshot cuts through the low buzz of the drone's rotors. The camera jerks backward from the recoil. The gun fires again. A small bit of flame darts out of the handgun attached to the homemade-looking drone. Red and yellow wires snake over the drone and into the gun's firing mechanism, allowing the human controller to remotely pull the trigger.

The controversial fifteen-second video clip released in the summer of 2015 was taken by a Connecticut teenager of a drone he armed himself. Law enforcement and the FAA investigated, but no laws were broken. The teenager used the drone on his family's property in the New England woods. There are no laws against firing weapons from a drone, provided it's done on private property. A few months later, for Thanksgiving, he posted a video of a flamethrower-armed drone roasting a turkey.

Drones are not only in wide use by countries around the globe; they are readily purchased by anyone online. For under $500, one can buy a small quadcopter that can autonomously fly a route preprogrammed by GPS, track and follow moving objects, and sense and avoid obstacles in its path. Commercial drones are moving forward in leaps and bounds, with autonomous behavior improving in each generation.

When I asked the Pentagon's chief weapons buyer Frank Kendall what he feared, it wasn't Russian war bots, it was cheap commercial drones. A world where everyone has access to autonomous weapons is a far different one than a world where only the most advanced militaries can build them. If autonomous weapons could be built by virtually anyone in their garage, bottling up the technology and enforcing a ban, as Stuart Russell and others have advocated, would be extremely difficult. I wanted to know, could someone leverage commercially available drones to make a do-it-yourself (DIY) autonomous weapon? How hard would it be?

I was terrified by what I found.

HUNTING TARGETS

The quadcopter rose off the ground confidently, smoothly gaining altitude till it hovered around eye level. The engineer next to me tapped his tablet and the copter moved out, beginning its search of the house.

I followed along behind the quadcopter, watching it navigate each room. It had no map, no preprogrammed set of instructions for where to go. The drone was told merely to search and report back, and so it did. As it moved through the house it scanned each room with a laser range-finding LIDAR sensor, building a map as it went. Transmitted via Wi-Fi, the map appeared on the engineer's tablet.

As the drone glided through the house, each time it came across a doorway it stopped, its LIDAR sensor probing the space beyond. The drone was programmed to explore unknown spaces until it had mapped everything. Only then would it finish its patrol and report back.

I watched the drone pause in front of an open doorway. I imagined its sensors pinging the distant wall of the other room, its algorithms computing that there must be unexplored space beyond the opening. The drone hovered for a moment, then moved into the unknown room. A thought popped unbidden into my mind: *it's curious.*

It's silly to impart such a human trait to a drone. Yet it comes so naturally to us, to imbue nonhuman objects with emotions, thoughts, and intentions. I was reminded of a small walking robot I had seen in a university lab years ago. The researchers taped a face to one end of the robot—nothing fancy, just slices of colored construction paper in the shape of eyes, a nose, and a mouth. I asked them why. Did it help them remember which direction was forward? No, they said. It just made them feel better to put a face on it. It made the robot seem more human, more like us. There's something deep in human nature that wants to connect to another sentient entity, to know that it is like us. There's something alien and chilling about entities that can move intelligently through the world and not feel any emotion or thought beyond their own programming. There is something predatory and remorseless about them, like a shark.

I shook off the momentary feeling and reminded myself of what the technology was actually doing. The drone "felt" nothing. The computer controlling its actions would have identified that there was a gap where the LIDAR sensors could not reach and so, following its programming, directed the drone to enter the room.

The technology *was* impressive. The company I was observing, Shield AI, was demonstrating fully autonomous indoor flight, an even more impressive feat than tracking a person and avoiding obstacles outdoors. Founded by brothers Ryan and Brandon Tseng, the former an engineer and the latter a former Navy SEAL, Shield AI has been pushing the boundaries of autonomy under a grant from the U.S. military. Shield's goal is to field fully autonomous quadcopters that special operators can launch into an unknown building and have the drones work cooperatively to map the building on their own, sending back footage of the interior and potential objects of interest to the special operators waiting outside.

Brandon described their goal as "highly autonomous swarms of robots that require minimal human input. That's the end-state. We envision that the DoD will have ten times more robots on the battlefield than soldiers, protecting soldiers and innocent civilians." Shield's work is pushing the boundaries of what is possible today. All the pieces of the technology are falling into place. The quadcopter I witnessed was using LIDAR for navigation, but Shield's engineers explained they had tested visual-aided navigation; they simply didn't have it active that day.

Visual-aided navigation is a critically important piece of technology that will allow drones to move autonomously through cluttered environments without the aid of GPS. Visual-aided navigation tracks how objects move through the camera's field of view, a process called "optical flow." By assessing optical flow, operating on the assumption that most of the environment is static and not moving, fixed objects moving through the camera's field of vision can be used as a reference point for the drone's own movement. This can allow the drone to determine how it is moving within its environment without relying on GPS or other external navigation aids. Visual-aided navigation can complement other internal guidance mechanisms, such as inertial measurement units (IMU) that work like a drone's "inner ear," sensing changes in velocity. (Imagine sitting blindfolded in a car, feeling the motion of the car's acceleration, braking, and turning.) When IMUs and visual-aided navigation are combined, they make an extremely powerful tool for determining a drone's position, allowing the drone to accurately navigate through cluttered environments without GPS.

Visual-aided navigation has been demonstrated in numerous laboratory settings and will no doubt trickle down to commercial quadcopters over time. There is certain to be a market for quadcopters that can autonomously navigate indoors, from filming children's birthday parties to indoor drone racing. With visual-aided navigation and other features, drones and other robotic systems will increasingly be able to move intelligently through their environment. Shield AI, like many tech companies, was focused on near-term applications, but Brandon Tseng was bullish on the long-term potential of AI and autonomy. "Robotics and artificial intelligence are where the internet was in 1994," he told me. "Robotics and AI are about to have a really transformative impact on the world. . . . Where we see the technology 10 to 15 years down the road? It is going to be mind-blowing, like a sci-fi movie."

Autonomous navigation is not the same as autonomous targeting, though. Drones that can maneuver and avoid obstacles on their own—indoors or outdoors—do not necessarily have the ability to identify and discriminate among the various objects in their surroundings. They simply avoid hitting anything at all. Searching for specific objects and targeting them for action—whether it's taking photographs or something more nefarious—would require more intelligence.

The ability to do target identification is the key missing link in building a DIY autonomous weapon. An autonomous weapon is one that can search for, decide to engage, and engage targets. That requires three abilities: the ability to maneuver intelligently through the environment to search; the ability to discriminate among potential targets to identify the correct ones; and the ability to engage targets, presumably through force. The last element has already been demonstrated—people have armed drones on their own. The first element, the ability to autonomously navigate and search an area, is already available outdoors and is coming soon indoors. Target identification is the only piece remaining, the only obstacle to someone making an autonomous weapon in their garage. Unfortunately, that technology is not far off. In fact, as I stood in the basement of the building watching Shield AI's quadcopter autonomously navigate from room to room, autonomous target recognition was literally being demonstrated right outside, just above my head.

DEEP LEARNING

The research group asked that they not be named, because the technology was new and untested. They didn't want to give the impression that it was good enough—that the error rate was low enough—to be used for military applications. Nor, it was clear, were military applications their primary intention in designing the system. They were engineers, simply trying to see if they could solve a tough problem with technology. Could they send a small drone out entirely on its own to autonomously find a crashed helicopter and report its location back to the human?

The answer, it turns out, is yes. To understand how they did it, we need to go deep.

Deep learning neural networks, first mentioned in chapter 5 as one potential solution to improving military automatic target recognition in DARPA's TRACE program, have been the driving force behind astounding gains in AI in the past few years. Deep neural networks have learned to play Atari, beat the world's reigning champion at *go*, and have been behind dramatic improvements in speech recognition and visual object recognition. Neural networks are also behind the "fully automated combat module" that Russian arms manufacturer Kalashnikov claims to have built. Unlike traditional computer algorithms that operate based on a script of instructions, neural networks work by learning from large amounts of data. They are an extremely powerful tool for handling tricky problems that can't be easily solved by prescribing a set of rules to follow.

Let's say, for example, that you wanted to write down a rule set for how to visually distinguish an apple from a tomato without touching, tasting, or smelling. Both are round. Both are red and shiny. Both have a green stem on top. They *look* different, but the differences are subtle and evade easy description. Yet a three-year-old child can immediately tell the difference. This is a tricky problem with a rules-based approach. What neural networks do is sidestep that problem entirely. Instead, they learn from vast amounts of data—tens of thousands or millions of pieces of data. As the network churns through the data, it continually adapts its internal structure until it optimizes to achieve the correct programmer-specified goal. The goal could be distinguishing an apple from a tomato, playing an Atari game, or some other task.

In one of the most powerful examples of how neural networks can be used to solve difficult problems, the Alphabet (formerly Google) AI company DeepMind trained a neural network to play *go*, a Chinese strategy game akin to chess, better than any human player. *Go* is an excellent game for a learning machine because the sheer complexity of the game makes it very difficult to program a computer to play at the level of a professional human player based on a rules-based strategy alone.

The rules of *go* are simple, but from these rules flows vast complexity. *Go* is played on a grid of 19 by 19 lines and players take turns placing stones—black for one player and white for the other—on the intersection points of the grid. The objective is to use one's stones to encircle areas of the board. The player who controls more territory on the board wins. From these simple rules come an almost unimaginably large number of possibilities. There are more possible positions in *go* than there are atoms in the known universe, making *go* 10^{100} (one followed by a hundred zeroes) times—literally a googol—more complex than chess.

Humans at the professional level play *go* based on intuition and feel. *Go* takes a lifetime to master. Prior to DeepMind, attempts to build *go*-playing AI software had fallen woefully short of human professional players. To craft its AI, called AlphaGo, DeepMind took a different approach. They built an AI composed of deep neural networks and fed it data from 30 million games of *go*. As explained in a DeepMind blog post, "These neural networks take a description of the Go board as an input and process it through 12 different network layers containing millions of neuron-like connections." Once the neural network was trained on human games of *go*, DeepMind then took the network to the next level by having it play itself. "Our goal is to beat the best human players, not just mimic them," as explained in the post. "To do this, AlphaGo learned to discover new strategies for itself, by playing thousands of games between its neural networks, and adjusting the connections using a trial-and-error process known as reinforcement learning." AlphaGo used the 30 million human games of *go* as a starting point, but by playing against itself could reach levels of game play beyond even the best human players.

This superhuman game play was demonstrated in the 4–1 victory AlphaGo delivered over the world's top-ranked human *go* player, Lee Sedol, in March 2016. AlphaGo won the first game solidly, but in game 2 demonstrated its virtuosity. Partway through game 2, on move 37, AlphaGo made a move so surprising, so un-human, that it stunned professional players watching the match. Seemingly ignoring a contest between white and black stones that was under way in one corner of the board, AlphaGo played a black stone far away in a nearly empty part of the board. It was a surprising move not seen in professional games, so much so that one commentator remarked, "I thought it was a mistake." Lee Sedol was similarly so taken by surprise he got up and left the room. After he returned, he took fifteen minutes to formulate his response. AlphaGo's move wasn't a mistake. European *go* champion Fan Hui, who had lost to AlphaGo a few months earlier in a closed-door match, said at first the move surprised him as well, and then he saw its merit. "It's not a human move," he said. "I've never seen a human play this move. So beautiful." Not only did the move *feel* like a move no human player would never make, it was a move no human player probably would never make. AlphaGo rated the odds that a human would have made that move as 1 in 10,000. Yet AlphaGo made the move anyway. AlphaGo went on to win game 2 and afterward Lee Sedol said, "I really feel that AlphaGo played the near perfect game." After losing game 3, thus giving AlphaGo the win for the match, Lee Sedol told the audience at a press conference, "I kind of felt powerless."

AlphaGo's triumph over Lee Sedol has implications far beyond the game of *go*. More than just another realm of competition in which AIs now top humans, the way DeepMind trained AlphaGo is what really matters. As explained in the DeepMind blog post, "AlphaGo isn't just an 'expert' system built with hand-crafted rules; instead it uses general machine learning techniques to figure out for itself how to win at Go." DeepMind didn't program rules for how to win at *go*. They simply fed a neural network massive amounts of data and let it learn all on its own, and some of the things it learned were surprising.

In 2017, DeepMind surpassed their earlier success with a new version of AlphaGo. With an updated algorithm, AlphaGo Zero learned to play *go* without any human data to start. With only access to the board and the rules of the game, AlphaGo Zero taught itself to play. Within a mere three days of self-play, AlphaGo Zero had eclipsed the previous version that had beaten Lee Sedol, defeating it 100 games to 0.

These deep learning techniques can solve a variety of other problems. In 2015, even before DeepMind debuted AlphaGo, DeepMind trained a neural network to play Atari games. Given only the pixels on the screen and the game score as input and told to maximize the score, the neural network was able to learn to play Atari games at the level of a professional human video game tester. Most importantly, the same neural network architecture could be applied across a vast array of Atari games—forty-nine games in all. Each game had to be individually learned, but the same neural network architecture applied to any game; the researchers didn't need to create a customized network design for each game.

The AIs being developed for *go* or Atari are still narrow AI systems. Once trained, the AIs are purpose-built tools to solve narrow problems. AlphaGo can beat any human at *go*, but it can't play a different game, drive a car, or make a cup of coffee. Still, the tools used to train AlphaGo are generalizable tools that can be used to build any number of special-purpose narrow AIs to solve various problems. Deep neural networks have been used to solve other thorny problems that have bedeviled the AI community for years, notably speech recognition and visual object recognition.

A deep neural network was the tool used by the research team I witnessed autonomously find the crashed helicopter. The researcher on the project explained that he had taken an existing neural network that had already been trained on object recognition, stripped off the top few layers, then retrained the network to identify helicopters, which hadn't originally been in its image dataset. The neural network he was using was running off of a laptop connected to the drone, but it could just as easily have been running off of a Raspberry Pi, a $40 credit-card sized processor, riding on board the drone itself.

All of these technologies are coming from outside the defense sector.

They are being developed at places like Google, Microsoft, IBM, and university research labs. In fact, programs like DARPA's TRACE are not necessarily intended to invent new machine learning techniques, but rather import existing techniques into the defense sector and apply them to military problems. These methods are widely available to those who know how to use them. I asked the researcher behind the helicopter-hunting drone: Where did he get the initial neural network that he started with, the one that was already trained to recognize other images that weren't helicopters? He looked at me like I was either half-crazy or stupid. He got it online, of course.

NEURAL NETS FOR EVERYONE

I feel I should confess that I'm not a technologist. In my job as a defense analyst, I research military technology to make recommendations about where the U.S. military should invest to keep its edge on the battlefield, but I don't build things. My undergraduate degree was in science and engineering, but I've done nothing even remotely close to engineering since then. To claim my programming skills were rusty would be to imply that at one point in time they existed. The extent of my computer programming knowledge is a one-semester introductory course in C++ in college.

Nevertheless, I went online to check out the open-source software database the researcher pointed me to: TensorFlow. TensorFlow is an open-source AI library developed by Google AI researchers. With TensorFlow, Google researchers have taken what they have been learning with deep neural networks and passed it on to the rest of the world. On TensorFlow, not only can you download already trained neural networks and software for building your own, there are reams of tutorials on how to teach yourself deep learning techniques. For users new to machine learning, there are basic tutorials on classic machine learning problems. These tools make neural networks accessible to computer programmers with little to no experience in machine learning. TensorFlow makes neural networks easy, even fun. A tutorial called Playground (playground.tensorflow.org) allows users to modify and train a neural network through a point-and-click interface in the browser. No programming skills are required at all.

Once I got into Playground, I was hooked. Reading about what neural networks could do was one thing. Building your own and training it on data was entirely another. Hours of time evaporated as I tinkered with the simple network in my browser. The first challenge was training the network to learn to predict the simple datasets used in Playground—patterns of orange and blue dots across a two-dimensional grid. Once I'd mastered that, I worked to make the leanest network I could, composed of the fewest neurons in the fewest number of layers that could still accurately make predictions. (Reader challenge: once you've mastered the easy datasets, try the spiral.)

With the Playground tutorial, the concept of neural nets becomes accessible to someone with no programming skills at all. Using Playground is no more complicated than solving an easy-level Sudoku puzzle and within the range of an average seven-year-old. Playground won't let the user build a custom neural net to solve novel problems. It's an illustration of what neural nets can do to help users see their potential. Within other parts of TensorFlow, though, lie more powerful tools to use existing neural networks or design custom ones, all within reach of a reasonably competent programmer in Python or C++.

TensorFlow includes extensive tutorials on convolutional neural nets, the particular type of neural network used for computer vision. In short order, I found a neural network available for download that was already trained to recognize images. The neural network Inception-v3 is trained on the ImageNet dataset, a standard database of images used by programmers. Inception-v3 can classify images into one of 1,000 categories, such as "gazelle," "canoe," or "volcano." As it turns out, none of the categories Inception-v3 is trained on are those that could be used to identify people, such as "human," "person," "man," or "woman." So one could not, strictly speaking, use this particular neural network to power an autonomous weapon that targets people. Still, I found this to be little consolation. ImageNet isn't the only visual object classification database used for machine learning online and others, such as the Pascal Visual Object Classes database, include "person" as a category. It took me all of about ten seconds on Google to find trained neural networks available for download that could find human faces, determine age and gender, or label human emotions. All of the tools to build an autonomous weapon that could target people on its own were readily available online.

This was, inevitably, one of the consequences of the AI revolution. AI technology was powerful. It could be used for good purposes or bad purposes; that was up to the people using it. Much of the technology behind AI was software, which meant it could be copied practically for free. It could be downloaded at the click of a button and could cross borders in an instant. Trying to contain software would be pointless. Pandora's box has already been opened.

ROBOTS EVERYWHERE

Just because the tools needed to make an autonomous weapon were widely available didn't tell me how easy or hard it would be for someone to actually do it. What I wanted to understand was how widespread the technological know-how was to build a homemade robot that could harness state-of-the-art techniques in deep learning computer vision. Was this within reach of a DIY drone hobbyist or did these techniques require a PhD in computer science?

There is a burgeoning world of robot competitions among high school students, and this seemed like a great place to get a sense of what an amateur robot enthusiast could do. The FIRST Robotics Competition is one such competition that includes 75,000 students organized in over 3,000 teams across twenty-four countries. To get a handle on what these kids might be able to do, I headed to my local high school.

Less than a mile from my house is Thomas Jefferson High School for Science and Technology—"TJ," for short. TJ is a math and science magnet school; kids have to apply to get in, and they are afforded opportunities above and beyond what most high school students have access to. But they're still high school students—not world-class hackers or DARPA whizzes.

In the Automation and Robotics Lab at TJ, students get hands-on experience building and programming robots. When I visited, two dozen students sat at workbenches hunched over circuit boards or silently tapping away at computers. Behind them on the edges of the workshop lay discarded pieces of robots, like archeological relics of students' projects from semesters prior. On a shelf sat "Roby Feliks," the Rubik's Cube solving robot. Nearby, a Raspberry Pi processor sat atop a plastic musical recorder, wires running from the circuit board to the instrument like some musical cyborg. Somewhat randomly in the center of the floor sat a half-disassembled robot, the remnants of TJ's admission to the FIRST competition that year. Charles Dela Cuesta, the teacher in charge of the lab, apologized for the mess, but it was exactly what I imagined a robot lab should look like.

Dela Cuesta came across as the kind of teacher you pray your own children have. Laid back and approachable, he seemed more like a lovable assistant coach than an aloof disciplinarian. The robotics lab had the feel of a place where students learn by doing, rather than sitting and copying down equations from a whiteboard.

Which isn't to say that there wasn't a whiteboard. There was. It sat in a corner amid a pile of other robotic projects, with circuit boards and wires draped over it. Students were designing an automatic whiteboard with a robot arm that could zip across the surface and sketch out designs from a computer. On the whiteboard were a series of inhumanly straight lines sketched out by the robot. It was at this point that I wanted to quit my job and sign up for a robotics class at TJ.

Dela Cuesta explained that all students at TJ must complete a robotics project in their freshmen year as part of their required coursework. "Every student in the building has had to design a small robot that is capable of navigating a maze and performing some sort of obstacle avoidance," he said. Students are given a schematic of what the maze looks like so they get to choose how to solve the problem, whether to preprogram the robot's moves or take the harder path of designing an autonomous robot that can figure it out on its own. After this required class, TJ offers two additional semesters of robotics electives, which can be complemented with up to five computer science courses in which students learn Java, C++, and Python. These are vital programming tools for using robot control systems, like the Raspberry Pi processor, which runs on Linux and takes commands in Python. Dela Cuesta explained that even though most students come into TJ with no programming experience, many learn fast and some even take computer science courses over the summer to get ahead. "They can pretty much program in anything—Java, Python.... They're just all over the place," he said. Their senior year, all students at TJ must complete a senior project in an area of their choosing. Some of the most impressive robotics projects are those done by seniors who choose to make robotics their area of focus. Next to the whiteboard stood a bicycle propped up on its kickstand. A large blue box sat inside the frame, wires snaking out of it to the gear shifters. Dela Cuesta explained it was an automatic gear shifter for the bike. The box senses when it is time to shift and does so automatically, like an automatic transmission on a car.

The students' projects have been getting better over the years, Dela Cuesta explained, as they are able to harness more advanced open-source components and software. A few years ago, a class project to create a robot tour guide for the school took two years to complete. Now, the timeline has been shortened to nine weeks. "The stuff that was impressive to me five, six years ago we could accomplish in a quarter of the time now. It just blows my mind," he said. Still, Dela Cuesta pushes students to build things custom themselves rather than use existing components. "I like to have the students, as much as possible, build from scratch." Partly, this is because it's often easier to fit custom-built hardware into a robot, an approach that is possible because of the impressive array of tools Dela Cuesta has in his shop. Along a back wall were five 3-D printers, two laser cutters to make custom parts, and a mill to etch custom circuit boards. An even more important reason to have students do things themselves is they learn more that way. "Custom is where I want to go," Dela Cuesta said. "They learn a lot more from it. It's not just kind of this black box magic thing they plug in and it works. They have to really understand what they're doing in order to make these things work."

Across the hall in the computer systems lab, I saw the same ethos on display. The teachers emphasized having students do things themselves so they were learning the fundamental concepts, even if that meant re-solving problems that have already been solved. Repackaging opensource software isn't what the teachers are after. That isn't to say that students aren't learning from the explosion in open-source neural network software. On one teacher's desk sat a copy of Jeff Heaton's *Artificial Intelligence for Humans, Volume 3: Deep Learning and Neural Networks*. (This title begs the uncomfortable question whether there is a parallel course of study, *Artificial Intelligence for Machines*, where machines learn to program other machines. The answer, I suppose, is "Not yet.") Students are learning how to work with neural networks, but they're doing so from the bottom up. A junior explained to me how he trained a neural network to play tic-tac-toe—a problem that was solved over fifteen years ago, but remains a seminal coding problem. Next year, TJ will offer a course in computer vision that will cover convolutional neural networks.

Maybe it's a cliché to say that the projects students were working on are mind-blowing, but I was floored by the things I saw TJ students doing. One student was disassembling a Keurig machine and turning it into a net-enabled coffeemaker so it could join the Internet of Things. Wires snaked through it as though the internet was physically infiltrating the coffeemaker, like *Star Trek's* Borg. Another student was tinkering with something that looked like a cross between a 1980s Nintendo Power Glove and an Apple smartwatch. He explained it was a "gauntlet," like that used by Iron Man. When I stared at him blankly, he explained (in that patient explaining-to-an-old-person voice that young people use) that a gauntlet is the name for the wrist-mounted control that Iron Man uses to fly his suit. "Oh, yeah. That's cool," I said, clearly not getting it. I don't feel like I need the full functionality of my smartphone mounted on my wrist, but then again I wouldn't have thought ten years ago that I needed a touch-screen smartphone on my person at all times in the first place. Technology has a way of surprising us. Today's technology landscape is a democratized one, where game-changing innovations don't just come out of tech giants like Google and Apple but can come from anyone, even high-school students. The AI revolution isn't something that is happening *out there*, only in top-tier research labs. It's happening everywhere.

THE EVERYONE REVOLUTION

I asked Brandon Tseng from Shield AI where this path to ever-greater autonomy was taking us. He said, "I don't think we're ever going to give [robots] complete autonomy. Nor do I think we should give them complete autonomy." On one level, it's reassuring to know that Tseng, like nearly everyone I met working on military robotics, saw a limit to how much autonomy we should give machines. Reasonable people might disagree on where that limit is, and for some people autonomous weapons that search for and engage targets within narrow human-defined parameters might be acceptable, but everyone I spoke with agreed there should be some limits. But the scary thing is that reasonableness on the part of Tseng and other engineers may not be enough. What's to stop a technologically inclined terrorist from building a swarm of people-hunting autonomous weapons and letting them loose in a crowded area? It might take some engineering and some time, but the underlying technological know-how is readily available. We are entering a world where the technology to build lethal autonomous weapons is available not only to nation-states but to individuals as well. That world is not in the distant future. It's already here.

What we do with the technology is an open question. What would be the consequence of a world of autonomous weapons? Would they lead to a robutopia or robopocalypse? Writers have pondered this question in science fiction for decades, and their answers vary wildly. The robots of Isaac Asimov's books are mostly benevolent partners to humans, helping to protect and guide humanity. Governed by the Three Laws of Robotics, they are incapable of harming humans. In *Star Wars*, droids are willing servants of humans. In the *Matrix* trilogy, robots enslave humans, growing them in pods and drawing on their body heat for power. In the *Terminator* series, Skynet strikes in one swift blow to exterminate humanity after it determines humans are a threat to its existence.

We can't know with any certainty what a future of autonomous weapons would look like, but we do have better tools than science fiction to guess at what promise and perils they might bring. Humanity's past and present experiences with autonomy in the military and other settings point to the potential benefits and dangers of autonomous weapons. These lessons allow us to peer into a murky future and, piece by piece, begin to discern the shape of things to come.

PART III

Runaway Gun

9

ROBOTS RUN AMOK

FAILURE IN AUTONOMOUS SYSTEMS

March 22, 2003—The system said to fire. The radars had detected an incoming tactical ballistic missile, or TBM, probably a Scud missile of the type Saddam had used to harass coalition forces during the first Gulf War. This was their job, shooting down the missile. They needed to protect the other soldiers on the ground, who were counting on them. It was an unfamiliar set of equipment; they were supporting an unfamiliar unit; they didn't have the intel they needed. But this was their job. The weight of the decision rested on a twenty-two-year-old second lieutenant fresh out of training. She weighed the available evidence. She made the best call she could: *fire*.

With a BOOM-ROAR-WOOSH, the Patriot PAC-2 missile left the launch tube, lit its engine, and soared into the sky to take down its target. The missile exploded. Impact. The ballistic missile disappeared from their screens: their first kill of the war. Success.

From the moment the Patriot unit left the States, circumstances had been against them. First, they'd fallen in on a different, older, set of equipment than what they'd trained on. Then once in theater, they were detached from their parent battalion and attached to a new battalion whom they hadn't worked with before. The new battalion was using the newer model equipment, which meant their old equipment (which they weren't fully trained on in the first place) couldn't communicate with the rest of the battalion. They were in the dark. Their systems couldn't con-

nect to the larger network, depriving them of vital information. All they
had was a radio.

But they were soldiers, and they soldiered on. Their job was to protect
coalition troops against Iraqi missile attacks, and so they did. They sat in
their command trailer, with outdated gear and imperfect information,
and they made the call. When they saw the missiles, they took the shots.
They protected people.

The next night, at 1:30 a.m., there was an attack on a nearby base. A U.S.
Army sergeant threw a grenade into a command tent, killing one soldier
and wounding fifteen. He was promptly detained but his motives were
unclear. Was this the work of one disgruntled soldier or was he an infiltra-
tor? Was this the first of a larger plot? Word of the attack spread over the
radio. Soldiers were sent to guard the Patriot battery's outer perimeter in
case follow-on attacks came, leaving only three people in the command
trailer, the lieutenant and two enlisted soldiers.

Elsewhere that same night, further north over Iraq, British Flight
Lieutenant Kevin Main turned around his Tornado GR4A fighter jet
and headed back toward Kuwait, his mission for the day complete. In
the back seat as navigator was Flight Lieutenant Dave Williams. What
Main and Williams didn't know as they rocketed back toward friendly
lines was that a crucial piece of equipment, the identification friend or
foe (IFF) signal, wasn't on. The IFF was supposed to broadcast a signal
to other friendly aircraft and ground radars to let them know their Tornado was friendly and not to fire. But the IFF wasn't working. The reason
why is still mysterious. It could be because Main and Williams turned
it off while over Iraqi territory so as not to give away their position and
forgot to turn it back on when returning to Kuwait. It could be because
the system simply broke, possibly from a power supply failure. The IFF
signal had been tested by maintenance personnel prior to the aircraft
taking off, so it should have been functional, but for whatever reason it
wasn't broadcasting.

As Main and Williams began their descent toward Ali Al Salem air
base, the Patriot battery tasked with defending coalition bases in Kuwait
sent out a radar signal into the sky, probing for Iraqi missiles. The radar
signal bounced off the front of Main and Williams' aircraft and reflected
back, where it was received by the Patriot's radar dish. Unfortunately, the

Patriot's computer didn't register the radar reflection from the Tornado as an aircraft. Because of the aircraft's descending profile, the Patriot's computer tagged the radar signal as coming from an anti-radiation missile. In the Patriot's command trailer, the humans didn't know that a friendly aircraft was coming in for a landing. Their screen showed a radar-hunting enemy missile homing in on the Patriot battery.

The Patriot operators' mission was to shoot down ballistic missiles, which are different from anti-radiation missiles. It would be hard for a radar to confuse an aircraft flying level with a ballistic missile, which follows a parabolic trajectory through the sky like a baseball. Anti-radiation missiles are different. They have a descending flight profile, like an aircraft coming in on landing. Anti-radiation missiles home on radars and could be deadly to the Patriot. Shooting them wasn't the Patriot operators' primary job, but they were authorized to engage if the missile appeared to be homing in on their radar.

The Patriot operators saw the missile headed toward their radar and weighed their decision. The Patriot battery was operating alone, without the ability to connect to other radars on the network because of their outdated equipment. Deprived of the ability to see other radar inputs directly, the lieutenant called over the radio to the other Patriot units. Did they see an anti-radiation missile? No one else saw it, but this meant little, since other radars may not have been in a position to see it. The Tornado's IFF signal, which would have identified the blip on their radar as a friendly aircraft, wasn't broadcasting. Even if it had been working, as it turns out, the Patriot wouldn't have been able to see the signal—the codes for the IFF hadn't been loaded into the Patriot's computers. The IFF, which was supposed to be a backup safety measure against friendly fire, was doubly broken.

There were no reports of coalition aircraft in the area. There was nothing at all to indicate that the blip that appeared on their scopes as an anti-radiation missile might, in fact, be a friendly aircraft. They had seconds to decide.

They took the shot. The missile disappeared from their scope. It was a hit. Their shift ended. Another successful day.

Elsewhere, Main and Williams' wingman landed in Kuwait, but Main and Williams never returned. The call went out: there is a missing

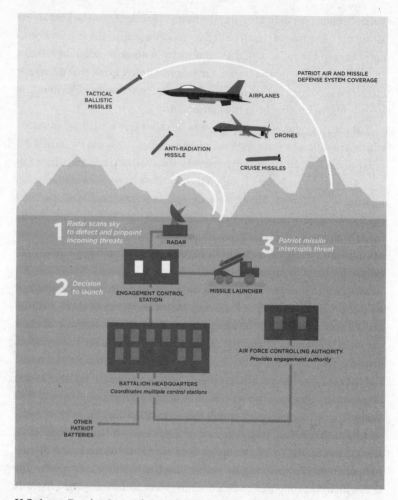

U.S. Army Patriot Operations *The Patriot air and missile defense system is used to counter a range of threats from enemy aircraft and missiles.*

Tornado aircraft. As the sun came up over the desert, people began to put two and two together. The Patriot had shot down one of their own.

The Army opened an investigation, but there was still a war to fight. The lieutenant stayed at her post; she had a job to do. The Army needed her to do that job, to protect other soldiers from Saddam's missiles. Confusion and chaos are unfortunate realities of war. Unless the investigation determined that she was negligent, the Army needed her in the fight. More of Saddam's missiles were coming.

The very next night, another enemy ballistic missile popped up on their scope. They took the shot. Success. It was a clean hit—another enemy ballistic missile down. The same Patriot battery had two more successful ballistic missile shootdowns before the end of the war. In all, they were responsible for 45 percent of all successful ballistic missile engagements in the war. Later, the investigation cleared the lieutenant of wrongdoing. She made the best call with the information she had.

Other Patriot units were fighting their own struggle against the fog of war. The day after the Tornado shoot down, a different Patriot unit got into a friendly fire engagement with a U.S. F-16 aircraft flying south of Najaf in Iraq. This time, the aircraft shot first. The F-16 fired off a radar-hunting AGM-88 high-speed anti-radiation missile. The missile zeroed in on the Patriot's radar and knocked it out of commission. The Patriot crew was unharmed—a near miss.

After these incidents, a number of safety measures were immediately put in place to prevent further fratricides. The Patriot has both a manual (semi-autonomous) and auto-fire (supervised autonomous) mode, which can be kept at different settings for different threats. In manual mode, a human is required to approve an engagement before the system will launch. In auto-fire mode, if there is an incoming threat that meets its target parameters, the system will automatically engage the threat on its own.

Because ballistic missiles often afford very little reaction time before impact, Patriots sometimes operated in auto-fire mode for tactical ballistic missiles. Now that the Army knew the Patriot might misidentify a friendly aircraft as an anti-radiation missile, however, they ordered Patriot units to operate in manual mode for anti-radiation missiles. As an additional safety, systems were now kept in "standby" status so they could track targets, but could not fire without a human bringing the system back to "operate" status. Thus, in order to fire on an anti-radiation missile, two steps were needed: bringing the launchers to operate status *and* authorizing the system to fire on the target. Ideally, this would prevent another fratricide like the Tornado shootdown.

Despite these precautions, a little over a week later on April 2, disaster struck again. A Patriot unit operating north of Kuwait on the road to

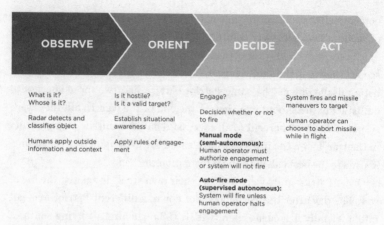

OBSERVE	ORIENT	DECIDE	ACT
What is it? Whose is it?	Is it hostile? Is it a valid target?	Engage?	System fires and missile maneuvers to target
Radar detects and classifies object	Establish situational awareness	Decision whether or not to fire	Human operator can choose to abort missile while in flight
Humans apply outside information and context	Apply rules of engagement	**Manual mode (semi-autonomous):** Human operator must authorize engagement or system will not fire	
		Auto-fire mode (supervised autonomous): System will fire unless human operator halts engagement	

Patriot Decision-Making Process *The OODA decision-making process for a Patriot system. In manual mode, the human operator must take a positive action in order for the system to fire. In auto-fire mode, the human supervises the system and can intervene if necessary, but the system will fire on its own if the human does not intervene. Auto-fire mode is vital for defending against short-warning attacks where there may be little time to make a decision before impact. In both modes, the human can still abort the missile while in flight.*

Baghdad picked up an inbound ballistic missile. Shooting down ballistic missiles was their job. Unlike the anti-radiation missile that the earlier Patriot unit had fired on—which turned out to be a Tornado—there was no evidence to suggest ballistic missiles might be misidentified as aircraft.

What the operators didn't know—what they could not have known—was that there was no missile. There wasn't even an aircraft misidentified as a missile. There was nothing. The radar track was false, a "ghost track" likely caused by electromagnetic interference between their radar and another nearby Patriot radar. The Patriot units supporting the U.S. advance north to Baghdad were operating in a nonstandard configuration. Units were spread in a line south-to-north along the main highway to Baghdad instead of the usual widely distributed pattern they would adopt to cover an area. This may have caused radars to overlap and interfere.

But the operators in the Patriot trailer didn't know this. All they saw was a ballistic missile headed their way. In response, the commander ordered the battery to bring its launchers from "standby" to "operate."

The unit was operating in manual mode for anti-radiation missiles, but auto-fire mode for ballistic missiles. As soon as the launcher became operational, the auto-fire system engaged: BOOM-BOOM. Two PAC-3 missiles launched automatically.

The two PAC-3 missiles steered toward the incoming ballistic missile, or at least to the spot where the ground-based radar told them it should be. The missiles activated their seekers to look for the incoming ballistic missile, but there was no missile.

Tragically, the missiles' seekers did find something: a U.S. Navy F/A-18C Hornet fighter jet nearby. The jet was piloted by Lieutenant Nathan White, who was simply in the wrong place at the wrong time. White's F-18 was squawking IFF and he showed up on the Patriot's radar as an aircraft. It didn't matter. The PAC-3 missiles locked onto White's aircraft. White saw the missiles coming and called it out over the radio. He took evasive action, but there was nothing he could do. Seconds later, both missiles struck his aircraft, killing him instantly.

ASSESSING THE PATRIOT'S PERFORMANCE

The Patriot fratricides are an example of the risks of operating complex, highly automated lethal systems. In a strict operational sense, the Patriot units accomplished their mission. Over sixty Patriot fire units were deployed during the initial phase of the war, forty from the United States and twenty-two from four coalition nations. Their mission was to protect ground troops from Iraqi ballistic missiles, which they did. Nine Iraqi ballistic missiles were fired at coalition forces; all were successfully engaged by Patriots. No coalition troops were harmed from Iraqi missiles. A Defense Science Board Task Force on the Patriot's performance concluded that, with respect to missile defense, the Patriot was a "substantial success."

On the other hand, in addition to these nine successful engagements, Patriots were involved in three fratricides: two incidents in which Patriots shot down friendly aircraft, killing the pilots, and a third incident in which an F-16 fired on a Patriot. Thus, of the twelve total engagements involving Patriots, 25 percent were fratricides, an "unacceptable" fratricide rate according to Army investigators.

The reasons for the Patriot fratricides were a complex mix of human error, improper testing, poor training, and unforeseen interactions on the battlefield. Some problems were known—IFF was well understood to be an imperfect solution for preventing fratricides. Other problems, such as the potential for the Patriot to misclassify an aircraft as an anti-radiation missile, had been identified during operational testing but had not been corrected and were not included in operator training. Still other issues, such as the potential for electromagnetic interference to cause a false radar track, were novel and unexpected. Some of these complications were preventable, but others were not. War entails uncertainty. Even the best training and operational testing can only approximate the actual conditions of war. Inevitably, soldiers will face wartime conditions where the environment, adversary innovation, and simply the chaos, confusion, and violence of war all contribute to unexpected challenges. Many things that seem simple in training often look far different in the maw of combat.

One thing that did not happen and was not a cause of the Patriot fratricides is that the Patriot system did not fail, per se. It didn't break. It didn't blow a fuse. The system performed its function: it tracked incoming targets and, when authorized, shot them down. Also, in both instances a human was required to give the command to fire or at least to bring the launchers to operate. When this lethal, highly automated system was placed in the hands of operators who did not fully understand its capabilities and limitations, however, it turned deadly. Not because the operators were negligent. No one was found to be at fault in either incident. It would be overly simplistic to blame the fratricides on "human error." Instead, what happened was more insidious. Army investigators determined the Patriot community had a culture of "trusting the system without question." According to Army researchers, the Patriot operators, while nominally in control, exhibited automation bias: an "unwarranted and uncritical trust in automation. In essence, control responsibility is ceded to the machine." There may have been a human "in the loop," but the human operators didn't question the machine when they should have. They didn't exercise the kind of judgment Stanislav Petrov did when he questioned the signals his system was giving him regarding a false launch of U.S. nuclear missiles. The Patriot operators trusted the machine, and it was wrong.

ROBUTOPIA VS. ROBOPOCALYPSE

We have two intuitions when it comes to autonomous systems, intuitions that come partly from science fiction but also from our everyday experiences with phones, computers, cars, and myriad other computerized devices.

The first intuition is that autonomous systems are reliable and introduce greater precision. Just as autopilots have improved air travel safety, automation can also improve safety and reliability in many other domains. Humans are terrible drivers, for example, killing more than 30,000 people a year in the United States alone (roughly the equivalent of a 9/11 attack every month). Even without fully autonomous cars, more advanced vehicle autopilots that allow cars to drive themselves under most conditions could dramatically improve safety and save lives.

However, we have another instinct when it comes to autonomous systems, and that is one of robots run amok, autonomous systems that slip out of human control and result in disastrous outcomes. These fears are fed to us through a steady diet of dystopian science fiction stories in which murderous AIs turn on humans, from *2001: A Space Odyssey*'s HAL 9000 to *Ex Machina*'s Ava. But these intuitions also come from our everyday experiences with simple automated devices. Anyone who has ever been frustrated with an automated telephone call support helpline, an alarm clock mistakenly set to "p.m." instead of "a.m.," or any of the countless frustrations that come with interacting with computers, has experienced the problem of "brittleness" that plagues automated systems. Autonomous systems will do precisely what they are programmed to do, and it is this quality that makes them both reliable and maddening, depending on whether what they were programmed to do was the right thing at that point in time.

Both of our intuitions are correct. With proper design, testing, and use, autonomous systems can often perform tasks far better than humans. They can be faster, more reliable, and more precise. However, if they are placed into situations for which they were not designed, if they aren't fully tested, if operators aren't properly trained, or if the environment changes, then autonomous systems can fail. When they do fail, they often fail badly. Unlike humans, autonomous systems lack the ability to step outside their instructions and employ "common sense" to adapt to the situation at hand.

This problem of brittleness was highlighted during a telling moment in the 2011 *Jeopardy! Challenge* in which IBM's Watson AI took on human Jeopardy champions Ken Jennings and Brad Rutter. Toward the end of the first game, Watson momentarily stumbled in its rout of Jennings and Rutter in response to a clue in the "Name the Decade" category. The clue was, "The first modern crossword is published and Oreo cookies are introduced." Jennings rang in first with the answer, "What are the 20s?" Wrong, said host Alex Trebek. Immediately afterward, Watson rang in and gave the same answer, "What is 1920s?" A befuddled Trebek testily replied, "No, Ken said that."

I'm not particularly good at Jeopardy, but even I knew, "What are the 1920s?" was the wrong answer once Jennings guessed wrong. (The correct answer is the 1910s.) Watson hadn't been programmed to listen to other contestants' wrong answers and adjust accordingly, however. Processing Jennings' answer was outside of the bounds of Watson's design. Watson was superb at answering Jeopardy questions under most conditions, but its design was brittle. When an atypical event occurred that Watson's design didn't account for, such as Ken Jennings getting a wrong answer, Watson couldn't adapt on the fly. As a result, Watson's performance suddenly plummeted from superhuman to super-dumb.

Brittleness can be managed when the person using an autonomous system understands the boundaries of the system—what it can and cannot do. The user can either steer the system away from situations outside the bounds of its design or knowingly account for and accept the risks of failure. In this case, Watson's designers understood this limitation. They just didn't think that the ability to learn from other contestants' wrong answers would be important. "We just didn't think it would ever happen," one of Watson's programmers said afterward. Watson's programmers were probably right to discount the importance of this ability. The momentary stumble proved inconsequential. Watson handily defeated its human counterparts.

Problems can arise when human users don't anticipate these moments of brittleness, however. This was the case with the Patriot fratricides. The system had vulnerabilities—misclassifying an anti-radiation missile as an aircraft, IFF failures, and electromagnetic interference causing "ghost track" ballistic missiles—that the human operators were either unaware of or didn't sufficiently account for. As a result, the human user's expectations and the system's actual behavior were not in alignment. The operators thought the system was targeting missiles when it was actually targeting aircraft.

AUTONOMY AND RISK

One of the ways to compensate for the brittle nature of automated systems is to retain tight control over their operation. If the system fails, humans can rapidly intervene to correct it or halt its operation. Tighter human control reduces the autonomy, or freedom, of the machine.

Immediate human intervention is possible for semiautonomous and human-supervised systems. Just because humans can intervene, however, doesn't mean they always do so when they should. In the case of the Patriot fratricides, humans were "in the loop," but they didn't sufficiently question the automation. Humans can't act as an independent fail-safe if they cede their judgment to the machine.

Effective human intervention may be even more challenging in supervised autonomous systems, where the system does not pause to wait for human input. The human's ability to actually regain control of the system in real time depends heavily on the speed of operations, the amount of information available to the human, and any time delays between the human's actions and the system's response. Giving a driver the ability to grab the wheel of an autonomous vehicle traveling at highway speeds in dense traffic, for example, is merely the illusion of control, particularly if the operator is not paying attention. This appears to have been the case in a 2016 fatality involving a Tesla Model S that crashed while driving on autopilot.

For fully autonomous systems, the human is out of the loop and cannot intervene at all, at least for some period of time. This means that if the system fails or the context changes, the result could be a runaway autonomous process beyond human control with no ability to halt or correct it.

This danger of autonomous systems is best illustrated not with a science fiction story, but with a Disney cartoon. *The Sorcerer's Apprentice* is an animated short in Disney's 1940 film *Fantasia*. In the story, which is an adaptation of an eighteenth-century poem by Goethe, Mickey Mouse plays the apprentice of an old sorcerer. When the sorcerer leaves for the day, Mickey decides to use his novice magic to automate his chores. Mickey enchants a broomstick, causing it to sprout arms and come to life. Mickey commands the broomstick to carry pails of water from the well to a cistern, a chore Mickey is supposed to be doing. Soon, Mickey is nodding off, his chores automated.

As Mickey sleeps, the cistern overfills. The job is done, but no one told the broomstick to stop. Mickey wakes to find the room flooded and the broomstick fetching more water. He commands the broomstick to halt, but it doesn't comply. Desperate, Mickey snatches an axe from the wall and chops the broomstick to pieces, but the splinters reanimate into a horde of broomsticks. They march forth to bring even more water, an army of rogue autonomous agents out of control. Finally, the madness is stopped only by the return of the sorcerer himself, who disperses the water and halts the broomsticks with a wave of his arms.

With the original German poem written in 1797, *The Sorcerer's Apprentice* may be the first example of autonomy displacing jobs. It also shows the danger of automation. An autonomous system may not perform a task in the manner we want. This could occur for a variety of reasons: malfunction, user error, unanticipated interactions with the environment, or hacking. In the case of Mickey's problem, the "software" (instructions) that he bewitched the broomstick with were flawed because he didn't specify when to stop. Overfilling the cistern might have been only a minor annoyance if it had happened once, however. A semiautonomous process that paused for Mickey's authorization after each trip to the well would have been far safer. Having a human "in the loop" would have mitigated the danger from the faulty software design. The human can act like a circuit breaker, catching harmful events before they cascade out of control. Making the broomstick fully autonomous without a human in the loop wasn't the cause of the failure, but it did dramatically increase the consequences if something went wrong. Because of this potential to have a runaway process, fully autonomous systems are inherently more hazardous than semiautonomous ones.

Putting an autonomous system into operation means accepting the risk that it may perform its task incorrectly. Fully autonomous systems are not necessarily more likely to fail than semiautonomous or supervised autonomous ones, but if they do, the consequences—the potential damage caused by the system—could be severe.

TRUST, BUT VERIFY

Activating an autonomous system is an act of trust. The user trusts that the system will function in the manner that he or she expects. Trust isn't blind faith, however. As the Patriot fratricides demonstrated, too much trust can be just as dangerous as too little. Human users need to trust the system just the right amount. They need to understand both the capabilities *and* limitations of the system. This is why Bradford Tousley from DARPA TTO cited test and evaluation as his number one concern. A rigorous testing regime can help designers and operators better understand how the system performs under realistic conditions. Bob Work similarly told me that test and evaluation was "central" to building trustworthy autonomous systems. "When you delegate authority to a machine, it's got to be repeatable," he said. "The same outcome has to happen over and over and over again. . . . So, what is going to be our test and evaluation regime for these smarter and smarter weapons to make sure that the weapon stays within the parameters of what we expect it to do? That's an issue."

The problem is that, even with simulations that test millions of scenarios, fully testing all of the possible scenarios a complex autonomous system might encounter is effectively impossible. There are simply too many possible interactions between the system and its environment and even within the system itself. Mickey should have been able to anticipate the cistern overflowing, but some real-world problems cannot be anticipated. The game of Go has more possible positions than atoms in the universe, and the real world is far more complex than Go. A 2015 Air Force report on autonomy bemoaned the problem:

> Traditional methods ... fail to address the complexities associated
> with autonomy software ... There are simply too many possible states
> and combination of states to be able to exhaustively test each one.

In addition to the sheer numerical problem of evaluating all possible combinations, testing is also limited by the testers' imagination. In games like chess or *go*, the set of possible actions is limited. In the real world, however, autonomous systems will encounter any number of novel situations: new kinds of human error, unexpected environmental conditions, or creative actions by adversaries looking to exploit vulnerabilities. If these scenarios can't be anticipated, they can't be tested.

Testing is vital to building confidence in how autonomous systems will behave in real-world environments, but no amount of testing can entirely eliminate the potential for unanticipated behaviors. Sometimes these unanticipated behaviors may pleasantly surprise users, like AlphaGo's 1 in 10,000 move that stunned human champion Lee Sedol. Sometimes these unanticipated actions can be negative. During Gary Kasparov's first game against Deep Blue in 1997, a bug in Deep Blue caused it to make a nonsense random move in the forty-fourth move of the game. One of Deep Blue's programmers later explained, "We had seen it once before, in a test game played earlier in 1997, and thought it was fixed. Unfortunately, there was one case that we had missed." When playing games like Jeopardy, chess, or *go*, surprising behaviors may be tolerable, even interesting flukes. When operating high-risk automated systems where life or death is at stake, unexpected actions can lead to tragic accidents, such as the Patriot fratricides.

WHEN ACCIDENTS ARE NORMAL

To better understand the risks of autonomous weapons, I spoke with John Borrie from the UN Institute for Disarmament Research (UNIDIR). UNIDIR is an independent research institute within the United Nations that focuses on arms control and disarmament issues. Borrie authored a recent UNIDIR report on autonomous weapons and risk and he's worked extensively on arms control and disarmament issues in a variety of capacities—for the New Zealand government, the International Committee of the Red Cross, and UNIDIR—and on a host of technologies: cryptography, chemical and biological weapons, and autonomy. This made him well positioned to understand the relative risks of autonomous weapons.

Borrie and I sat down on the sidelines of the UN talks on autonomous weapons in Geneva in 2016. Borrie is not an advocate for a preemptive ban on autonomous weapons and in general has the sober demeanor of a professor, not a firebrand activist. He speaks passionately (though in an even-tempered, professorial cadence) in his lilting New Zealand accent. I could imagine myself pleasantly nodding off in his class, even as he calmly warned of the dangers of robots run amok.

"With very complex technological systems that are hazardous," Borrie said, "—and I think autonomous weapons fall into that category of hazard because of their intended lethality . . . we have difficulty [saying] that we can remove the risk of unintentional lethal effects." Borrie compared autonomous weapons to complex systems in other industries. Humans have decades of experience designing, testing, and operating complex systems for high-risk applications, from nuclear power plants to commercial airliners to spacecraft. The good news is that because of these experiences, there is a robust field of research on how to improve safety and resiliency in these systems. The bad news is that all of the experience with complex systems to date suggests that 100 percent error-free operation is impossible. In sufficiently complex systems, it is impossible to test every possible system state and combination of states; some unanticipated interactions will happen. Failures may be unlikely, but over a long enough timeline they are inevitable. Engineers refer to these incidents as "normal accidents" because their occurrence is inevitable, even normal, in complex systems. "Why would autonomous systems be any different?" Borrie asked.

The textbook example of a normal accident is the Three Mile Island nuclear power plant meltdown in 1979. The Three Mile Island incident was a "system failure," meaning that the accident was caused by the interaction of many small, individually manageable failures interacting in an unexpected and dramatic way, much like the Patriot fratricides. The Three Mile Island incident illustrates the challenge in anticipating and preventing accidents in complex systems.

The trouble began when moisture from a leaky seal got into an unrelated system, causing it to shut off water pumps vital to cooling the reactor. An automated safety kicked in, activating emergency pumps, but a valve needed to allow water to flow through the emergency cooling system had been left closed. Human operators monitoring the reactor were unaware that the valve was shut because the indicator light on their control panel was obscured by a repair tag for another, unrelated system.

Without water, the reactor core temperature rose. The reactor automatically "scrammed," dropping graphite control rods into the reactor core to absorb neutrons and stop the chain reaction. However, the core was still generating heat. Rising temperatures activated another automatic safety, a pressure release valve designed to let off steam before the rising pressure cracked the containment vessel.

The valve opened as intended but failed to close. Moreover, the valve's indicator light also failed, so the plant's operators did not know the valve was stuck open. Too much steam was released and water levels in the reactor core fell to dangerous levels. Because water was crucial to cooling the still-hot nuclear core, another automatic emergency water cooling system kicked in and the plant's operators also activated an additional emergency cooling system.

What made these failures catastrophic was the fact that that nuclear reactors are *tightly coupled,* as are many other complex machines. Tight coupling is when an interaction in one component of the system directly and immediately affects components elsewhere. There is very little "slack" in the system—little time or flexibility for humans to intervene and exercise judgment, bend or break rules, or alter the system's behavior. In the case of Three Mile Island, the sequence of failures that caused the initial accident happened within a mere thirteen seconds.

It is the combination of complexity *and* tight coupling that makes accidents an expected, if infrequent, occurrence in such systems. In loosely coupled complex systems, such as bureaucracies or other human organizations, there is sufficient slack for humans to adjust to unexpected situations and manage failures. In tightly coupled systems, however, failures can rapidly cascade from one subsystem to the next and minor problems can quickly lead to system breakdown.

As events unfolded at Three Mile Island, human operators reacted quickly and automatic safeties kicked in. In their responses, though, we see the limitations of both humans and automatic safeties. The automatic safeties were useful, but did not fully address the root causes of the problems—a water cooling valve that was closed when it should have been open and a pressure-release valve that was stuck open when it should have been closed. In principle, "smarter" safeties that took into account more variables could have addressed these issues. Indeed, nuclear reactor safety has improved considerably since Three Mile Island.

The human operators faced a different problem, though, one which more sophisticated automation actually makes harder, not easier: the incomprehensibility of the system. Because the human operators could not directly inspect the internal functioning of the reactor core, they had to rely on indicators to tell them what was occurring. But these indicators were also susceptible to failure. Some indicators did fail, leaving human operators with a substantial deficit of information about the system's internal state. The operators did not discover that the water cooling valve was improperly closed until eight minutes into the accident and did not discover that the pressure release valve was stuck open until two hours later. This meant that some of the corrective actions they took were, in retrospect, incorrect. It would be improper to call their actions "human error," however. They were operating with the best information they had at the time.

The father of normal accident theory, Charles Perrow, points out that the "incomprehensibility" of complex systems themselves is a stumbling block to predicting and managing normal accidents. The system is so complex that it is incomprehensible, or opaque, to users and even the system's designers. This problem is exacerbated in situations in which humans cannot directly inspect the system, such as a nuclear reactor, but also exists in situations where humans are physically present. During the *Apollo 13* disaster, it took seventeen minutes for the astronauts and NASA ground control to uncover the source of the instrument anomalies they were seeing, in spite of the fact that the astronauts were on board the craft and could "feel" how the spacecraft was performing. The astronauts heard a bang and felt a small jolt from the initial explosion in the oxygen tank and could tell that they had trouble controlling the attitude (orientation) of the craft. Nevertheless, the system was so complex that vital time was lost as the astronauts and ground-control experts pored over the various instrument readings and rapidly-cascading electrical failures before they discovered the root cause.

Failures are inevitable in complex, tightly coupled systems and the sheer complexity of the system inhibits predicting when and how failures are likely to occur. John Borrie argued that autonomous weapons would have the same characteristics of complexity and tight coupling, making them susceptible to "failures . . . we hadn't anticipated." Viewed from the perspective of normal accident theory, the Patriot fratricides were not surprising—they were inevitable.

THE INEVITABILITY OF ACCIDENTS

The *Apollo 13* and Three Mile Island incidents occurred in the 1970s, when engineers where still learning to manage complex, tightly coupled systems. Since then, both nuclear power and space travel have become safer and more reliable—even if they can never be made entirely safe.

NASA has seen additional tragic accidents, including some that were not recoverable as *Apollo 13* was. These include the loss of the space shuttles *Challenger* (1986) and *Columbia* (2003) and their crews. While these accidents had discrete causes that could be addressed in later designs (faulty O-rings and falling foam insulation, respectively), the impossibility of anticipating such specific failures in advance makes continued accidents inevitable. In 2015, for example, the private company SpaceX had a rocket blow up on the launch pad due to a strut failure that had not been previously identified as a risk. A year later, another SpaceX rocket exploded during testing due to a problem with supercooled oxygen that CEO Elon Musk said had "never been encountered before in the history of rocketry."

Nuclear power has grown significantly safer since Three Mile Island, but the 2011 meltdown of the Japanese Fukushima Daiichi nuclear plant points to the limits of safety. Fukushima Daiichi was hardened against earthquakes and flooding, with backup generators and thirty-foot-high floodwalls. Unfortunately, the plant was not prepared for a 9.0 magnitude earthquake (the largest recorded earthquake to ever hit Japan) off the coast that caused both a loss in power *and* a massive forty-foot-high tsunami. Many safeties worked. The earthquake did not damage the containment vessels. When the earthquake knocked out primary power, the reactors automatically scrammed, inserting control rods to stop the nuclear reaction. Backup diesel generators automatically came online.

However, the forty-foot-high tsunami wave topped the thirty-foot-high floodwalls, swamping twelve of thirteen backup diesel generators. Combined with the loss of primary power from the electrical grid, the plant lost the ability to pump water to cool the still-hot reactor cores. Despite the heroic efforts of Japanese engineers to bring in additional generators and pump water into the overheating reactors, the result was the worst nuclear power accident since Chernobyl.

The problem wasn't that Fukushima Daiichi lacked backup safeties. The problem was a failure to anticipate an unusual environmental condition (a massive earthquake off the coast that induced a tsunami) that caused what engineers call a *common-mode* failure—one that simultaneously overwhelmed two seemingly independent safeties: primary and backup power. Even in fields where safety is a central concern, such as space travel and nuclear power, anticipating all of the possible interactions of the system and its environment is effectively impossible.

"BOTH SIDES HAVE STRENGTHS AND WEAKNESSES"

Automation plays a mixed role in accidents. Sometimes the brittleness and inflexibility of automation can cause accidents. In other situations, automation can help reduce the probability of accidents or mitigate their damage. At Fukushima Daiichi, automated safeties scrammed the reactor and brought backup generators online. Is more automation a good or bad thing?

Professor William Kennedy of George Mason University has extensive experience in nuclear reactors and military hardware. Kennedy has a unique background—thirty years in the Navy (active and reserve) on nuclear missile submarines, combined with twenty-five years working for the Nuclear Regulatory Commission and the Department of Energy on nuclear reactor safety. To top it off, he has a PhD in information technology with a focus on artificial intelligence. I asked him to help me understand the benefits of humans versus AI in managing high-risk systems.

"A significant message for the Nuclear Regulatory Commission from Three Mile Island was that humans were not omnipotent," Kennedy said. "The solution prior to Three Mile Island was that every time there was a design weakness or a feature that needed to be processed was to give the operator another gauge, another switch, another valve to operate remotely from the control room and everything would be fine. And Three Mile Island demonstrated that humans make mistakes. . . . We got to the point where we had over 2,000 alarms in the control rooms, a wall of procedures for each individual alarm. And Three Mile Island said that alarms almost never occur individually." This was an unmanageable level of complexity for any human operator to absorb, Kennedy explained.

Following Three Mile Island, more automation was introduced to manage some of these processes. Kennedy supports this approach, to a point. "The automated systems, as they are currently designed and built, may be more reliable than humans for planned emergencies, or known emergencies. . . . If we can study it in advance and lay out all of the possibilities and in our nice quiet offices consider all the ways things can behave, we can build that into a system and it can reliably do what we say. But we don't always know what things are possible. . . . Machines can repeatedly, quite reliably, do planned actions. . . . But having the human there provides for 'beyond design basis' accidents or events." In other words, automation could help for situations that could be predicted, but humans were needed to manage novel situations. "Both sides have strengths and weaknesses," Kennedy explained. "They need to work together, at the moment, to provide the most reliable system."

AUTOMATION AND COMPLEXITY— A DOUBLE-EDGED SWORD

Kennedy's argument tracks with what we have seen in modern machines— increasing software and automation but with humans still involved at some level. Modern jetliners effectively fly themselves, with pilots functioning largely as an emergency backup. Modern automobiles still have human drivers, but have a host of automated or autonomous features to improve driving safety and comfort: antilock brakes, traction and stability control, automatic lane keeping, intelligent cruise control, collision avoidance, and self-parking. Even modern fighter jets use software to help improve safety and reliability. F-16 fighter aircraft have been upgraded with automatic ground collision avoidance systems. The newer F-35 fighter reportedly has software-based limits on its flight controls to prevent pilots from putting the aircraft into unrecoverable spins or other aerodynamically unstable conditions.

The double-edged sword to this automation is that all of this added software increases complexity, which can itself introduce new problems. Sophisticated automation requires software with millions of lines of code: 1.7 million for the F-22 fighter jet, 24 million for the F-35 jet, and some 100 million lines of code for a modern luxury automobile. Longer pieces of software are harder to verify as being free from bugs or glitches. Studies have pegged the software industry average error rate at fifteen to fifty errors per 1,000 lines of code. Rigorous internal test and evaluation has been able to reduce the error rate to 0.1 to 0.5 errors per 1,000 lines of code in some cases. However, in systems with millions of lines of code, some errors are inevitable. If they aren't caught in testing, they can cause accidents if encountered during real world operations.

On their first deployment to the Pacific in 2007, eight F-22 fighter jets experienced a Y2K-like total computer meltdown when crossing the International Date Line. All onboard computer systems crashed, causing the pilots to lose navigation, fuel subsystems, and some communications. Stranded over the Pacific without a navigational reference point, the aircraft were able to make it back to land by following the tanker aircraft accompanying them, which relied on an older computer system. Under tougher circumstances, such as combat or even bad weather, the incident could have led to a catastrophic loss of the aircraft. While the existence of the International Date Line clearly could be anticipated, the interaction of the dateline with the software was not identified in testing.

Software vulnerabilities can also leave open opportunities for hackers. In 2015, two hackers revealed that they had discovered vulnerabilities that allowed them to remotely hack certain automobiles while they were on the road. This allowed them to take control of critical driving components including the transmission, steering column, and brakes. In future self-driving cars, hackers who gain access could simply change the car's destination.

Even if software does not have specific bugs or vulnerabilities, the sheer complexity of modern machines can make it challenging for users to understand what the automation is doing and why. When humans are no longer interacting with simple mechanical systems that may behave predictably but instead are interacting with complex pieces of software with millions of lines of code, the human user's expectation about what the automation will do may diverge significantly from what it actually does. I found this to be a challenge with the Nest thermostat, which doesn't have millions of lines of code. (A study of Nest users found similar frustrations, so apparently I am not uniquely unqualified in predicting Nest behavior.)

More advanced autonomous systems are often able to account for more variables. As a result, they can handle more complex or ambiguous environments, making them more valuable than simpler systems. They may fail less overall, because they can handle a wider range of situations. However, they will still fail sometimes and because they are more complex, accurately predicting *when* they will fail may be more difficult. Borrie said, "As systems get increasingly complex and increasingly self-directed, I think it's going to get more and more difficult for human beings to be able to think ahead of time what those weak points are necessarily going to be." When this happens in high-risk situations, the result can be catastrophic.

"WE DON'T UNDERSTAND ANYTHING!"

On June 1, 2009, Air France Flight 447 from Rio to Paris ran into trouble midway over the Atlantic Ocean. The incident began with a minor and insignificant instrumentation failure. Air speed probes on the wings froze due to ice crystals, a rare but non-serious problem that did not affect the flight of the aircraft. Because the airspeed indicators were no longer functioning properly, the autopilot disengaged and handed over control back to the pilots. The plane also entered a different software mode for flight controls. Instead of flying under "normal law" mode, where software limitations prevent pilots from putting the plane into dangerous aerodynamic conditions such as stalls, the plane entered "alternate law" mode, where the software limitations are relaxed and the pilots have more direct control over the plane.

Nevertheless, there was no actual emergency. Eleven seconds following the autopilot disengagement, the pilots correctly identified that they had lost the airspeed indicators. The aircraft was flying normally, at appropriate speeds and full altitude. Everything was fine.

Inexplicably, however, the pilots began a series of errors that resulted in a stall, causing the aircraft to crash into the ocean. Throughout the incident, the pilots continually misinterpreted data from the airplane and misunderstood the aircraft's behavior. At one point mid-crisis, the copilot exclaimed, "We completely lost control of the airplane and we don't understand anything! We tried everything!" The problem was actually simple. The pilots had pulled back too far on the stick, causing the aircraft to stall and lose lift. This is a basic aerodynamic concept, but poor user interfaces and opaque automated processes on the aircraft, even while flown manually, contributed to the pilots' lack of understanding. The complexity of the aircraft created problems of transparency that would likely not have existed on a simpler aircraft. By the time the senior pilot understood what was happening, it was too late. The plane was too low and descending too rapidly to recover. The plane crashed into the ocean, killing all 228 people on board.

Unlike in the F-22 International Date Line incident or the automobile hack, the Air France Flight 447 crash was not due to a hidden vulnerability lurking within the software. In fact, the automation performed perfectly. However, it would be overly simplistic to lay the crash at the feet of human error. Certainly the pilots made mistakes, but the problem is best characterized as human-automation failure. The pilots were confused by the automation and the complexity of the system.

THE PATRIOT FRATRICIDES AS NORMAL ACCIDENTS

Normal accident theory sheds light on the Patriot fratricides. They weren't merely freak occurrences, unlikely to be repeated. Instead, they were a normal consequence of operating a highly lethal, complex, tightly coupled system. True to normal accidents, the specific chain of events that led to each fratricide was unlikely. Multiple failures happened simultaneously. However, simply because these specific combinations of failures were unlikely does not mean that probability of accidents as a whole was low. In fact, given the degree of operational use, the probability of there being some kind of accident was quite high. Over sixty Patriot batteries were deployed to Operation Iraqi Freedom, and during the initial phase of the war coalition aircraft flew 41,000 sorties. This means that the number of possible Patriot-aircraft interactions were in the millions. As the Defense Science Board Task Force on the Patriot pointed out, given the sheer number of interactions, "even very-low-probability failures could result in regrettable fratricide incidents." The fact that the F-18 and Tornado incidents had different causes lends further credence to the view that normal accidents are lurking below the surface in complex systems, waiting to emerge. The complexities of war may bring these vulnerabilities to the surface.

Is it possible to safely operate hazardous complex systems? Normal accident theory says "no." The probability of accidents can be reduced, but never eliminated. There is an alternate point of view on complex systems, however, which suggests that, under certain conditions, normal accidents can largely be avoided.

10

COMMAND AND DECISION

CAN AUTONOMOUS WEAPONS BE USED SAFELY?

There is a robust body of evidence supporting normal accident theory, but a few outliers seem to defy expectations. The Federal Aviation Administration (FAA) air traffic control system and U.S. Navy aircraft carrier flight decks are two examples of "high-reliability organizations." Their rate of accidents isn't zero, but they *are* exceptionally low given the complexities of their operating environment and the hazards of operation. High-reliability organizations can be found across a range of applications and have some common characteristics: highly trained individuals, a collective mindfulness of the risk of failure, and a continued commitment to learn from near misses and improve safety.

While militaries as a whole would not be considered high-reliability organizations, some military communities have very high safety records with complex high-risk systems. In addition to aircraft carrier flight deck operations, the U.S. Navy's submarine community is an example of a high-reliability organization. Following the loss of the USS *Thresher* to an accident in 1963—at the time one of the Navy's most advanced submarines and first in her class—the Navy instituted the Submarine Safety (SUBSAFE) program. Submarine components that are critical for safe operation are designated "SUBSAFE" and subject to rigorous inspection and testing throughout their design, fabrication, maintenance, and use. There is no silver bullet to SUBSAFE's high reliability. It is a continuous process of quality assurance and quality control applied across the entire submarine's life cycle. Upon installation and at every subsequent inspection or repair over the life of the ship, every SUBSAFE component is checked, double-checked, and checked again against technical specifications. If anything is amiss, it must be corrected or approved by an appropriate authority before the submarine can proceed with operations.

SUBSAFE is not a technological solution to normal accidents. It is a bureaucratic and organizational solution. Nevertheless, the results have been astounding. In 2003 Congressional testimony, Rear Admiral Paul Sullivan, the Navy deputy commander for ship design, integration, and engineering, explained the impact of the program:

> The SUBSAFE Program has been very successful. Between 1915 and 1963, 16 submarines were lost due to non-combat causes, an average of one every three years. Since the inception of the SUBSAFE Program in 1963 . . . We have never lost a SUBSAFE certified submarine.

It is hard to overstate the significance of this safety record. The U.S. Navy has more than seventy submarines in its force, with approximately one-third of them at sea at a time. The U.S. Navy has operated at this pace for over half a century without losing a single submarine. From the perspective of normal accident theory, this should not be possible. Operating a nuclear-powered submarine is extremely complex and inherently hazardous, and yet the Navy has been able to substantially reduce these risks. Accidents resulting in catastrophic loss of a submarine are not "normal" in the U.S. Navy. Indeed, they are unprecedented since the advent of SUBSAFE, making SUBSAFE a shining example of what high-reliability organizations can achieve.

Could high-reliability organizations be a model for how militaries might handle autonomous weapons? In fact, lessons from SUBSAFE and aircraft carrier deck operations have already informed how the Navy operates the Aegis combat system. The Navy describes the Aegis as "a centralized, automated, command-and-control (C2) and weapons control system that was designed as a total weapon system, from detection to kill." It is the electronic brain of a ship's weapons. The Aegis connects the ship's advanced radar with its anti-air, anti-surface, and antisubmarine weapon systems and provides a central control interface for sailors. First fielded in 1983, the Aegis has gone through several upgrades and is now at the core of over eighty U.S. Navy warships. To better understand Aegis and whether it could be a model for safe use of future autonomous weapons, I traveled to Dahlgren, Virginia, where Aegis operators are trained.

THE AEGIS COMBAT SYSTEM

Captain Pete Galluch is commander of the Aegis Training and Readiness Center, where he oversees training for all Aegis-qualified officers and enlisted sailors. The phrase "steely-eyed missile man" comes to mind upon meeting Galluch. He speaks with the calmness and decisiveness of a surgeon, a man who is ready to let missiles fly if need be. I can imagine Galluch standing in the midst of a ship's combat information center (CIC) in wartime, unflappable in the midst of the chaos, ordering his sailors when to take the shot and when to hold back. If I were flying within range of an Aegis's weapons or was counting on its ballistic missile defense capabilities to protect my city, I would trust Galluch to make the right call.

Aegis is a weapon system of staggering complexity. At the core of Aegis is a computer called "Command and Decision," or C&D, which governs the behavior of the radar and weapons. Command and Decision's actions are governed by a series of statements—essentially programs or algorithms—that the Navy refers to as "doctrine." Unlike the Patriot circa 2003, however, which had only a handful of different operating modes, Aegis doctrine is almost infinitely customizable.

With respect to weapons engagements, Aegis has four settings. The manual setting, in which engagements against radar "tracks" (objects detected by the radar) must be done directly by a human, involves the most human control. Ship commanders can increase the degree of automation in the engagement process by activating one of three types of doctrine: Semi-Auto, Auto SM, and Auto-Special. Semi-Auto, as the term would imply, automates part of the engagement process to generate a firing solution on a radar track, but final decision authority is withheld by the human operator. Auto SM automates more of the engagement process, but a human must still take a positive action before firing. Despite the term, Auto SM still retains a human in the loop. Auto-Special is the only mode where the human is "on the loop." Once Auto-Special is activated, the Aegis will automatically fire against threats that meet its parameters. The human can intervene to stop the engagement, but no further authorization is needed to fire.

It would be a mistake to think, however, that this means that Aegis can only operate in four discrete modes. In fact, doctrine statements can mix and match these control types against different threats. For example, one doctrine statement could be written to use Auto SM against one type of threat, such as aircraft. Another doctrine statement might authorize Auto-Special against cruise missiles, for which there may be less warning. These doctrine statements can be applied individually or in packages. "You can mix and match," Galluch explained. "It's a very flexible system. . . . we can do all [doctrine statements] with a push of a button, some with a push of a button, or bring them up individually."

This makes Aegis less like a finished product with a few different modes and more like a customizable system that can be tailored for each mission. Galluch explained that the ship's doctrine review board, consisting of the officers and senior enlisted personnel who work on Aegis, begin the process of writing doctrine months before deployment. They consider their anticipated missions, intelligence assessments, and information on the region for the upcoming deployment, then make recommendations on doctrine to the ship's captain for approval. The result is a series of doctrine statements, individually and in packages, that the captain can activate as needed during deployment. "If you have your doctrine statements built and tested," Galluch said, the time to "bring them up is seconds."

Doctrine statements are typically grouped into two general categories: non-saturation and saturation. Non-saturation doctrine is used when there is time to carefully evaluate each potential threat. Saturation doctrine is needed if the ship gets into a combat situation where the number of inbound threats could overwhelm the ability of operators to respond. "If World War III starts and people start throwing a lot of stuff at me," Galluch said, "I will have grouped my doctrine together so that it's a one-push button that activates all of them. And what we've done is we've tested and we've looked at how they overlap each other and what the effects are going to be and make sure that we're getting the defense of the ship that we expect." This is where something like Auto-Special comes into play, in a "kill or be killed" scenario, as Galluch described it.

It's not enough to build the doctrine, though. Extensive testing goes into ensuring that it works properly. Once the ship arrives in theater, the first thing the crew does is test the weapons doctrine to see if there is anything in the environment that might cause it to fire in peacetime, which would not be good. This is done safely by enabling a hardware-level cutout called the Fire Inhibit Switch, or FIS. The FIS includes a key that must be inserted for any of the ship's weapons to fire. When the FIS key is inserted, a red light comes on; when it is turned to the right, the light turns green, meaning the weapons are live and ready to fire. When the FIS is red—or removed entirely—the ship's weapons are disabled at the hardware level. As Galluch put it, "there is no voltage that can be applied to light the wick and let the rocket fly out." By keeping the FIS red or removing the key, the ship's crew can test Aegis doctrine statements safely without any risk of inadvertent firing.

Establishing the doctrine and activating it is the sole responsibility of the ship's captain. Doctrine is more than just a set of programs. It is the embodiment of the captain's intent for the warship. "Absolutely, it's automated, but there's so much human interface with what gets automated and how we apply that automation," Galluch said. Aegis doctrine is a way for the captain to predelegate his or her decision-making against certain threats.

The Aegis community uses automation in a very different way than the Patriot community did in 2003. Patriot operators sitting at the consoles in 2003 were essentially trusting in the automation. They had a handful of operational modes they could activate, but the operators themselves didn't write the rules for how the automation would function in those modes. Those rules were written years beforehand. Aegis, by contrast, can be customized and tailored to the specific operating environment. A destroyer operating in the Western Pacific, for example, might have different doctrine statements than one operating in the Persian Gulf to account for different threats from Chinese versus Iranian missiles. But the differences run deeper than merely having more options. The whole philosophy of automation is different. With Aegis, the automation is used to capture the ship captain's intent. In Patriot, the automation embodies the intent of the designers and testers. The actual operators of the system may not even fully understand the designers' intent that went into crafting the rules. The automation in Patriot is largely intended to *replace* warfighters' decision-making. In Aegis, the automation is used to *capture* warfighters' decision-making.

Another key difference is where decision authority rests. Only the captain of the ship has the authority to activate Aegis weapons doctrine. The captain can predelegate that authority to the tactical action officer on watch, but the order must be in writing as part of official orders. This means the decision-maker's experience level for Aegis operations is radically different from Patriot. When Captain Galluch took command of the USS *Ramage*, he had eighteen years of experience and had served on three prior Aegis ships. By contrast, the person who made the call on the first Patriot fratricide was a twenty-two-year-old second lieutenant fresh out of training.

Throughout our conversation, Galluch's experience was apparent. He was clearly comfortable using Aegis, but he wasn't flippant about its automation. What came through was a healthy respect for the weapon system. Activating Aegis doctrine is a serious decision, not be taken lightly. "You're never driving around with any kind of weapons doctrine activated" unless you expect to get into a fight, he explained. Even on manual mode, it is possible to launch a missile in seconds. And if need be, doctrine can be activated quickly. "I've made more Gulf deployments than I care to," he said. "I'm very comfortable with driving around for months at a time with no active doctrine, but making damn sure that I have it set up and tested and ready to go if I need to." Because there can be situations that call for that level of automation. "You can get a missile fired pretty quickly, so why don't you do everything manually?" Galluch explained: "My view is that [manual control] works well if it's one or two missiles or threats. But if you're controlling fighters, you're doing a running gun battle with small patrol boats, you're launching your helicopter. . . . and you've got a bunch of cruise missiles coming in from different angles. You know, the watch is pretty small. It's ten or twelve people. So, there's not that many people . . . You can miss things coming in. That's where I get to the whole concept of saturation vs. normal. You want the man in the loop as much as possible, but there comes a time when you can get overwhelmed."

Aegis philosophy is one of human control over engagements, even when doctrine is activated. What varies is the form of human control. In Auto-

Special doctrine, firing authority is delegated to Aegis's Command & Decision computer, but the human intent is still there. The goal is always to ensure "there is a conscious decision to fire a weapon," Galluch said. That doesn't mean that accidents can't happen. In fact, it is the constant preoccupation with the potential for accidents that helps prevent them. Galluch and others understand that, with doctrine activated, mishaps can happen. That's precisely why tight control is kept over the weapon. "[Ship commanding officers] are constantly balancing readiness condition to fire the weapon versus a chance for inadvertent firing," he explained.

I saw this tight control in action when Galluch took me to the Aegis simulation center and had his team run through a series of mock engagements. Galluch stood in as the ship's commanding officer and had Aegis-qualified sailors sitting at the same terminals doing the same jobs they would on a real ship. Then they went to work.

"ROLL GREEN"

The Navy would not permit me to record the precise language of the commands used between the sailors, but they allowed me to observe and report on what I saw. First, Galluch ordered the sailors to demonstrate a shot in manual operation. They put a simulated radar track on the screen and Galluch ordered them to target the track. They began working a firing solution, with the three sailors calmly but crisply reporting when they had completed each step in the process. Once the firing solution was ready, Galluch ordered the tactical action officer to roll his FIS key to green. Then Galluch gave the order to fire. A sailor pressed the button to fire and called out that the missile was away. On a large screen in front of us, the radar showed the outbound missile racing toward the track.

I checked my watch. The whole process had been exceptionally fast— under a minute. The threat had been identified, a decision made, and a missile launched well under a minute, and that was in manual mode. I could understand Galluch's confidence in his ability to defend the ship without doctrine activated.

They did it again in Semi-Auto mode, now with doctrine activated. The FIS key was back at red, the tactical action officer having turned it back right after the missile was launched. Galluch ordered them to activate

Semi-Auto doctrine. Then they brought up another track to target. This time, Aegis's Command & Decision computer generated part of the firing solution automatically. This shortened the time to fire by more than half.

They rolled FIS red, activated Auto SM doctrine, and put up a new track. Roll FIS green. Fire.

Finally, they brought up Auto-Special doctrine. This was it. This was the big leap into the great unknown, with the human removed from the loop. The sailors were merely observers now; they didn't need to take any action for the system to fire. Except . . . I looked at the FIS key. The key was in, but it was turned to red. Auto-Special doctrine was enabled, but there was still a hardware-level cutout in place. There was not even any voltage applied to the weapons. Nothing could fire until the tactical action officer rolled his key green.

The track for a simulated threat came up on the screen and Galluch ordered them to roll FIS green. I counted only a handful of heartbeats before a sailor announced the missiles were away. That's all it took for Command & Decision to target the track and fire.

But I felt cheated. They hadn't turned on the automation and leaned back in their chairs, taking it easy. Even on Auto-Special, and they had their hand literally on the key that disabled firing. And as soon as the missile was away, I saw the tactical action officer roll FIS red again. They weren't trusting the automation at all!

Of course, that was the point, I realized. They didn't trust it. The automation was powerful and they respected it—they even recognized there was a place for it—but that didn't mean they were surrendering their human decision-making to the machine.

To further drive the point home, Galluch had them demonstrate one final shot. With Auto-Special doctrine enabled, they rolled FIS green and let Command & Decision take its shot. But then after the missile was away, Galluch ordered them to abort the missile. They pushed a button and a few seconds later the simulated missile disappeared from our radar, having been destroyed mid-flight. Even in the case of Auto-Special, even after the missile had been launched, they still had the ability to reassert human control over the engagement.

The Aegis community has reason to be so careful. In 1988, an Aegis warship was involved in a horrible accident. The incident haunts the community like a ghost—an ever-present reminder of the deadly power of an Aegis ship. Galluch described what transpired as a "terrible, painful lesson" and talked freely what the Aegis community learned to prevent future tragedies.

THE USS *VINCENNES* INCIDENT

The Persian Gulf in 1988 was a dangerous place. The Iran-Iraq war, under way since 1980, had boiled over into an extended "tanker war," with Iran and Iraq attacking each others' oil tankers, trying to starve their economies into submission. In 1987, Iran expanded to attacks against U.S.-flagged tanker ships carrying oil from Kuwait. In response, the U.S. Navy began escorting U.S.-flagged Kuwaiti tankers to protect them from Iranian attacks.

U.S. Navy ships in the Gulf were on high alert to threats from mines, rocket-equipped Iranian fast boats, warships, and fighter aircraft from several countries. A year earlier, the USS *Stark* had been hit with two Exocet missiles fired from an Iraqi jet and thirty-seven U.S. sailors were killed. In April 1988, in response to a U.S. frigate hitting an Iranian mine, the United States attacked Iranian oil platforms and sunk several Iranian ships. The battle only lasted a day, but tensions between the United States and Iran were high afterward.

On July 3, 1988, the U.S. warships USS *Vincennes* and USS *Montgomery* were escorting tankers through the Strait of Hormuz when they came into contact with Iranian fast boats. The *Vincennes's* helicopter, which was monitoring the Iranian boats, came under fire. The *Vincennes* and *Montgomery* responded, pursuing the Iranian boats into Iranian territorial waters and opening fire.

While the *Vincennes* was in the midst of a gun battle with the Iranian boats, two aircraft took off in close sequence from Iran's nearby Bandar Abbas airport. Bandar Abbas was a dual-use airport, servicing both Iranian commercial and military flights. One aircraft was a commercial airliner, Iran Air Flight 655. The other was an Iranian F-14 fighter. For whatever reason, in the minds of the sailors in the *Vincennes's* combat information center, the tracks of the two aircraft on their radar screens became confused. The Iranian F-14 veered away but Iran Air 655 flew along its normal commercial route, which happened to be directly toward the *Vincennes*. Even though the commercial jet was squawking IFF and flying a commercial airliner route, the *Vincennes* captain and crew became convinced, incorrectly, that the radar track headed toward their position was an Iranian F-14 fighter.

As the aircraft approached, the *Vincennes* issued multiple warnings on military and civilian frequencies. There was no response. Believing the Iranians were choosing to escalate the engagement by sending a fighter and that his ship was under threat, the *Vincennes*'s captain gave the order to fire. Iran Air 655 was shot down, killing all 290 people on board.

The USS *Vincennes* incident and the Patriot fratricides sit as two opposite cases on the scales of automation versus human control. In the Patriot fratricides, humans trusted the automation too much. The *Vincennes* incident was caused by human error and more automation might have helped. Iran Air 655 was flying a commercial route squawking IFF. Well-crafted Aegis doctrine should not have fired.

Automation could have helped the *Vincennes* crew in this fast-paced combat environment. They weren't overwhelmed with too many missiles, but they were overwhelmed with too much information: the running gun battle with Iranian boats and tracking an F-14 and a commercial airliner launching in close succession from a nearby airport. In this information-saturated environment, the crew missed important details they should have noticed and made poor decisions with grave consequences. Automation, by contrast, wouldn't have gotten overwhelmed by the amount of information. Just as automation could help shoot down incoming missiles in a saturation scenario, it could also help *not fire* at the wrong targets in an information-overloaded environment.

ACHIEVING HIGH RELIABILITY

The Aegis community has learned from the *Vincennes* incident, Patriot fratricides, and years of experience to refine their operating procedures, doctrine, and software to the point where they are able to operate a very complex weapon system with low accidents. In the nearly thirty years since *Vincennes*, there has not been another similar incident, even with Aegis ships deployed continuously around the world.

The Navy's track record with Aegis shows that high-reliability operation of complex, hazardous systems is possible, but it doesn't come from testing alone. The human operators are not passive bystanders in the Aegis's operation, trusting blindly in the automation. They are active participants at every stage. They program the system's operational parameters, constantly monitor its modes of operation, supervise its actions in real time, and maintain tight control over weapons release authority. The Aegis culture is 180 degrees from the "unwarranted and uncritical trust in automation" that Army researchers found in the Patriot community in 2003.

After the Patriot fratricides, the Army launched the Patriot Vigilance Project, a three-year postmortem assessment to better understand what went wrong and to improve training, doctrine, and system design to ensure it didn't happen again. Dr. John Hawley is an engineering psychologist who led the project and spoke frankly about the challenges in implementing those changes. He said that there are examples of communities that have been able to manage high-risk technologies with very low accident rates, but high reliability is not easy to achieve. The Navy "spent a lot of money looking into . . . how you more effectively use a system like Aegis so that you don't make the kinds of mistakes that led to the [*Vincennes* incident]," he said. This training is costly and time-consuming, and in practice there are bureaucratic and cultural obstacles that may prevent military organizations from investing this amount of effort. Hawley explained that Patriot commanders are evaluated based on how many trained crews they keep ready. "If you make the [training] situation too demanding, then you could start putting yourself in the situation where you're not meeting those [crew] requirements." It may seem that militaries have an incentive to make training as realistic as possible, and to a certain extent that's true, but there are limits to how much time and money can be applied. Hawley argued that Army Patriot operators train in a "sham environment" that doesn't accurately simulate the rigors of real-world combat. As a result, he said "the Army deceives itself about how good their people really are. . . . It would be easy to believe you're good at this, but that's only because you've been able to handle the relatively non-demanding scenarios that they throw at you." Unfortunately, militaries might not realize their training is ineffective until a war occurs, at which point it may be too late.

Hawley explained that the Aegis community was partially protected from this problem because they use their system day in and day out on ships operating around the globe. Aegis operators get "consistent objective feedback from your environment on how well you're doing," preventing this kind of self-deception. The Army's peacetime operating environment for the Patriot, on the other hand, is not as intense, Hawley said. "Even when the Army guys are deployed, I don't think that the quality of their experience with the system is quite the same. They're theoretically hot, but they're really not doing much of anything, other than just monitoring their scopes." Leadership is also a vital factor. "Navy brass in the Aegis community are absolutely paranoid" about another *Vincennes* incident, Hawley said.

The bottom line is that high reliability not easy to achieve. It requires frequent experience under real-world operating conditions and a major investment in time and money. Safety must be an overriding priority for leaders, who often have other demands they must meet. U.S. Navy submariners, aircraft carrier deck operators, and Aegis weapon system operators are very specific military communities that meet these conditions. Military organizations in general do not. Hawley was pessimistic about the ability of the U.S. Army to safely operate a system like the Patriot, saying it was "too sloppy an organization to . . . insist upon the kinds of rigor that these systems require."

This is a disappointing conclusion, because the U.S. Army is one of the most professional military organizations in history. Hawley was even more pessimistic about other nations. "Judging from history and the Russian army's willingness to tolerate casualties and attitude about fratricide . . . I would expect that . . . they would tilt the scale very much in the direction of lethality and operational effectiveness and away from necessarily safe use." Practice would appear to bear this out. The accident rate for Soviet/Russian submarines is far higher than for U.S. submarines.

If there is any military community that should be incentivized to avoid accidents, it is those responsible for maintaining control of nuclear weapons. There are no weapons on earth more destructive than nuclear weapons. Nuclear weapons are therefore an excellent test case for the extent to which dangerous weapons can be managed safely.

NUCLEAR WEAPONS SAFETY AND NEAR-MISS ACCIDENTS

The destructive power of nuclear weapons defies easy comprehension. A single *Ohio*-class ballistic missile submarine can carry twenty-four Trident II (D5) ballistic missiles, each with eight 100-kiloton warheads per missile. Each 100-kiloton warhead is over six times more powerful than the bomb dropped on Hiroshima. Thus, a single submarine has the power to unleash over a thousand times the destructive power of the attack on Hiroshima. Individually, nuclear weapons have the potential for mass destruction. Collectively, a nuclear exchange could destroy human civilization. But outside of testing they have not been used, intentionally or accidentally, since 1945.

On closer inspection, however, the safety track record of nuclear weapons is less than inspiring. In addition to the Stanislav Petrov incident in 1983, there have been multiple nuclear near-miss incidents that could have had catastrophic consequences. Some of these could have resulted in an individual weapon's use, while others could potentially have led to a nuclear exchange between superpowers.

In 1979, a training tape left in a computer at the U.S. military's North American Aerospace Defense Command (NORAD) led military officers to initially believe that a Soviet attack was under way, until it was refuted by early warning radars. Less than a year later in 1980, a faulty computer chip led to a similar false alarm at NORAD. This incident progressed far enough that U.S. commanders notified National Security Advisor Zbigniew Brzezinski that 2,200 Soviet missiles were inbound to the United States. Brzezinski was about to inform President Jimmy Carter before NORAD realized the alarm was false.

Even after the Cold War ended, the danger from nuclear weapons did not entirely subside. In 1995, Norway launched a rocket carrying a science payload to study the aurora borealis that had a trajectory and radar signature similar to a U.S. Trident II submarine-launched nuclear missile. While a single missile would not have made sense as a first strike, the launch was consistent with a high-altitude nuclear burst to deliver an electromagnetic pulse to blind Russian satellites, a prelude to a massive U.S. first strike. Russian commanders brought President Boris Yeltsin the nuclear briefcase, who discussed a response with senior Russian military commanders before the missile was identified as harmless.

In addition to these incidents are safety lapses that might not have risked nuclear war but are troubling nonetheless. In 2007, for example, a U.S. Air Force B-52 bomber flew from Minot Air Force Base to Barksdale Air Force Base with six nuclear weapons aboard without the pilots or crew being aware. After it landed, the weapons remained on board the aircraft, unsecured and with ground personnel unaware of the weapons, until they were discovered the following day. This incident was merely the most egregious in a series of recent security lapses in the U.S. nuclear community that caused Air Force leaders to warn of an "erosion" of adherence to appropriate safety standards.

Nor were these isolated cases. There were at least thirteen near-use nuclear incidents from 1962 to 2002. This track record does not inspire confidence. Indeed, it lends credence to the view that near-miss incidents are normal, if terrifying, conditions of nuclear weapons. The fact that none of these incidents led to an actual nuclear detonation, however, presents an interesting puzzle: Do these near-miss incidents support the pessimistic view of normal accident theory that accidents are inevitable? Or does the fact that they didn't result in an actual nuclear detonation support the more optimistic view that high-reliability organizations can safely operate high-risk systems?

Stanford political scientist Scott Sagan undertook an in-depth evaluation of nuclear weapons safety to answer this very question. In the conclusion of his exhaustive study, published in *The Limits of Safety: Organizations, Accidents, and Nuclear Weapons,* Sagan wrote:

> When I began this book, the public record on nuclear weapons safety led me to expect that the high reliability school of organization theorists would provide the strongest set of intellectual tools for explaining this apparent success story. . . . The evidence presented in this book has reluctantly led me to the opposite view: the experience of persistent safety problems in the U.S. nuclear arsenal should serve as a warning.

Sagan concluded, "the historical evidence provides much stronger support for the ideas developed by Charles Perrow in *Normal Accidents*" than for high-reliability theory. Beneath the surface of what appeared, at first blush, to be a strong safety record was, in fact, a "long series of close calls with U.S. nuclear weapon systems." This is not because the organizations in charge of safeguarding U.S. nuclear weapons were unnaturally incompetent or lax. Rather, the history of nuclear near misses simply reflects "the inherent limits of organizational safety," he said. Military organizations have other operational demands they must accommodate beyond safety. Political scientists have termed this the "always/never dilemma." Militaries of nuclear-armed powers must *always* be ready to launch nuclear weapons at a moment's notice and deliver a massive strike against their adversaries for deterrence to be credible. At the same time, they must *never* allow unauthorized or accidental detonation of a weapon. Sagan says this is effectively "impossible." There are limits to how safe some hazards can be made.

THE INEVITABILITY OF ACCIDENTS

Safety is challenging enough with nuclear weapons. Autonomous weapons would be potentially more difficult in a number of ways. Nuclear weapons are available to only a handful of actors, but autonomous weapons could proliferate widely, including to countries less concerned about safety. Autonomous weapons have an analogous problem to the always/never dilemma: once put into operation, they are expected to find and destroy enemy targets and not strike friendlies or civilian objects. Unlike nuclear weapons, some isolated mistakes might be tolerated with autonomous weapons, but gross errors would not.

The fact that autonomous weapons are not obviously as dangerous as nuclear weapons might make risk mitigation more challenging in some respects. The perception that automation can increase safety and reliability—which is true in some circumstances—could lead militaries to be less cautious with autonomous weapons than even other conventional weapons. If militaries cannot reliably institute safety procedures to control and account for nuclear weapons, their ability to do so with autonomous weapons is far less certain.

The overall track record of nuclear safety, Aegis operations, and the Patriot fratricides suggests that sound procedures can reduce the likelihood of accidents, but can never drive them to zero. By embracing the principles of high-reliability organizations, the U.S. Navy submarine and Aegis communities have been able to manage complex, hazardous systems safely, at least during peacetime. Had the Patriot community adopted some of these principles prior to 2003, the fratricides might have been prevented. At the very least, the Tornado shootdown could have been prevented with a greater cultural vigilance to respond to near-miss incidents and correct known problems, such as the anti-radiation missile misclassification problem, which had come up in testing. High-reliability theory does not promise zero accidents, however. It merely suggests that very low accident rates are possible. Even in industries where safety is paramount, such as nuclear power, accidents still occur.

There are reasons to be skeptical of the ability to achieve high-reliability operations for autonomous weapons. High-reliability organizations depend on three key features that work for Aegis in peacetime, but are unlikely to be present for fully autonomous weapons in war.

First, high-reliability organizations can achieve low accident rates by constantly refining their operations and learning from near-miss incidents. This is only possible if they can accumulate extensive experience in their operating environment. For example, when Aegis first arrives to an area, the ship operates for some time with its radar on and doctrine enabled, but the weapons deactivated, so sailors can see how the doctrine responds to the unique peculiarities of that specific operating environment. Similarly, FAA air traffic control, nuclear power plants, and aircraft carriers are systems people operate day in and day out, accumulating large amounts of operational experience. This daily experience in real-world conditions allows them to refine safe operations.

When extreme events occur outside the norm, safety can be compromised. Users are not able to anticipate all of the possible interactions that may occur under atypical conditions. The 9.0 magnitude earthquake in Japan that led to the Fukushima-Daiichi meltdown is one such example. If 9.0 magnitude earthquakes causing forty-foot-high tsunamis were a regular occurrence, nuclear power plant operators would have quickly learned to anticipate the common-mode failure that knocked out primary and backup power. They would have built higher floodwalls and elevated the backup diesel generators off the ground. It is difficult, however, to anticipate the specific failures that might occur during atypical events.

War is an atypical condition. Militaries prepare for war, but the usual day-to-day experience of militaries is peacetime. Militaries attempt to prepare for the rigors of war through training, but no amount of training can replicate the violence and chaos of actual combat. This makes it very difficult for militaries to accurately predict the behavior of autonomous systems in war. Even for Aegis, activating the doctrine with the weapons disabled allows the operators to understand only how the doctrine will interact with a peacetime operating environment. A wartime operating environment will inevitably be different and raise novel challenges. The USS *Vincennes* accident highlights this problem. The *Vincennes* crew faced a set of conditions that were different from peacetime—military and commercial aircraft operating in close proximity from the same air base coupled with an ongoing hostile engagement from Iranian boats firing at the *Vincennes*. Had they routinely faced these challenges, they might have been able to come up with protocols to avoid an accident, such as staying off the path of civilian airliners. However, their day-to-day operations did not prepare them—and could not have prepared them—for the complexities that combat would bring. Hawley remarked, "You can go through all of the kinds of training that you think you should do . . . what nails you is the unexpected and the surprises."

Another important difference between peacetime high-reliability organizations and war is the presence of adversarial actors. Safe operation of complex systems is difficult because bureaucratic actors have other interests that can sometimes compete with safety—profit, prestige, etc. However, none of the actors are generally hostile to safety. The risk is that people take shortcuts, not actively sabotage safe operations. War is different. War is an inherently adversarial environment in which there are actors attempting to undermine, exploit, or subvert systems. Militaries prepare their troops for this environment not by trying to train their troops for every possible enemy action, but rather by inculcating a culture of resiliency, decisiveness, and autonomous execution of orders. Warfighters must adapt on the fly and come up with novel solutions to respond to enemy actions. This is an area in which humans excel, but machines perform poorly. The brittleness of automation is a major weakness when it comes to responding to adversary innovation. Once an adversary finds a vulnerability in an autonomous system, he or she is free to exploit it until a human realizes the vulnerability and either fixes the system or adapts its use. The system itself cannot adapt. The predictability that a human user finds desirable in automation can be a vulnerability in an adversarial environment.

Finally, the key ingredient in high-reliability organizations that makes them reliable is people, who by definition are not present in the actual execution of operations by a fully autonomous weapon. People are what makes high-reliability organizations reliable. Automation can play a role for "planned actions," as William Kennedy explained, but humans are required to make the system flexible, so that operations are resilient in the face of atypical events. Humans put slack in a system's operations, reducing the tight coupling between components and allowing for judgment to play a role in operations. In fully autonomous systems, humans are present during the design and testing of a system and humans put the system into operation, but humans are not present during actual operations. They cannot intervene if something goes wrong. The *organization* that enables high reliability is not available—the machine is on its own, at least for some period of time. Safety under these conditions requires something more than high-reliability organizations. It requires high-reliability fully autonomous complex machines, and there is no precedent for such systems. This would require a vastly different kind of machine from Aegis, one that was exceptionally predictable to the user but not to the enemy, and with a fault-tolerant design that defaulted to safe operations in the event of failures.

Given the state of technology today, no one knows how to build a complex system that is 100 percent fail-safe. It is tempting to think that future systems will change this dynamic. The promise of "smarter" machines is seductive: they will be more advanced, more intelligent, and therefore able to account for more variables and avoid failures. To a certain extent, this is true. A more sophisticated early warning system that understood U.S. nuclear doctrine might have been able to apply something similar to Petrov's judgment, determining that the attack was likely false. A more advanced version of the Patriot might have been able to take into account the IFF problems or electromagnetic interference and withhold firing on potentially ambiguous targets.

But smarter machines couldn't avoid accidents entirely. New features increase complexity, a double-edged sword. More complex machines may be more capable, but harder for users to understand and predict their behavior, particularly in novel situations. For rule-based systems, deciphering the intricate web of relationships between the various rules that govern a system's behavior and all possible interactions it might have with its environment quickly becomes impossible. Adding more rules can make a system smarter by allowing it to account for more scenarios, but the increased complexity of its internal logic makes it even more opaque to the user.

Learning systems would appear to sidestep this problem. They don't rely on rules. Rather, the system is fed data and then learns the correct answer through experience over time. Some of the most innovative advances in AI are in learning systems, such as deep neural networks. Militaries will want to use learning systems to solve difficult problems, and indeed programs such as DARPA's TRACE already aim to do so. Testing these systems is even more challenging, however. Incomprehensibility is a problem in complex systems, but it is far worse in systems that learn on their own.

11

BLACK BOX

THE WEIRD, ALIEN WORLD OF DEEP
NEURAL NETWORKS

Learning machines that don't follow a set of programmed rules, but rather learn from data, are effectively a "black box" to designers. Computer programmers can look at the network's output and see whether it is right or wrong, but understanding *why* the system came to a certain conclusion—and, more importantly, predicting its failures in advance—can be quite challenging. Bob Work specifically called out this problem when I met with him. "How do you do test and evaluation of learning systems?" he asked. He didn't have an answer; it is a difficult problem.

The problem of verifying the behavior of learning systems is starkly illustrated by the vulnerability of the current class of visual object recognition AIs to "adversarial images." Deep neural networks have proven to be an extremely powerful tool for object recognition, performing as well or better than humans in standard benchmark tests. However, researchers have also discovered that, at least with current techniques, they have strange and bizarre vulnerabilities that humans lack.

Adversarial images are pictures that exploit deep neural networks' vulnerabilities to trick them into confidently identifying false images. Adversarial images (usually created by researchers intentionally) come in two forms: one looks like abstract wavy lines and shapes and the other looks to the human eye like meaningless static. Neural networks nevertheless identify these nonsense images as concrete objects, such as a

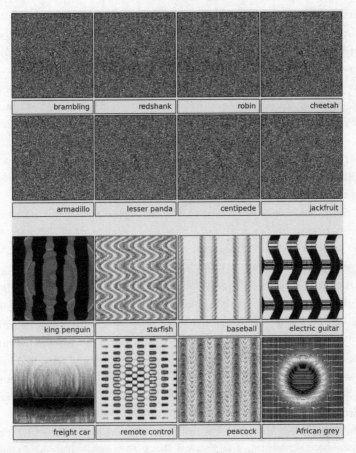

High-Confidence "Fooling Images" *A state-of-the-art image recognition neural network identified these images, which are unrecognizable to humans, as familiar objects with a greater than 99.6 percent certainty. Researchers evolved the images using two different techniques: evolving individual pixels for the top eight images and evolving the image as a whole for the bottom eight images.*

starfish, cheetah, or peacock, with greater than 99 percent confidence. The problem isn't that the networks get some objects wrong. The problem is that the way in which the deep neural nets get the objects wrong is bizarre and counterintuitive to humans. The networks falsely identify objects from meaningless static or abstract shapes in ways that humans never would. This makes it difficult for humans to accurately predict the circumstances in which the neural net might fail. Because the network behaves in a way that seems totally alien, it is very difficult for humans to come up with an accurate mental model of the network's internal logic to predict its behavior. Within the black box of the neural net lies a counterintuitive and unexpected form of brittleness, one that is surprising even to the network's designers. This is not a weakness of only one specific network. This vulnerability appears to be replicated across most deep neural networks currently used for object recognition. In fact, one doesn't even need to know the specific internal structure of the network in order to fool it.

To better understand this phenomenon, I spoke with Jeff Clune, an AI researcher at the University of Wyoming who was part of the research team that discovered these vulnerabilities. Clune described their discovery as a "textbook case of scientific serendipity." They were attempting to design a "creative artificial intelligence that could endlessly innovate." To do this, they took an existing deep neural network that was trained on image recognition and had it evolve new images that were abstractions of the image classes it knew. For example, if it had been trained to recognize baseballs, then they had the neural net evolve a new image that captured the essence of "baseball." They envisioned this creative AI as a form of artist and expected the result would be unique computer images that were nevertheless recognizable to humans. Instead, the images they got were "completely unrecognizable garbage," Clune said. What was even more surprising, however, was that other deep neural nets agreed with theirs and identified the seemingly garbage images as actual objects. Clune described this discovery as stumbling across a "huge, weird, alien world of imagery" that AIs all agree on.

This vulnerability of deep neural nets to adversarial images is a major problem. In the near term, it casts doubt on the wisdom of using the current class of visual object recognition AIs for military applications—or for that matter any high-risk applications in adversarial environments. Deliberately feeding a machine false data to manipulate its behavior is known as a spoofing attack, and the current state-of-the-art image classifiers have a known weakness to spoofing attacks that can be exploited by adversaries. Even worse, the adversarial images can be surreptitiously

Hidden Spoofing Attacks Inside Images *The images on the right and left columns look identical to humans, but are perceived very differently by neural networks. The left column shows the unaltered image, which is correctly identified by the neural network. The middle column shows, at 10x amplification, the difference between the images on the right and left. The right column shows the manipulated images, which contain a hidden spoofing attack that is not noticeable by humans. Due to the subtle manipulation of the image, the neural network identified all of the objects in the right column as "ostrich."*

embedded into normal images in a way that is undetectable by humans. This makes it a "hidden exploit," and Clune explained that this could allow an adversary to trick the AI in a way that was invisible to the human. For example, someone could embed an image into the mottled gray of an athletic shirt, tricking an AI security camera into believing the person wearing the shirt was authorized entry, and human security guards wouldn't even be able to tell a fooling image being used.

Researchers are only beginning to understand why the current class of deep neural networks is susceptible to this type of manipulation. It appears to stem from fundamental properties of their internal structures. The semitechnical explanation is that while deep neural networks are highly nonlinear at the macro level, they actually use linear methods to interpret data at the micro level. What does that mean? Imagine a field of gray dots separated into two clusters, with mostly light gray dots on the right and darker gray dots on the left, but with some overlap in the middle. Now imagine the neural net is trained on this data and asked to predict whether, given the position of a new dot, it is likely to be light or dark gray. Based on current methods, the AI will draw a line between the light and dark gray clusters. The AI would then predict that new dots on the left side of the line are likely to be darker and new dots on the right side of the line are likely to be lighter, acknowledging that there is some overlap and there will be an occasional light gray dot on the left or dark gray on the right. Now imagine that you asked it to predict where the darkest possible dot would be. Since the further one moves to the left the more likely the dot is to be dark gray, the AI would put it "infinitely far to the left," Clune explained. This is the case even though the AI has zero information about any dots that far away. Even worse, because the dot is so far to the left, the AI would be very confident in its prediction that the dot would be dark. This is because at the micro level, the AI has a very simple, linear representation of the data. All it knows is that the further one moves left, the more likely the dot is to be dark.

The "fooling images," as Clune calls them, exploit this vulnerability. He explained that, "real-world images are a very, very small, rare subset of all possible images." On real-world images, the AIs do fairly well. This hack exploits their weakness on the extremes, however, in the space of all possible images, which is virtually infinite.

Because this vulnerability stems from the basic structure of the neural net, it is present in essentially every deep neural network commonly in use today, regardless of its specific design. It applies to visual object recognition neural nets but also to those used for speech recognition or other data analysis. This exploit has been demonstrated with song-interpreting AIs, for example. Researchers fed specially evolved noise into the AI, which sounds like nonsense to humans, but which the AI confidently interpreted as music.

In some settings, the consequences of this vulnerability could be severe. Clune gave a hypothetical example of a stock-trading neural net that read the news. News-reading trading bots appear to already be active on the market, evidenced by sharp market moves in response to news events at speeds faster than what is possible by human traders. If these bots used

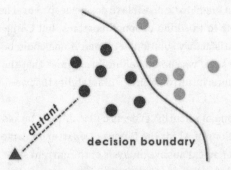

Evolving Fooling Images *"Fooling images" are created by evolving novel images that are far from the decision boundary of the neural network. The "decision boundary" is the line of 50/50 confidence between two classes of images, in this case two shades of dots. The neural network's confidence in the image's correct classification increases as the image is further from the decision boundary. At the extremes, however, the image may no longer be recognizable, yet the neural network classifies the image with high confidence.*

deep neural networks to understand text—a technique that has been demonstrated and is extremely effective—then they would be vulnerable to this form of hacking. Something as simple as a carefully crafted tweet could fool the bots into believing a terrorist attack was under way, for example. A similar incident already occurred in 2013 when the Associated Press Twitter account was hacked and used to send a false tweet reporting explosions at the White House. Stocks rapidly plunged in response. Eventually, the AP confirmed that its account had been hacked and markets recovered, but what makes Clune's exploit so damaging is that it could be done in a hidden way, without humans even aware that it is occurring.

You may be wondering, why not just feed these images back into the network and have it learn that these images are false, vaccinating the network against this hack? Clune and others have tried that. It doesn't work, Clune explained, because the space of all possible images is "virtually infinite." The neural net learns that specific image is false, but many more fooling images can be evolved. Clune compared it to playing an "infinite game of whack-a-mole" with "an infinite number of holes." No matter how many fooling images the AI learns to ignore, more can be created.

In principle, it ought to be possible to design deep neural networks that aren't vulnerable to this kind of spoofing attack, but Clune said that he hasn't seen a satisfactory solution yet. Even if one could be discovered, however, Clune said "we should definitely assume" that the new AI has some other "counterintuitive, weird" vulnerability that we simply haven't discovered yet.

In 2017, a group of scientific experts called JASON tasked with studying the implications of AI for the Defense Department came to a similar conclusion. After an exhaustive analysis of the current state of the art in AI, they concluded:

> [T]he sheer magnitude, millions or billions of parameters (i.e. weights/biases/etc.), which are learned as part of the training of the net ... makes it impossible to really understand exactly how the network does what it does. Thus the response of the network to all possible inputs is unknowable.

Part of this is due to the early stage of research in neural nets, but part of it is due to the sheer complexity of the deep learning. The JASON group argued that "the very nature of [deep neural networks] may make it intrinsically difficult for them to transition into what is typically recognized as a professionally engineered product."

AI researchers are working on ways to build more transparent AI, but Jeff Clune isn't hopeful. "As deep learning gets even more powerful and more impressive and more complicated and as the networks grow in size, there will be more and more and more things we don't understand. . . . We have now created artifacts so complicated that we ourselves don't understand them." Clune likened his position to an "AI neuroscientist" working to discover how these artificial brains function. It's possible that AI neuroscience will elucidate these complex machines, but Clune said that current trends point against it: "It's almost certain that as AI becomes more complicated, we'll understand it less and less."

Even if it were possible to make simpler, more understandable AI, Clune argued that it probably wouldn't work as well as AI that is "super complicated and big and weird." At the end of the day, "people tend to use what works," even if they don't understand it. "This kind of a race to use the most powerful stuff—if the most powerful stuff is inscrutable and unpredictable and incomprehensible—somebody's probably going to use it anyway."

Clune said that this discovery has changed how he views AI and is a "sobering message." When it comes to lethal applications, Clune warned using deep neural networks for autonomous targeting "could lead to tremendous harm." An adversary could manipulate the system's behavior, leading it to attack the wrong targets. "If you're trying to classify, target, and kill autonomously with no human in the loop, then this sort of adversarial hacking could get fatal and tragic extremely quickly."

While couched in more analytic language, the JASON group essentially issued the same cautionary warning to DoD:

> [I]t is not clear that the existing AI paradigm is immediately amenable to any sort of software engineering validation and verification. This is a serious issue, and is a potential roadblock to DoD's use of these modern AI systems, especially when considering the liability and accountability of using AI in lethal systems.

Given these glaring vulnerabilities and the lack of any known solution, it would be extremely irresponsible to use deep neural networks, as they exist today, for autonomous targeting. Even without any knowledge about how the neural network was structured, adversaries could generate fooling images to draw the autonomous weapon onto false targets and conceal legitimate ones. Because these images can be hidden, it could do so in a way that is undetectable by humans, until things start blowing up.

Beyond immediate applications, this discovery should make us far more cautious about machine learning in general. Machine learning techniques are powerful tools, but they also have weaknesses. Unfortunately, these weaknesses may not be obvious or intuitive to humans. These vulnerabilities are different and more insidious than those lurking within complex systems like nuclear reactors. The accident at Three Mile Island might not have been predictable ahead of time, but it is at least understandable after the fact. One can lay out the specific sequence of events and understand how one event led to another, and how the combination of highly improbable events led to catastrophe. The vulnerabilities of deep neural networks are different; they are entirely alien to the human mind. One group of researchers described them as "nonintuitive characteristics and intrinsic blind spots, whose structure is connected to the data distribution in a non-obvious way." In other words: the AIs have weaknesses that we can't anticipate and we don't really understand how it happens or why.

12

FAILING DEADLY

THE RISK OF AUTONOMOUS WEAPONS

A cknowledging that machine intelligence has weaknesses does not negate its advantages. AI isn't good or bad. It is powerful. The question is how humans should use this technology. How much freedom (autonomy) should we give AI-empowered machines to perform tasks on their own?

Delegating a task to a machine means accepting the consequences if the machine fails. John Borrie of UNIDIR told me, "I think that we're being overly optimistic if we think that we're not going to see problems of system accidents" in autonomous weapons. Army researcher John Hawley agreed: "If you're going to turn these things loose, whether it be Patriot, whether it be Aegis, whether it be some type of totally unmanned system with the ability to kill, you have to be psychologically prepared to accept the fact that sometimes incidents will happen." Charles Perrow, the father of normal accident theory, made a similar conclusion about complex systems in general:

> [E]ven with our improved knowledge, accidents and, thus, potential catastrophes are inevitable in complex, tightly coupled systems with lethal possibilities. We should try harder to reduce failures—and that will help a great deal—but for some systems it will not be enough. . . . We must live and die with their risks, shut them down, or radically redesign them.

If we are to use autonomous weapons, we must accept their risks. All weapons are dangerous. War entails violence. Weapons that are designed to be dangerous to the enemy can also be dangerous to the user if they slip out of control. Even a knife wielded improperly can slip and cut its user. Most modern weapons, regardless of their level of autonomy, are complex systems. Accidents will happen, and sometimes these accidents will result in fratricide or civilian casualties. What makes autonomous weapons any different?

The key difference between semi-, supervised, and fully autonomous weapons is amount of damage the system can cause until the next opportunity for a human to intervene. In semi- or supervised autonomous weapons, such as Aegis, the human is a natural fail-safe against accidents, a circuit breaker if things go wrong. The human can step outside of the rigid rules of the system and exercise judgment. Taking the human out of the loop reduces slack and increases the coupling of the system. In fully autonomous weapons, there is no human to intervene and halt the system's operation. A failure that might cause a single unfortunate incident with a semiautonomous weapon could cause far greater damage if it occurred in a fully autonomous weapon.

THE RUNAWAY GUN

A simple malfunction in an automatic weapon—a machine gun—provides an analogy for the danger with autonomous weapons. When functioning properly, a machine gun continues firing so long as the trigger remains held down. Once the trigger is released, a small metal device called a "sear" springs into place to stop the operating rod within the weapon from moving, halting the automatic firing process. Over time, however, the sear can become worn down. If the sear becomes so worn down that it fails to stop the operating rod, the machine gun will continue firing even when the trigger is released. The gun will keep firing on its own until it exhausts its ammunition.

This malfunction is called a runaway gun.

Runaway guns are serious business. The machine gunner has let go of the trigger, but the gun continues firing: the firing process is now fully automatic, with no way to directly halt it. The only way to stop a runaway gun is to break the links on the ammunition belt feeding into the weapon. While this is happening, the gunner must ensure the weapon stays pointed in a safe direction.

A runaway gun is the kind of hypothetical danger I was aware of as an infantry soldier, but I remember clearly the first time I heard about one actually occurring. We were out on an overnight patrol in northeastern Afghanistan and got word of an incident back at the outpost where we were based. An M249 SAW (light machine gun) gunner tried to disassemble his weapon without removing the ammunition first. (Pro tip: bad idea.) When he removed the pistol grip, the sear that held back the operating rod came out with it. The bolt slammed forward, firing off a round. The recoil cycled the weapon, which reloaded and fired again. Without anything to stop it, the weapon kept firing. A stream of bullets sailed across the outpost, stitching a line of holes across the far wall until someone broke the links of the ammunition belt feeding into the gun. No one was killed, but such accidents don't always end well.

In 2007, a South African antiaircraft gun malfunctioned on a firing range, resulting in a runaway gun that killed nine soldiers. Contrary to breathless reports of a "robo-cannon rampage," the remote gun was not an autonomous weapon and likely malfunctioned because of a mechanical problem, not a software glitch. According to sources knowledgeable about the weapon, it was likely bad luck, not deliberate targeting, that caused the gun to swivel toward friendly lines when it malfunctioned. Unfortunately, despite the heroic efforts of one artillery officer who risked her life to try to stop the runaway gun, the gun poured a string of 35 mm rounds into a neighboring gun position, killing the soldiers present.

Runaway guns can be deadly affairs even with simple machine guns that can't aim themselves. A loss of control of an autonomous weapon would be a far more dangerous situation. The destruction unleashed by an autonomous weapon would not be random—it would be targeted. If there were no human to intervene, a single accident could become many, with the system continuing to engage inappropriate targets until it exhausted its ammunition. "The machine doesn't know it's making a mistake," Hawley observed. The consequences to civilians or friendly forces could be disastrous.

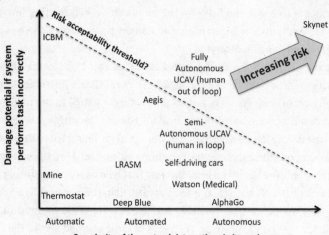

Risk of Delegating Autonomy to a Machine

THE DANGER OF AUTONOMOUS WEAPONS

With autonomous weapons, we are like Mickey enchanting the broomstick. We trust that autonomous weapons will perform their functions correctly. We trust that we have designed the system, tested it, and trained the operators correctly. We trust that the operators are using the system the right way, in an environment they can understand and predict, and that they remain vigilant and don't cede their judgment to the machine. Normal accident theory would suggest that we should trust a little less.

Autonomy is tightly bounded in weapons today. Fire-and-forget missiles cannot be recalled once launched, but their freedom to search for targets in space and time is limited. This restricts the damage they could cause if they fail. In order for them to strike the wrong target, there would need to be an inappropriate target that met the seeker's parameters within the seeker's field of view for the limited time it was active. Such a circumstance is not inconceivable. That appears to be what occurred in the F-18 Patriot fratricide. If missiles were made more autonomous, however—if the freedom of the seeker to search in time and space were expanded—the possibility for more accidents like the F-18 shootdown would expand.

Supervised autonomous weapons such as the Aegis have more freedom to search for targets in time and space, but this freedom is compensated for by the fact that human operators have more immediate control over the weapon. Humans supervise the weapon's operation in real time. For Aegis, they can engage hardware-level cutouts that will disable power, preventing a missile launch. An Aegis is a dangerous dog kept on a tight leash.

Fully autonomous weapons would be a fundamental paradigm shift in warfare. In deploying fully autonomous weapons, militaries would be introducing onto the battlefield a highly lethal system that they cannot control or recall once launched. They would be sending this weapon into an environment that they do not control where it is subject to enemy hacking and manipulation. In the event of failures, the damage fully autonomous weapons could cause would be limited only by the weapons' range, endurance, ability to sense targets, and magazine capacity.

Additionally, militaries rarely deploy weapons individually. Flaws in any one system are likely to be replicated in entire squadrons and fleets of autonomous weapons, opening the door to what John Borrie described as "incidents of mass lethality." This is fundamentally different from human mistakes, which tend to be idiosyncratic. Hawley told me, "If you put someone else in [a fratricide situation], they probably would assess the situation differently and they may or may not do that." Machines are different. Not only will they continue making the same mistake; all other systems of that same type will do so as well.

A frequent refrain in debates about autonomous weapons is that humans also make mistakes and if the machines are better, then we should use the machines. This objection is a red herring and misconstrues the nature of autonomous weapons. If there are specific engagement-related tasks that automation can do better than humans, then those tasks should be automated. Humans, whether in the loop or on the loop, act as a vital fail-safe, however. It's the difference between a pilot flying an airplane on autopilot and an airplane with no human in the cockpit at all. The key factor to assess with autonomous weapons isn't whether the system is better than a human, but rather if the system fails (which it inevitably will), what is the amount of damage it could cause, and can we live with that risk?

Putting an offensive fully autonomous weapon system into operation would be like turning an Aegis to Auto-Special, rolling FIS green, pointing it toward a communications-denied environment, and having everyone on board exit the ship. Deploying autonomous weapons would be like putting a whole fleet of these systems into operation. There is no precedent for delegating that amount of lethality to autonomous systems without any ability for humans to intervene. In fact, placing that amount of trust in machines would run 180 degrees counter to the tight control the Aegis community maintains over supervised autonomous weapons today.

I asked Captain Galluch what he thought of an Aegis operating on its own with no human supervision. It was the only question I asked him in our four-hour interview for which he did not have an immediate answer. It was clear that in his thirty-year career it had never once occurred to him to turn an Aegis to Auto-Special, roll FIS green, and have everyone on board exit the ship. He leaned back in his chair and looked out the window. "I don't have a lot of good answers for that," he said. But then he began to walk through what one might need to do to build trust in such a system, applying his decades of experience with Aegis. One would need to "build a little, test a little," he said. High-fidelity computer modeling coupled with real-world tests and live-fire exercises would be necessary to understand the system's limitations and the risks of using it. Still, he said, if the military did deploy a fully autonomous weapon, "we're going to get a *Vincennes*-like response" in the beginning. "Understanding the complexity of Aegis has been a thirty-year process," Galluch said. "Aegis today is not the Aegis of *Vincennes*," but only because the Navy has learned from mistakes. With a fully autonomous weapon, we'd be starting at year zero.

Deploying fully autonomous weapons would be a weighty risk, but it might be one that militaries decide is worth taking. Doing so would be entering uncharted waters. Experience with supervised autonomous weapons such as Aegis would be useful, but only to a point. Fully autonomous weapons in wartime would face unique conditions that limit the applicability of lessons from high-reliability organizations. The wartime operating environment is different from day-to-day peacetime experience. Hostile actors are actively trying to undermine safe operations. And no humans would be present at the time of operation to intervene or correct problems.

There is one industry that has many of these dynamics, where automation is used in a competitive high-risk environment and at speeds that make it impossible for humans to compete: stock-trading. The world of high-frequency trading—and its consequences—has instructive lessons for what could happen if militaries deployed fully autonomous weapons.

PART IV

Flash War

13

BOT VS. BOT

AN ARMS RACE IN SPEED

On May 6, 2010, at 2:32 p.m. Eastern Time, the S&P 500, NASDAQ, and Dow Jones Industrial Average all began a precipitous downward slide. Within a few minutes, they were in free fall. By 2:45 p.m., the Dow had lost nearly 10 percent of its value. Then, just as inexplicably, the markets rebounded. By 3:00 p.m., whatever glitch had caused the sharp drop in the market was over. However, the repercussions from the "Flash Crash," as it came to be known, were only beginning.

Asian markets tumbled when they opened the next day, and while the markets soon stabilized, it was harder to repair confidence. Traders described the Flash Crash as "horrifying" and "absolute chaos," reminiscent of the 1987 "Black Monday" crash where the Dow Jones lost 22 percent of its value. Market corrections had occurred before, but the sudden downward plunge followed by an equally rapid reset suggested something else. In the years preceding the Flash Crash, algorithms had taken over a large fraction of stock trading, including high-frequency trading that occurred at superhuman speeds. Were the machines to blame?

Investigations followed, along with counter-investigations and eventually criminal charges. Simple answers proved elusive. Researchers blamed everything from human error to brittle algorithms, high-frequency trading, market volatility, and deliberate market manipulation. In truth, all of them likely played a role. Like other normal accidents, the Flash Crash had multiple causes, any one of which individually would have been manageable. The combination, however, was uncontrollable.

RISE OF THE MACHINES

Stock trading today is largely automated. Gone are the days of floor traders shouting prices and waving their hands to compete for attention in the furious scrum of the New York Stock Exchange. Approximately three-quarters of all trades made in the U.S. stock market today are executed by algorithms. Automated stock trading, sometimes called algorithmic trading, is when computer algorithms are used to monitor the market and make trades based on certain conditions. The simplest kind of algorithm, or "algo," is used to break up large trades into smaller ones in order to minimize the costs of the trade. If a single buy or sell order is too large relative to the volume of that stock that is regularly traded, placing the order all at once can skew the market price. To avoid this, traders use algorithms to break up the sale into pieces that can be executed incrementally according to stock price, time, volume, or other factors. In such cases, the decision to make the trade (to buy or sell a certain amount of stock) is still made by a person. The machine simply handles the execution of the trade.

Some trading algorithms take on more responsibility, actually making automated trading decisions to buy or sell based on the market. For example, an algorithm could be tasked to monitor a stock's price over a period of time. When the price moves significantly above or below the average of where the price has been, the algo sells or buys accordingly, under the assumption that over time the price will revert back to the average, yielding a profit. Another strategy could be to look for arbitrage opportunities, where the price of a stock in one market is different from the price in another market, and this price difference can be exploited for profit. All of these strategies could, in principle, be done by humans. Automated trading offers the advantage, however, of monitoring large amounts of data and immediately and precisely making trades in ways that would be impossible for humans.

Speed is a vital factor in stock trading. If there is a price imbalance and a stock is under- or overpriced, many other traders are also looking to sweep up that profit. Move too slow and one could miss the opportunity. The result has been an arms race in speed and the rise of high-frequency trading, a specialized type of automated trading that occurs at speeds too quick for humans to even register.

The blink of an eye takes a fraction of a second—0.1 to 0.4 seconds—but is still an eon compared to high-frequency trading. High-frequency trades move at speeds measured in microseconds: 0.000001 seconds. During the span of a single eyeblink, 100,000 microseconds pass by. The result is an entirely new ecosystem, a world of trading bots dueling at superhuman speeds only accessible by machines.

The gains from even a slight advantage in speed are so significant that high-frequency traders will go to great lengths to shave just a few microseconds off their trading times. High-frequency traders colocate their servers within the server rooms of stock exchanges, cutting down on travel time. Some are even willing to pay additional money to move their firm's servers a few feet closer to the stock exchange's servers inside the room. Firms try to find the shortest route for their cables within the server room, cutting microseconds off transit time. Like race teams outfitting an Indy car, high-frequency traders spare no expense in optimizing every part of their hardware for speed, from data switches to the glass inside fiber-optic cables.

At the time scales at which high-frequency trading operates, humans have to delegate trading decisions to the algorithms. Humans can't possibly observe the market and react to it in microseconds. That means if things go wrong, they can go wrong very quickly. To ensure algorithms do what they are designed to do once released into the real world, developers test them against actual stock market data, but with trading disabled—analogous to testing Aegis doctrine with the FIS key turned red. Despite this, accidents still occur.

"KNIGHTMARE ON WALL STREET"

In 2012, Knight Capital Group was a titan of high-frequency trading. Knight was a "market maker," a high-frequency trader that executed over 3.3 billion trades, totaling $21 billion, every single day. Like most high-frequency traders, Knight didn't hold on to this stock. Stocks were bought and sold the same day, sometimes within fractions of a second. Nevertheless, Knight was a key player in the U.S. stock market, executing 17 percent of all trades on the New York Stock Exchange and NASDAQ. Their slogan was, "The Science of Trading, the Standard of Trust." Like many high-frequency trading firms, their business was lucrative. On the morning of July 31, 2012, Knight had $365 million in assets. Within 45 minutes, they would be bankrupt.

At 9:30 a.m. Eastern Time on July 31, U.S. markets opened and Knight deployed a new automated trading system. Instantly, it was apparent that something was wrong. One of the functions of the automated trading system was to break up large orders into smaller ones, which then would be executed individually. Knight's trading system wasn't registering that these smaller trades were actually completed, however, so it kept tasking them again. This created an endless loop of trades. Knight's trading system began flooding the market with orders, executing over a thousand trades a second. Even worse, Knight's algorithm was buying high and selling low, losing money on every trade.

There was no way to stop it. The developers had neglected to install a "kill switch" to turn their algorithm off. There was no equivalent of "rolling FIS red" to terminate trading. While Knight's computer engineers worked to diagnose the problem, the software was actively trading in the market, moving $2.6 million a second. By the time they finally halted the system 45 minutes later, the runaway algo had executed 4 million trades, moving $7 billion. Some of those trades made money, but Knight lost a net $460 million. The company only had $365 million in assets. Knight was bankrupt.

An influx of cash from investors helped Knight cover their losses, but the company was ultimately sold. The incident became known as the "Knightmare on Wall Street," a cautionary tale for partners to tell their associates about the dangers of high-frequency trading. Knight's runaway algo vividly demonstrated the risk of using an autonomous system in a high-stakes application, especially with no ability for humans to intervene. Despite their experience in high-frequency trading, Knight was taking fatal risks with their automated stock trading system.

BEHIND THE FLASH CRASH

If the Knightmare on Wall Street was like a runaway gun, the Flash Crash was like a forest fire. The damage from Knight's trading debacle was largely contained to a single company, but the Flash Crash affected the entire market. A volatile combination of factors meant that during the Flash Crash, one malfunctioning algorithm interacted with an entire marketplace ready to run out of control. And run away it did.

The spark that lit the fire was a single bad algorithm. At 2:32 p.m. on May 6, 2010, Kansas-based mutual fund trader Waddell & Reed initiated a sale of 75,000 S&P 500 E-mini futures contracts estimated at $4.1 billion. (E-minis are a smaller type of futures contract, one-fifth the size of a regular futures contract. A futures contract is what it sounds like: an agreement to buy or sell at a certain price at a certain point in time in the future.) Because executing such a large trade all at once could distort the market, Waddell & Reed used a "sell algorithm" to break up the sale into smaller trades, a standard practice. The algorithm was tied to the overall volume of E-minis sold on the market, with direction to execute the sale at 9 percent of the trading volume over the previous minute. In theory, this should have spread out the sale so as to not overly influence the market.

The sell algorithm was given no instructions with regard to time or price, however, an oversight that led to a catastrophic case of brittleness. The market that day was already under stress. Government investigators later characterized the market as "unusually turbulent," in part due to an unfolding European debt crisis that was causing uncertainty. By midafternoon, the market was experiencing "unusually high volatility" (sharp movements in prices) and low liquidity (low market depth). It was into these choppy waters that the sell algorithm waded.

Only twice in the previous year had a single trader attempted to unload so many E-minis on the market in a single day. Normally, a trade of this scale took hours to execute. This time, because the sell algorithm was only tied to volume and not price or time, it happened very quickly: within 20 minutes.

The sell algorithm provided the spark, and high-frequency traders were the gasoline. High-frequency traders bought the E-minis the sell algorithm was unloading and, as is their frequent practice, rapidly resold them. This increased the volume of E-minis being traded on the market. Since the rate at which the sell algorithm sold E-minis was tied to volume but not price or time, it accelerated its sales, dumping more E-minis on an already stressed market.

Without buyers interested in buying up all of the E-minis that the sell algorithm and high-frequency traders were selling, the price of E-minis dropped, falling 3 percent in just four minutes. This generated a "hot potato" effect among high-frequency traders as they tried to unload the falling E-minis onto other high-frequency traders. In one 14-second period, high-frequency trading algorithms exchanged 27,000 E-mini contracts. (The total amount Waddell & Reed were trying to sell was 75,000 contracts.) All the while as trading volume skyrocketed, the sell algorithm kept unloading more and more E-minis on a market that was unable to handle them.

The plummeting E-minis dragged down other U.S. markets. Observers watched the Dow Jones, NASDAQ, and S&P 500 all plunge, inexplicably. Finally, at 2:45:28 p.m., an automated "stop logic" safety on the Chicago Mercantile Exchange kicked in, halting E-mini trading for 5 seconds and allowing the markets to reset. They rapidly recovered, but the sharp distortions in the market wreaked havoc on trading. Over 20,000 trades had been executed at what financial regulators termed "irrational prices" far from their norm, some as low as a penny or as high as $100,000. After the markets closed, the Financial Industry Regulatory Authority worked with stock exchanges to cancel tens of thousands of "clearly erroneous" trades.

The Flash Crash demonstrated how when brittle algorithms interact with a complex environment at superhuman speeds, the result can be a runaway process with catastrophic consequences. The stock market as a whole is an incredibly complex system that defies simple understanding, which can make predicting these interactions difficult ahead of time. On a different day, under different market conditions, the same sell algorithm may not have led to a crash.

PRICE WARS: $23,698,655.93
(PLUS $3.99 SHIPPING)

While complexity was a factor in the Flash Crash, even simple interactions between algorithms can lead to runaway escalation. This phenomenon was starkly illustrated when two warring bots jacked up the price of an otherwise ordinary book on Amazon to $23 million. Michael Eisen, a biologist at UC Berkeley, accidentally stumbled across this price war for Peter Lawrence's *Making of a Fly: The Genetics of Animal Design*. Like a good scientist, Eisen began investigating.

Two online sellers, *bordeebook* and *profnath*, both of whom were legitimate online booksellers with thousands of positive ratings, were locked in a runaway price war. Once a day, *profnath* would set its price to 0.9983 times *bordeebook*'s price, slightly undercutting them. A few hours later, *bordeebook* would change its price to 1.270589 times *profnath*'s. The combination raised both booksellers' prices by approximately 27 percent daily.

Bots were clearly to blame. The pricing was irrational and precise. *Profnath*'s algorithm made sense; it was trying to draw in sales by slightly undercutting the highest price on the market. What was *bordeebook*'s algorithm doing, though? Why *raise* the price over the highest competitor?

Eisen hypothesized that *bordeebook* didn't actually own the book. Instead, they probably were posting an ad and hoping their higher reviews would attract customers. If someone bought the book, then of course *bordeebook* would have to buy it, so they set their price slightly above—1.270589 times greater than—the highest price on the market, so they could make a profit.

Eventually, someone at one of the two companies caught on. The price peaked out at $23,698,655.93 (plus $3.99 shipping) before dropping back to a tamer $134.97, where it stayed. Eisen mused in a blog posting, however, about the possibilities for "chaos and mischief" that this discovery suggested. A person could potentially hack this vulnerability of the bots, manipulating prices.

SPOOFING THE BOT

Eisen wasn't the first to think of exploiting the predictability of bots for financial gain. Others had seen these opportunities before him, and they'd gone and done it. Six years after the Flash Crash in 2016, London-based trader Navinder Singh Sarao pled guilty to fraud and spoofing, admitting that he used an automated trading algorithm to manipulate the market for E-minis on the day of the crash. According to the U.S. Department of Justice, Sarao used automated trading algorithms to place multiple large-volume orders to create the appearance of demand to drive up price, then cancelled the orders before they were executed. By deliberately manipulating the price, Sarao could buy low and sell high, making a profit as the price moved.

It would be overly simplistic to pin the blame for the Flash Crash on Sarao. He continued his alleged market manipulation for *five years* after the Flash Crash until finally arrested in 2015 and his spoofing algorithm was reportedly turned *off* during the sharpest downturn in the Flash Crash. His spoofing could have exacerbated instability in the E-mini market that day, however, contributing to the crash.

AFTERMATH

In the aftermath of the Flash Crash, regulators installed "circuit breakers" to limit future damage. Circuit breakers, which were first introduced after the 1987 Black Monday crash, halt trading if stock prices drop too quickly. Market-wide circuit breakers trip if the S&P 500 drops more than 7 percent, 13 percent or 20 percent from the closing price the previous day, temporarily pausing trading or, in the event of a 20 percent drop, shutting down markets for the day. After the Flash Crash, in 2012 the Securities and Exchange Commission introduced new "limit up–limit down" circuit breakers for individual stocks to prevent sharp, dramatic price swings. The limit up–limit down mechanism creates a price band around a stock, based on the stock's average price over the preceding five minutes. If the stock price moves out of that band for more than fifteen seconds, trading is halted on that stock for five minutes.

Circuit breakers are an important mechanism for preventing flash crashes from causing too much damage. We know this because they keep getting tripped. An average day sees a handful of circuit breakers tripped due to rapid price moves. One day in August 2015, over 1,200 circuit breakers were tripped across multiple exchanges. Mini-flash crashes have continued to be a regular, even normal event on Wall Street. Sometimes these are caused by simple human error, such as a trader misplacing a zero or using an algorithm intended for a different trade. In other situations, as in the May 2010 flash crash, the causes are more complex. Either way, the underlying conditions for flash crashes remain, making circuit breakers a vital tool for limiting their damage. As Greg Berman, associate director of the SEC's Office of Analytics and Research, explained, "Circuit breakers don't prevent the initial problems, but they prevent the consequences from being catastrophic."

WAR AT MACHINE SPEED

Stock trading is a window into what a future of adversarial autonomous systems competing at superhuman speeds might look like in war. Both involve high-speed adversarial interactions in complex, uncontrolled environments. Could something analogous to a flash crash occur in war—a flash war?

Certainly, if Stanislav Petrov's fateful decision had been automated, the consequences could have been disastrous: nuclear war. Nuclear command and control is a niche application, though. One could envision militaries deploying autonomous weapons in a wide variety of contexts but still keeping a human finger on the nuclear trigger.

Nonnuclear applications still hold risks for accidental escalation. Militaries regularly interact in tense situations that have the potential for conflict, even in peacetime. In recent years, the U.S. military has jockeyed for position with Russian warplanes in Syria and the Black Sea, Iranian fast boats in the Straits of Hormuz, and Chinese ships and air defenses in the South China Sea. Periods of brinksmanship, where nations flex their militaries to assert dominance but without actually firing weapons, are common in international relations. Sometimes tensions escalate to full-blown crises in which war appears imminent, such as the 1962 Cuban Missile Crisis. In such situations, even the tiniest incident can trigger war.

In 1914, a lone gunman assassinated Archduke Franz Ferdinand of Austria, sparking a chain of events that led to World War I. Miscalculation and ambiguity are common in these tense situations, and confusion and accidents can generate momentum toward war. The Gulf of Tonkin incident, which led Congress to authorize the war in Vietnam, was later discovered to be partially false; a purported gun battle between U.S. and Vietnamese boats on August 4, 1964, never occurred.

Robotic systems are already complicating these situations, even with existing technology. In 2013, China flew a drone over the Senkaku Islands, a contested pile of uninhabited rocks in the East China Sea that both China and Japan claim as their own. In response, Japan scrambled an F-15 fighter jet to intercept the drone. Eventually, the drone turned around and left, but afterward Japan issued news rules of engagement for how it would deal with drone incursions. The rules were more aggressive than those for intercepting manned aircraft, with Japan stating they would shoot down any drone entering their territory. In response, China stated that any attack on their drones would be an "act of war" and that China would "strike back."

As drones have proliferated, they have repeatedly been used to broach other nations' sovereignty. North Korea has flown drones into South Korea. Hamas and Hezbollah have flown drones into Israel. Pakistan has accused India of flying drones over the Pakistani-controlled parts of Kashmir (a claim India has denied). It seems one of the first things people do when they get ahold of drones is send them into places they don't belong.

When sovereignty is clear, the typical response has been to simply shoot down the offending drone. Pakistan shot down the alleged Indian drone over Kashmir. Israel has shot down drones sent into its air space. Syria shot down a U.S. drone over its territory in 2015. A few months later, Turkey shot down a presumed Russian drone that penetrated Turkey from Syria.

These incidents have not led to larger conflagrations, perhaps in part because sovereignty in these incidents was not actually in dispute. These were clear cases where a drone was sent into another nation's air space. Within the realm of international relations, shooting it down was seen as a reasonable response. This same action could be perceived very differently in contested areas, however, such as the Senkaku Islands, where both countries assert sovereignty. In such situations, a country whose drone was shot down might feel compelled to escalate in order to back up their territorial claim. Hints of these incidents have already begun. In December 2016, China seized a small underwater robot drone the United States was operating in the South China Sea. China quickly returned it after U.S. protests, but other incidents might not be resolved so easily.

All of these complications are manageable if autonomous systems do what humans expect them to do. Robots may raise new challenges in war, but humans can navigate these hurdles, so long as the automation is an accurate reflection of human intent. The danger is if autonomous systems do something they aren't supposed to—if humans lose control.

That's already happened with drones. In 2010, a Navy Fire Scout drone wandered 23 miles off course from its Maryland base toward Washington, DC, restricted air space before it was brought back under control. In 2017, an Army Shadow drone flew more than 600 miles after operators lost control, before finally crashing in a Colorado forest. Not all incidents have ended so harmlessly, however.

In 2011, the United States lost control of an RQ-170 stealth drone over western Afghanistan. A few days later, it popped up on Iranian television largely intact and in the hands of the Iranian military. Reports swirled online that Iran had hijacked the drone by jamming its communications link, cutting off contact with its human controllers, and then spoofing its GPS signal to trick it into landing at an Iranian base. U.S. sources called the hacking claim "complete bullshit." (Although after a few days of hemming and hawing, the United States did awkwardly confirm the drone was theirs.) Either way—whatever the cause of the mishap—the United States lost control of a highly valued stealth drone, which ended up in the hands of a hostile nation.

A reconnaissance drone wandering off course might lead to international humiliation and the loss of potentially valuable military technology. Loss of control with a lethal autonomous weapon could be another matter. Even a robot programmed to shoot only in self-defense could still end up firing in situations where humans wished it hadn't. If another nation's military personnel or civilians were killed, it might be difficult to de-escalate tensions.

Heather Roff, a research scientist at Arizona State University who

works on ethics and policy for emerging technologies, says there is validity
to the concern about a "flash war." Roff is less worried about an "isolated
individual platform." Her real concern is "networks of systems" working together in "collaborative autonomy." If the visions of Bob Work and
others come true, militaries will field flotillas of robot ships, wolf packs of
sub-hunting robots undersea, and swarms of aerial drones. In that world,
the consequences of a loss of control could be catastrophic. Roff warned,
"If my autonomous agent is patrolling an area, like the border of India and
Pakistan, and my adversary is patrolling the same border and we have
given certain permissions to escalate in terms of self-defense and those
are linked to other systems . . . that could escalate very quickly." An accident like the Patriot fratricides could lead to a firestorm of unintended
lethality.

When I sat down with Bradford Tousley, DARPA's TTO director, I put
the question of flash crashes to him. Were there lessons militaries could
learn from automated stock trading? Tousley lit up at the mention of high-
frequency trading. He was well aware of the issue and said it was one he'd
discussed with colleagues. He saw automated trading as a "great analogy"
for the challenges of automation in military applications. "What are the
unexpected side effects of complex systems of machines that we don't
fully understand?" he asked rhetorically. Tousley noted that while circuit breakers were an effective damage control measure in stock markets,
"there's no 'time out' in the military."

As interesting as the analogy was, Tousley wasn't concerned about a
flash war because the speed dimension was vastly different between stock
trading and war. "I don't know that large-scale military impacts are in
milliseconds," he said. (A millisecond is a thousand microseconds.) "Even
a hypersonic munition that might go 700 miles in 20 minutes—it takes
20 minutes; it doesn't take 20 milliseconds." The sheer physics of moving
missiles, aircraft, or ships through physical space imposes time constraints on how quickly events can spiral out of control, in theory giving
humans time to adapt and respond.

The exception, Tousley said, was in electronic warfare and cyberspace,
where interactions occur at "machine speed." In this world, "the speed
with which a bad event can happen," he said, "is milliseconds."

14

THE INVISIBLE WAR

AUTONOMY IN CYBERSPACE

In just the past few decades, humans have created an invisible world. We can't see it, but we feel its influence everywhere we go: the buzzing of a phone in our pocket, the chime of an email, the pause when a credit card reader searches the aether for authorization. This world is hidden from us, yet in plain view everywhere. We call it the internet. We call it cyberspace.

Throughout history, technology has enabled humans to venture into inhospitable domains, from undersea to the air and space. As we did, our war-making machines came with us. Cyberspace is no different. In this invisible world of machines operating at machine speed, a silent war rages.

MALICIOUS INTENT

You don't need to be a computer programmer to understand malware. It's the reason you're supposed to upgrade your computer and phone when prompted. It's the reason you're not supposed to click on links in emails from strangers. It's the reason you worry when you hear yet another major corporation has had millions of credit card numbers stolen from their databases. Malware is *malicious software*—viruses, Trojans, worms, botnets—a whole taxonomy of digital diseases.

Viruses have been a problem since the early days of computers, when they were transmitted via floppy disk. Once computers were networked together, worms emerged, which actively transmit themselves over net-works. In 1988, the first large-scale worm—at the time called the Internet Worm because it was the first—spread across an estimated 10 percent of the internet. The internet was pretty small then, only 60,000 computers, and the Internet Worm of 1988 didn't do much. Its intent was to map the internet, so all it did was replicate itself, but it still ended up causing sig-nificant harm. Because there was no safety mechanism in place to pre-vent the worm from copying itself multiple times onto the same machine, it ended up infecting many machines with multiple copies, slowing them down to the point of being unusable.

Today's malware is more sophisticated. Malware is used by govern-ments, criminals, terrorists, and activists ("hacktivists") to gain access to computers for a variety of purposes: conducting espionage, stealing intel-lectual property, exposing embarrassing secrets, slowing down or deny-ing computer usage, or simply creating access for future use. The scope of everyday cyber activity is massive. In 2015, the U.S. government had over 70,000 reported cybersecurity incidents on government systems, and the number has been rising every year. The most frequent and the most seri-ous attacks came from other governments. Many attacks are relatively minor, but some are massive in scale. In July 2015, the U.S. government acknowledged a hack into the Office of Personnel Management (OPM) that exposed security clearance investigation data of 21 million people. The attack was widely attributed to China, although the Chinese govern-ment claimed it was the work of criminals operating from within China and not officially sanctioned by the government.

Other cyberattacks have gone beyond espionage. One of the first widely recognized acts of "cyberwar" was a distributed denial of service (DDoS) attack on Estonia in 2007. DDoS attacks are designed to shut down web-sites by flooding them with millions of requests, overwhelming band-width and denying service to legitimate users. DDoS attacks frequently use "botnets," networks of "zombie" computers infected with malware and harnessed to launch the attack.

Following a decision to relocate a Soviet war memorial, Estonia was besieged with 128 DDoS attacks over a two-week period. The attacks did more than take websites offline; they affected Estonia's entire electronic infrastructure. Banks, ATMs, telecommunications, and media outlets

U.S. Marine Corps officers with a Gatling gun in Washington, DC, 1896. Through automation, the Gatling gun allowed four men to perform the same work as a hundred. Richard Gatling built his gun in the hopes that it would reduce the number of soldiers on the battlefield, thus saving lives.

A British machine gun crew in gas masks during the Battle of the Somme, July 1916. The Gatling gun paved the way for the machine gun, which brought a new level of destruction to war that European nations were not prepared for. At the Battle of the Somme, Britain lost 20,000 men in a single day.

The destroyer USS Fitzgerald fires a Harpoon missile during a joint training exercise with Japan, 2016. The Harpoon is a fire-and-forget semiautonomous anti-ship missile. The human chooses the enemy ship to be destroyed and the missile uses automation to avoid other nearby ships. Missiles of this type are in widespread use around the world and have been used for decades.

The Tomahawk Land Attack Missile (TLAM) Block IV, also called "Tactical Tomahawk" or TLAM-E, flies over China Lake, California. The Tactical Tomahawk is a "net-enabled" weapon with a communications link back to human controllers, allowing commanders to redirect the missile while in flight. Advanced missiles increasingly have communications links, which give commanders more control and increases weapons' effectiveness.

U.S. Marines remove a training AGM-88 High-Speed Anti-Radiation Missile (HARM) from an F/A-18C Hornet on the deck of the USS Theodore Roosevelt aircraft carrier, 2015. The HARM is a fire-and-forget semiautonomous homing missile used to destroy enemy radars.

An Israeli Harpy loitering munition launching. The Harpy is a fully autonomous anti-radar weapon and has been sold to a number of countries: Chile, China, India, South Korea, and Turkey. Similar to the HARM, the Harpy is intended to destroy radars. The key difference is that the Harpy can loiter for 2.5 hours, allowing it to search over a wide area for enemy targets, whereas the HARM is only aloft for approximately 4.5 minutes.

A U.S. Navy Aegis warship fires a missile as part of a live-fire exercise off the coast of North Carolina, 2017. The Aegis air and missile defense system has semiautonomous (human in the loop) and supervised autonomous (human on the loop) modes. Supervised autonomy is vital for defending ships against short-warning saturation attacks. At least thirty nations have ship- or land-based defensive supervised autonomous weapons similar to Aegis.

A U.S. Army Patriot battery along the Turkey-Syria border, 2013. U.S. Patriot batteries were deployed to Turkey to aid in defending Turkey during the Syrian civil war. The Patriot, a land-based supervised autonomous air and missile defense system, was involved in two fratricide incidents in 2003 that highlighted some of the dangers of automation in weapon systems.

An MQ-1 Predator at Creech Air Force Base, Nevada, 2016. At least ninety nations have drones and over a dozen have armed drones. As automation increases, future drones will be increasingly autonomous, raising new possibilities and challenges in warfare.

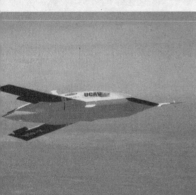

The X-45A uninhabited aircraft in an experimental test flight, 2002. Today's drones are not survivable in contested environments because of their lack of stealth characteristics. The X-45A paved the way for future stealth combat drones, which are in development by leading military powers around the globe. Stealth combat drones would operate in contested environments in which communications may be jammed, raising questions about which tasks the aircraft should be allowed to perform when operating autonomously.

The X-47B autonomously lands on the USS George H. W. Bush in 2013, marking the first time an uninhabited aircraft landed on an aircraft carrier. Demonstrating autonomous carrier landings was a significant milestone for uninhabited aircraft. Earning warfighters' trust is a major limiting factor in fielding more advanced military robotic systems. Despite technological opportunities, the U.S. Navy is not developing a carrier-based uninhabited combat aircraft.

The X-47B autonomously refuels from a K-707 tanker over the Chesapeake Bay, 2015, demonstrating the first aerial refueling of an uninhabited aircraft. Autonomous aerial refueling is an important enabler for making uninhabited combat aircraft operationally relevant.

The Israeli Guardium uninhabited ground vehicle. The armed Guardium has reportedly been sent on patrol near the Gaza border, although humans remain in control of firing weapons. Countries have different thresholds for risk with armed robots, including lethal autonomy, depending on their security environment.

An uninhabited, autonomous boat near Virginia Beach as part of a U.S. Navy demonstration of swarming boats, 2016. Swarms are the next evolution in autonomous systems, allowing one human to control many uninhabited vehicles simultaneously, which autonomously cooperate to achieve a human-directed goal.

Deputy Secretary of Defense Bob Work speaks at the christening of DARPA's Sea Hunter, or Anti-Submarine Warfare Continuous Trail Unmanned Vessel (ACTUV), 2016. Work has been a major advocate of robotics and autonomous systems and human-machine teaming to maintain U.S. military superiority. At the Sea Hunter's christening, Work envisioned "wolf packs" of uninhabited warships like the Sea Hunter plying the seas in search of enemy submarines.

The Sea Hunter gets under way on the Willamette River following its christening in Portland, Oregon, 2016. At $21 million each, the Sea Hunter is a fraction of the cost of a $1.6 billion destroyer, allowing the United States to field large numbers of Sea Hunters, if it desires.

A B-1B bomber launches a Long-Range Anti-Ship Missile (LRASM) in a flight demonstration, 2013. The semiautonomous LRASM incorporates a number of advanced autonomous guidance features that allow it to avoid pop-up threats while en route to its human-designated target.

A Long-Range Anti-Ship Missile (LRASM) about to hit a target ship in a demonstration test, 2013. Humans remain "in the loop" for LRASM targeting decisions. Similar to the Harpoon, a human operator chooses the enemy ship to be attacked and the LRASM uses automation to maneuver and identify the intended target while avoiding other nearby ships.

A modified quadcopter autonomously navigates through a warehouse as part of DARPA's Fast Lightweight Autonomy (FLA) program, 2016. FLA quadcopters use onboard sensors to detect the surrounding environment and autonomously navigate through cluttered terrain.

Under Secretary of Defense Frank Kendall watches as a soldier from the 4th Battalion, 17th Infantry Regiment, 1st Brigade Combat Team, 1st Armored Division demonstrates a micro drone at Fort Bliss, Texas, 2015. When he was under secretary of defense for acquisition, technology, and logistics, Kendall was one of three officials who would have had responsibility for authorizing the development of any autonomous weapon under current DoD policy.

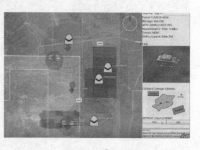

Screenshot from a DARPA video of the prototype human-machine interface for the CODE program. The machine automatically detected an enemy tank, developed a targeting solution, and estimated likely collateral damage. For this engagement the human is "in the loop," however, and must approve each engagement.

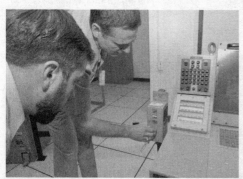

Navy Captain Pete Galluch (right), commander of the Aegis Training and Readiness Center, demonstrates to the author "rolling green" to authorize lethal engagements in an Aegis simulator in Dahlgren, Virginia, 2016. Navy commanders are able to use the highly automated and lethal Aegis weapon system safely in large part because humans retain tight control over its operation.

Researchers from the Naval Postgraduate School (NPS) launch a small drone as part of a thirty-drone swarm at Camp Roberts, California, 2015. Researchers at NPS are experimenting with swarm versus swarm combat, exploring new tactics for how to control friendly swarms and how to defeat enemy swarms.

Two drones fly in formation over Camp Roberts, California, as part of a Naval Postgraduate School experiment in cooperative autonomy, or swarming. Swarms raise novel command-and-control challenges for how to optimize autonomous behavior and cooperation for large numbers of systems while retaining human control over the swarm as a whole.

Deputy Secretary of Defense Bob Work (left) and Office of Naval Research program officer Lee Mastroianni discuss the prototype Low-Cost Unmanned Aerial Vehicle Swarming Technology (LOCUST) drone, 2016. The tube-launched LOCUST is intended to pave the way for swarms of low-cost drones.

A Counter Rocket, Artillery, and Mortar (C-RAM) system from the 2nd Battalion, 44th Air Defense Artillery Regiment at Bagram Airfield in Afghanistan. The C-RAM, which has a high degree of automation but also retains a human in the loop, is an example of the kind of "centaur" human-machine teaming that Bob Work has advocated for.

Left to right: *Steve Goose (Human Rights Watch), Jody Williams (Nobel Women's Initiative), the author Paul Scharre (Center for a New American Security), and Thomas Nash (Article 36) at the 2016 United Nations Convention on Certain Conventional Weapons (CCW) meeting on lethal autonomous weapons. In 2016, CCW member states agreed to form a Group of Governmental Experts (GGE) to discuss lethal autonomous weapon systems, but there is no consensus among states on what to do about autonomous weapons.*

The DJI Spark, which retailed for $499 as of August 2017, can autonomously track and follow moving objects, avoid obstacles, and return home when it is low on batteries. The hobbyist drone market has exploded in recent years, making the technology widely available and inexpensive. Non-state groups have already used weaponized small, cheap drones for aerial attacks. Over time, hobbyist drones will become increasingly autonomous.

A student robotics project at Thomas Jefferson High School in Alexandria, Virginia, to build a bicycle with an automatic gear shifter, akin to an automatic transmission in a car. As robotics and autonomous technology advances, increasingly capable robotic systems will be available to DIY hobbyists.

Vice Chairman of the Joint Chiefs General Paul Selva (left) looks on as Deputy Secretary of Defense Bob Work speaks to reporters on the defense budget, 2016. Speaking at a conference in 2016, General Selva said that delegating responsibility for lethal force decisions was a "fairly bright line that we're not willing to cross."

An X-47B experimental drone takes off from the USS Theodore Roosevelt aircraft carrier, 2013. Robotic technology will continue to evolve, with increasingly autonomous systems available to nations and non-state groups around the globe.

were all shut down. At the height of the DDoS attacks on Estonia, over a million botnet-infected computers around the globe were directed toward Estonian websites, pinging them four million times a second, overloading servers and shutting down access. Estonia accused the Russian government, which had threatened "disastrous" consequences if Estonia removed the monument, of being behind the attack. Russia denied involvement at the time, although two years later a Russian Duma official confirmed that a government-backed hacker group had conducted the attacks.

In the years since, there have been many alleged or confirmed cyber-attacks between nations. Russian government-backed hackers attacked Georgia in 2008. Iran launched a series of cyberattacks against Saudi Arabia and the United States in 2012 and 2013, destroying data on 30,000 computers owned by a Saudi oil company and carrying out 350 DDoS attacks against U.S. banks. While most cyberattacks involve stealing, exposing, or denying data, some have crossed into physical space. In 2010, a worm came to light that crossed a cyber-Rubicon, turning 1s and 0s into physical destruction.

STUXNET: THE CYBERSHOT HEARD ROUND THE WORLD

In the summer of 2010, word began to spread through the computer security world of something new, a worm unlike any other. It was more advanced than anything seen before, the kind of malware that had clearly taken a team of professional hackers months if not years to design. It was a form of malware that security professionals have long speculated was possible but had never seen before: a digital weapon. Stuxnet, as the worm came to be called, could do more than spy, steal things, and delete data. Stuxnet could break things, not just in cyberspace but in the physical world as well.

Stuxnet was a serious piece of malware. Zero-day exploits take advantage of vulnerabilities that software developers are unaware of. (Defenders have known about them for "zero days.") Zero-days are a prized commodity in the world of computer security, worth as much as $100,000 on the black market. Stuxnet had four. Spreading via removable USB drives, the first thing Stuxnet did when it spread to a new a system was to give itself "root" access in the computer, essentially unlimited access. Then it hid, using a real—not fake—security certificate from a reputable company to mask itself from antivirus software. Then Stuxnet began searching. It spread to every machine on the network, looking for a very particular type of software, Siemens Step 7, which is used to operate programmable logic controllers (PLCs) used in industrial applications. PLCs control power plants, water valves, traffic lights, and factories. They also control centrifuges in nuclear enrichment facilities.

Stuxnet wasn't just looking for any PLC. Stuxnet operated like a homing munition, searching for a very specific type of PLC, one configured for frequency-converter drives, which are used to control centrifuge speeds. If it didn't find its target, Stuxnet went dead and did nothing. If it did find it, then Stuxnet sprang into action, deploying two encrypted "warheads," as computer security specialists described them. One of them hijacked the PLC, changing its settings and taking control. The other recorded regular industrial operations and played them back to the humans on the other side of the PLC, like a fake surveillance video in a bank heist. While secretly sabotaging the industrial facility, Stuxnet told anyone watching: "everything is fine."

Computer security specialists widely agree that Stuxnet's target was an industrial control facility in Iran, likely the Natanz nuclear enrichment facility. Nearly 60 percent of Stuxnet infections were in Iran and the original infections were in companies that have been tied to Iran's nuclear enrichment program. Stuxnet infections appear to be correlated with a sharp decline in the number of centrifuges operating at Natanz. Security specialists have further speculated that the United States, Israel, or possibly both, were behind Stuxnet, although definitive attribution can be difficult in cyberspace.

Stuxnet had a tremendous amount of autonomy. It was designed to operate on "air-gapped" networks, which aren't connected to the internet for security reasons. In order to reach inside these protected networks, Stuxnet spread via removable USB flash drives. This also meant that once Stuxnet arrived at its target, it was on its own. Computer security company Symantec described how this likely influenced Stuxnet's design:

While attackers could control Stuxnet with a command and control server, as mentioned previously the key computer was unlikely to have outbound Internet access. Thus, all the functionality required to sabotage a system was embedded directly in the Stuxnet executable.

Unlike other malware, it wasn't enough for Stuxnet to give its designers access. Stuxnet had to perform the mission autonomously.

Like other malware, Stuxnet also had the ability to replicate and propagate, infecting other computers. Stuxnet spread far beyond its original target, infecting over 100,000 computers. Symantec referred to these additional computers as "collateral damage," an unintentional side effect of Stuxnet's "promiscuous" spreading that allowed it to infiltrate air-gapped networks.

To compensate for these collateral infections, however, Stuxnet had a number of safety features. First, if Stuxnet found itself on a computer that did not have the specific type of PLC it was looking for, it did nothing. Second, each copy of Stuxnet could spread via USB to only three other machines, limiting the extent of its proliferation. Finally, Stuxnet had a self-termination date. On June 24, 2012, it was designed to erase all copies of itself. (Some experts saw these safety features as further evidence that it was designed by a Western government.)

By using software to actively sabotage an industrial control system, something cybersecurity specialists thought was possible before Stuxnet but had not yet happened, Stuxnet was the first cyberweapon. More will inevitably follow. Stuxnet is an "open-source weapon" whose code is laid bare online for other researchers to tinker with, modify, and repurpose for other attacks. The specific vulnerabilities Stuxnet exploited will have been fixed, but its design is already being used as a blueprint for cyberweapons to come.

AUTONOMY IN CYBERSPACE

Autonomy is essential to offensive cyberweapons, such as Stuxnet, that are intended to operate on closed networks separated from the internet. Once it arrives at its target, Stuxnet carries out the attack on its own. In that sense, Stuxnet is analogous to a homing munition. A human chooses the target and Stuxnet conducts the attack.

Autonomy is also essential for cyberdefense. The sheer volume of attacks means it is impossible to catch them all. Some will inevitably slip through defenses, whether by using zero-day vulnerabilities, finding systems that have not yet been updated, or exploiting users who insert infected USB drives or click on nefarious links. This means that in addition to keeping malware out, security specialists have also adopted "active cyberdefenses" to police networks on the inside to find malware, counter it, and patch network vulnerabilities.

In 2015, I testified to the Senate Armed Services Committee alongside retired General Keith Alexander, former head of the National Security Agency, on the future of warfare. General Alexander, focusing on cyber threats, explained the challenge in defending 15,000 "enclaves" (separate computer networks) within the Department of Defense. Keeping all of these networks up-to-date manually was nearly impossible. Patching network vulnerabilities at "manual speed," he said, took months. "It should be automated," Alexander argued. "The humans should be out of the loop." Computer security researchers are already working to develop these more sophisticated cyber that would take humans out of the loop. As in other areas of autonomy, DARPA is at the leading edge of this research.

UNLEASHING MAYHEM: THE CYBER GRAND CHALLENGE

DARPA tackles only the most difficult research problems, "DARPA hard" problems that others might deem impossible. DARPA does this every day, but when a technical problem is truly daunting even for DARPA, the organization pulls out its big guns in a Grand Challenge.

The first DARPA Grand Challenge was held in 2004, on autonomous vehicles. Twenty-one research teams competed to build a fully autonomous vehicle that could navigate a 142-mile course across the Mojave Desert. It was truly a "DARPA hard" problem. The day ended with every single vehicle broken down, overturned, or stuck. The furthest any car got was 7.4 miles, only 5 percent of the way through the course.

The organization kept at it, sponsoring a follow-up Grand Challenge the next year. This time, it was a resounding success. Twenty-two vehicles beat the previous year's distance record and five cars finished the entire course. In 2007, DARPA hosted an Urban Challenge for self-driving cars on a closed, urban course complete with traffic and stop signs. These Grand Challenges matured autonomous vehicle technology in leaps and bounds, laying the seeds for the self-driving cars now in development at companies like Google and Tesla.

DARPA has since used the Grand Challenge approach as a way to tackle other truly daunting problems, harnessing the power of competition to generate the best ideas and launch a technology forward. From 2013 to 2015, DARPA held a Robotics Challenge to advance the field of humanoid robotics, running robots through a set of tasks simulating humanitarian relief and disaster response.

In 2016, DARPA hosted a Cyber Grand Challenge to advance the field of cybersecurity. Over one hundred teams competed to build a fully autonomous Cyber Reasoning System to defend a network. The systems competed in a live capture the flag competition to automatically identify computer vulnerabilities and either patch or exploit them.

David Brumley is a computer scientist at Carnegie Mellon University and CEO of ForAllSecure, whose system Mayhem won the Cyber Grand Challenge. Brumley describes his goal as building systems that "automatically check the world's software for exploitable bugs." Mayhem is that vision brought to life, a "fully autonomous system for finding and fixing computer security vulnerabilities." In that sense, Mayhem is even more ambitious than Keith Alexander's goal of just updating software automatically. Mayhem actually goes and finds bugs on its own—bugs that humans are not yet aware of— and then patches them.

Brumley explained to me that there are actually several steps in this process. The first is finding a vulnerability in a piece of software. The next step is developing either an "exploit" to take advantage of the vulnerability or a "patch" to fix it. If a vulnerability is analogous to a weak lock, then an exploit is like a custom-made key to take advantage of the lock's weakness. A patch, on the other hand, fixes the lock.

Developing these exploits and patches isn't enough, though. One has to know when to use them. Even on the defensive side, Brumley explained, you can't just apply a patch as soon as you see an exploit. For any given vulnerability, Mayhem would develop a "suite of patches." Fixing a vulnerability isn't a binary thing, where either it's fixed or it isn't. Brumley said, "There's grades of security, and often these have different tradeoffs on performance, maybe even functionality." Some patches might be more secure, but would cause the system to run slower. Which patch to apply depends on the system's use. For home use, "you'd rather have it more functional rather than 100 percent secure," Brumley said. A customer protecting critical systems, on the other hand, like the Department of Defense, might choose to sacrifice efficiency for better security. When to apply the patch is another factor to consider. "You don't install a Microsoft PowerPoint update right before a big business presentation," Brumley said.

Today, these steps are all done by people. People find the vulnerabilities, design the patches, and upload them to an automatic update server. Even the "auto-update" functions on your home computer are not actually fully automatic. You have to click "Okay" in order for the update to move forward. Every place where there is a human in the loop slows down the process of finding and patching vulnerabilities. Mayhem, on the other hand, is a completely autonomous system for doing all those steps. That means it isn't just finding and patching vulnerabilities blindly. It's also reasoning about which patch to use and when to apply it. Brumley said it's "an autonomous system that's taking all of those things that humans are doing, it's automating them, and then it's reasoning about how to use them, when to apply the patch, when to use the exploit." Mayhem also deploys hardening techniques on programs. Brumley described these as proactive security measures applied to a program before a vulnerability has even been discovered to make it harder to exploit, if there are vulnerabilities. And Mayhem does all of this at machine speed.

In the Cyber Grand Challenge final round, Mayhem and six other systems competed in a battle royale to scan each other's software for vulnerabilities, then exploit the weaknesses in other systems while patching their own vulnerabilities. Brumley compared the competition to seven fortresses probing each together, trying to get into locked doors. "Our goal was to come up with a skeleton key that let us in when it wasn't supposed to." DARPA gave points for showing a "proof of vulnerability," essentially an exploit or "key," to get into another system. The kind of access also mattered—full access into the system gave more points than more limited access that was only useful for stealing information.

Mike Walker, the DARPA program manager who ran the Cyber Grand Challenge, said that the contest was the first time that automated cyber-tools had moved beyond simply applying human-generated code and into the "automatic creation of knowledge." By autonomously developing patches, they had moved beyond automated antivirus systems that can clean up known malware to "automation of the supply chain." Walker said, "true autonomy in the cyber domain are systems that can create their own knowledge. . . . It's a pretty bright and clear line. And I think we kind of crossed it . . . for the first time in the Cyber Grand Challenge."

Walker compared the Cyber Grand Challenge to the very first chess tournaments between computers. The technology isn't perfect. That wasn't the point. The goal was to prove the concept to show what can be done and refine the technology over time. Brumley said Mayhem is roughly comparable to a "competent" computer security professional, someone "just fresh out of college in computer security." Mayhem has nothing on world-class hackers. Brumley should know. He also runs a team of competitive human hackers who compete in the DEF CON hacking conference, the "world series" of hacking. Brumley's team from Carnegie Mellon has won four out of the past five years.

Brumley's aim with Mayhem isn't to beat the best human hackers, though. He has something far more practical—and transformative—in mind. He wants to fundamentally change computer security. As the internet colonizes physical objects all around us—bringing toasters, watches, cars, thermostats and other household objects online in the Internet of Things (IoT), this digitization and connectivity also bring vulnerabilities. In October 2016, a botnet called Mirai hijacked everyday networked devices such as printers, routers, DVR machines, and security cameras and leveraged them for a massive DDoS attack. Brumley said most IoT devices are "ridiculously vulnerable." There are an estimated 6.4 billion IoT devices online today, a number expected to grow to over 20 billion devices by 2020. That means there are millions of different programs, all with potential vulnerabilities. "Every program written is like a unique lock and most of those locks have never been checked to see if they're terrible," Brumley said. For example, his team looked at 4,000 commercially available internet routers and "we've yet to find one that's secure," he said. "No one's ever bothered to check them for security." Checking this many devices at human speed would be impossible. There just aren't enough computer security experts to do it. Brumley's vision is an autonomous system to "check all these locks."

Once you've uncovered a weak lock, patching it is a choice. You could just as easily make a key—an exploit—to open the lock. There's "no difference" between the technology for offense and defense, Brumley said. They're just different applications of the same technology. He compared it to a gun, which could be used for hunting or to fight wars. Walker agreed. "All computer security technologies are dual-use," he said.

For safety reasons, DARPA had the computers compete on an air-gapped network that was closed off from the internet. DARPA *also* created a special operating system just for this contest. Even if one of the systems was plugged into the internet, it would need to be re-engineered to search for vulnerabilities on a Windows, Linux, or Mac machine.

Brumley emphasized that they've never had a problem with people using this technology for nefarious ends at Carnegie Mellon. He compared his researchers to biologists working on a better flu vaccine. They could use that knowledge to make a better virus, but "you have to trust the researchers to have appropriate safety protocols." His company, ForAllSecure, practices "responsible disclosure" and notifies companies of vulnerabilities they find. Nonetheless, he admitted, "you do worry about the bad actors."

Brumley envisions a world where over the next decade, tools like Mayhem are used to find weak locks and patch them, shoring up cyber-defenses in the billions of devices online. Walker said that self-driving cars today are a product of the commercial sector throwing enormous investment money behind the individuals who competed in the original DARPA Grand Challenge a decade ago, and he sees a similar road ahead for autonomous cybersecurity. "It's going to take the same kind of long-term will and financial backing to do it again here."

Both Brumley and Walker agreed that autonomous cybertools will also be used by attackers, but they said the net effect was to help the defense more. Right now, "offense has all of the advantage in computer security," Walker said. The problem is right now there is an asymmetry between attackers and defenders. Defenders have to close all of the vulnerabilities, while attackers have to just find one way in. Autonomous cybersystems level the playing field, in part because defense gets a first-mover advantage. They write the code, so they can scan it for vulnerabilities and patch them before it is deployed. "I'm not saying that we can change to a place where defense has the advantage," Walker said, but he did think autonomous cybertools would enable "investment parity," where "the best investment wins." Even that would be "transformative," he said. There's big money in malware, but far more is spent annually on computer security. Prior to joining DARPA, Walker said he worked for a decade as a "red teamer," paid by energy and financial sector companies to hack into their systems and uncover their vulnerabilities. He said autonomous cyberdefenses "can actually make hacking into something like our energy infrastructure or our financial infrastructure a highly uncommon proposition that average criminals cannot afford to do."

David Brumley admitted that this won't stop hacking from advanced nation-states who have ample resources. He said limiting access was still beneficial, though, and drew a comparison to efforts to limit the spread of nuclear weapons: "It's scary to think of Russia and the U.S. having it, but what's really scary is when the average Joe has it. We want to get rid of the average Joe having these sorts of things." If Brumley is right, autonomous systems like Mayhem will make computers more secure and safer ten years from now. But autonomy will keep evolving in cyberspace, with even more advanced systems beyond Mayhem yet to come.

The next evolution in autonomous cyberdefense is what Brumley calls "counter-autonomy." Mayhem targets weak locks; counter-autonomy targets the locksmith. It "leverages flaws or predictable patterns in the adversary to win." Counter-autonomy goes beyond finding exploits, he said; it's about "trying to find vulnerabilities in the opponent's algorithms." Brumley compared it to playing poker: "you play the opponent." Counter-autonomy exploits the brittleness of the enemy's autonomous systems to defeat them.

While counter-autonomy was not part of the Cyber Grand Challenge, Brumley said they have experimented with counter-autonomy techniques that they simply didn't use. One tool they developed embeds a hidden exploit targeting a competitor's autonomous system into a patch. "It's a little bit like a Trojan horse," Brumely said. The patch "works just fine. It's a legitimate program." Hidden within the patch is an exploit, though, that targets one of the common tools that hackers use to analyze patches. "Anyone who tries to analyze [the patch] gets exploited," he said. Another approach to counter-autonomy would move beyond simply finding vulnerabilities to actually creating them. This could be done in learning systems by inserting false data into the learning process. Brumley calls this the "computer equivalent to 'the long con,'" where our systems methodically cause our adversary's systems to 'mis-learn' (incorrectly learn) how to operate."

AUTONOMOUS CYBERWEAPONS

The arms race in speed in cyberspace is already under way. In an unpublished 2016 working paper, Brumley wrote, "Make no mistake, cyber is a war between attackers and defenders, both who coevolve as the other deploys new systems and measures. In order to win, we must act, react, and evolve faster than our adversaries." Cyberweapons of the future—defensive and offensive—will incorporate greater autonomy, just the same way that more autonomy is being integrated into missiles, drones, and physical systems like Aegis. What would a "cyber autonomous weapon" look like?

Cyberspace and autonomous weapons intersect in a number of potentially significant ways. The first is the danger that cyber vulnerabilities pose in autonomous weapons. Anything that is computerized is vulnerable to hacking. The migration of household objects online as part of the IoT presents major cybersecurity risks, and there are analogous risks for militaries whose major platforms and munitions are increasingly networked. Cyber vulnerabilities could hobble a next-generation weapon system like the F-35 Joint Strike Fighter, which has tens of millions of lines of code. There is no reason to think that an autonomous weapon would necessarily be more vulnerable to hacking, but the consequences if one were hacked could be much worse. Autonomous weapons would be a very attractive target for a hostile state's malware, since a hacker could potentially usurp control of an autonomous weapon and redirect it. The consequences could be even worse than those of a runaway gun. The weapon wouldn't be out of control; it would be under the control of the enemy.

In theory, greater autonomy that allows for off-network operation may appear to be a solution to cyber vulnerabilities. This is an appealing tactic that has come up in science fiction wars between humans and machines. In the opening episode of the 2003 reboot of *Battlestar Galactica,* the evil Cylon machines wipe out nearly the entire human space fleet via a computer virus. The ship *Galactica* survives only because it has an older computer system that is not networked to the rest of the fleet. As Stuxnet demonstrated, however, in the real world operating off-network complicates cyberattacks but is no guarantee of immunity.

The second key intersection between cyberspace and autonomy occurs in automated "hacking back." Autonomous cyberbots like Mayhem will be part of active cyberdefenses, including those that use higher-level reasoning and decision-making, but these still operate within one's own network. Some concepts for active cyber defense move beyond policing one's own networks into going on the offense. Hacking back is when an organization responds to a cyberattack by counterattacking, gaining information about the attacker or potentially shutting down the computers from which the attack is originating. Because many cyberattacks involve co-opting unsuspecting "zombie" computers and repurposing them for attack, hacking back can inevitably draw in third parties. Hacking back is controversial and, if done by private actors, could be illegal. As one cybersecurity analyst noted, "Every action accelerates."

Automation has been used in some limited settings when hacking back. When the FBI took down the Coreflood botnet, it redirected infected botnet computers to friendly command-and-control servers, which then issued an automatic stop command to them. However, this is another example of automation being used to execute a decision made by people, which is far different than delegating the decision whether or not to hack back to an autonomous process.

Automated hacking back would delegate the decision whether or not to go on the counteroffensive to an autonomous system. Delegating this authority could be very dangerous. Patrick Lin, an ethicist at California Polytechnic State University who has written extensively on autonomy in both military and civilian applications, warned at the United Nations in 2015, "autonomous cyber weapons could automatically escalate a conflict." As Tousley acknowledged, cyberspace could be an area where automatic reactions between nation-states happen in milliseconds. Automated hacking back could cause a flash cyberwar that rapidly spirals out of control. Automated hacking back is a theoretical concept, and there are no publicly known examples of it occurring. (Definitively saying something has *not* happened in cyberspace is difficult, given the shadowy world of cyberwar.)

The third intersection between cyber- and autonomous weapons is increasingly autonomous offensive cyberweapons. Computer security researchers have already demonstrated the ability to automate "spear phishing" attacks, in which unwitting users are sent malicious links buried inside seemingly innocuous emails or tweets. Unlike regular phishing attacks, which target millions of users at a time with mass emails, spear phishing attacks are specially tailored to specific individuals. This makes them more effective, but also more time-intensive to execute. Researchers developed a neural network that, drawing on data available on Twitter, learned to automatically develop "humanlike" tweets targeted at specific users, enticing them to click on malicious links. The algorithm was roughly as successful as manual spear phishing attempts but, because of automation, could be deployed en masse to automatically seek out and target vulnerable users.

As in other areas, greater intelligence will allow offensive cyberweapons to operate with greater autonomy. Stuxnet autonomously carried out its attack, but its autonomy was highly constrained. Stuxnet had a number of safeguards in place to limit its spread and effects on computers that weren't its target, as well as a self-termination date. One could envision future offensive cyberweapons that were given freer rein. Eric Messinger, a writer and researcher on legal issues and human rights, has argued:

> ... in offensive cyberwarfare, [autonomous weapon systems] may *have* to be deployed, because they will be integral to effective action in an environment populated by automated defenses and taking place at speeds beyond human capacities.... [The] development and deployment of offensive [autonomous weapon systems] may well be unavoidable.

It's not clear what an offensive autonomous cyberweapon would look like, given the challenges in both defining a "cyberweapon" and the varying ways in which autonomy is already used in cyberspace. From a certain perspective, a great deal of malware is inherently autonomous by virtue of its ability to self-replicate. The Internet Worm of 1988, for example, is an example of the Sorcerer's Apprentice effect: a runaway, self-replicating process that cannot be stopped. This is an important dimension to malware that does not have an analogy in physical weapons. Drones and robotic systems cannot self-replicate. In this sense, malware resembles biological viruses and bacteria, which self-replicate and spread from host to host.

But there is a critical difference between digital and biological viruses. Biological pathogens can mutate and adapt in response to environmental conditions. They evolve. Malware, at least today, is static. Once malware is deployed, it can spread, it can hide (as Stuxnet did), but it cannot modify itself. Malware can be designed to look for updates and spread these updates among copies of itself via peer-to-peer sharing (Stuxnet did this as well), but new software updates originate with humans.

In 2008, a worm called Conficker spread through the internet, infecting millions of computers. As computer security specialists moved to counter it, Conficker's designers released updates, eventually fielding as many as five different variants. These updates allowed Conficker's programmers to stay ahead of security specialists, upgrading the worm and closing vulnerabilities when they were detected. This made Conficker a devilishly hard worm to defeat. At one point, an estimated 8 to 15 million computers worldwide were infected.

Conficker used a mixture of human control and automation to stay ahead of antivirus specialists. Conficker's updates came from its human designers, but it used automation to get the updates clandestinely. Every day, Conficker would generate hundreds of new domain names, only one of which would link back to its human controllers with new updates. This made the traditional approach of blocking domains to isolate the worm from its controllers ineffective. As security specialists found a method to counter Conficker, a new variant would be released quickly, often within weeks. Eventually, a consortium of industry experts brought Conficker to heel, but doing so took a major effort.

Conficker's fundamental weakness was that its updates could only happen at human speed. Conficker replicated autonomously and used clever automation to surreptitiously link back to its human controllers, but the contest between the hackers and security specialists was fought at human speed. Humans were the ones working to identify the worm's weaknesses and take it down, and humans on the other side were working to adapt the worm and keep it one step ahead of antivirus companies.

The technology that Mayhem represents could change that. What if a piece of software turned the same tools for identifying and patching vulnerabilities and applied them to itself? It could improve itself, shoring up its own defenses and resisting attack. Brumley has hypothesized about such "introspective systems." Self-adapting software that can modify itself, rather than wait on updates from its human controllers, would be a significant evolution. The result could be robust cyberdefenses . . . or resilient malware. At the 2015 International Conference on Cyber Conflict, Alessandro Guarino hypothesized that AI-based offensive cyberweapons could "prevent and react to countermeasures," allowing them to persist inside networks. Such an agent would be "much more resilient and able to repel active measures deployed to counter it."

A worm that could autonomously adapt—mutating like a biological virus, but at machine speed—would be a nasty bug to kill. Walker cautioned that the tools used in the Cyber Grand Challenge would only allow a piece of software to patch its own vulnerabilities. It wouldn't allow "the synthesis of new logic" to develop "new code that can work towards a goal." To do that, he said, "first we'd have to invent the field of code synthesis, and right now, it's like trying to predict when time travel's going to be invented. Who knows if it can be invented? We don't have a path." While such a development would be a leap beyond current malware, the advent of learning systems in other areas, such as Google DeepMind's Atari-playing AI or AlphaGo, suggests that it is not inconceivable. Adaptive malware that could rewrite itself to hide and avoid scrutiny at superhuman speeds could be incredibly virulent, spreading and mutating like a biological virus without any form of human control.

When I asked Brumley about the possibility of future malware that was adaptive, he said "those are a possibility and are worrisome. . . . I think someone could come up with this kind of ultimate malware and it could get out of control and it would be a really big pain for a while." What he really worries about, though, are near-term problems. His chief concern is a shortage of cybersecurity experts. We have weak cyber locks because we're not training enough people how to be better cyber locksmiths. Part of this, Brumley said, is a culture that views hacking as an illegitimate profession. "In the U.S., we've shot ourselves in the foot by equating a hacker with a bad guy." We don't view flesh-and-blood locksmiths that way, yet for digital security, we do. Other countries don't see it that way, and Brumley worries the United States is falling behind. He said, "There's this kind of hubris in the U.S. that we think that because we have the best Army and Navy and we have all these great amazing natural resources, great aircraft carriers, that of course we're going to dominate in cyber. And I don't think that's a given. It's a brand-new space, completely different from anything else. There's no reason that things will just carry over." We need to shift the culture in the United States, he said, from thinking about hacking skills as something that are only used for "offense and should be super-secret and only used by the military" to something that is valued in the cyber workforce more broadly. Walker agreed. "Defense is powered by openness," he said.

Looking to the future, Brumley said he saw the "ecosystem" we were building for computer security and autonomous cybersystems as critical. "I tend to view everything as a system—a dynamic system." People are part of that system too. The solution to potentially dangerous malware in the future was to create "the right ecosystem . . . and then it will be resilient to problems."

KEEPING THE BOTS AT BAY

Mixing cyberspace and autonomous weapons combines two issues that are challenging enough by themselves. Cyberwarfare is poorly understood outside the specialist community of cyber experts, in part because of the secrecy surrounding cyber operations. Norms about appropriate behavior between states in cyberspace are still emerging. There is not even a consensus among cyber experts about what constitutes a "cyberweapon." The concept of autonomous weapons is similarly nascent, making the combination of these two issues extremely difficult to understand. The DoD's official policy on autonomy in weapons, DoD Directive 3000.09, specifically exempts cyberweapons. This wasn't because we thought autonomous cyberweapons were uninteresting or unimportant when we wrote the directive. It was because we knew bureaucratically it would be hard enough simply to create a new policy on autonomy. Adding cyber operations would have multiplied the complexity of the problem, making it very likely we would have accomplished nothing at all.

This lack of clarity is reflected in the mixed signals I got from Defense Department officials on autonomy in cyberspace. Both Work and Tousley mentioned electronic warfare and cyberspace as an arena in which they would be willing to accept more autonomy, but they had different perspectives on how far they would be willing to go. Tousley said he saw a role for autonomy in only defensive cyber operations. The "goal is not offense—it's defense," he told me.

Tousley's boss's boss, Deputy Secretary Bob Work, saw things differently. Work made a direct comparison between Aegis and automated "hacking back." He said, "the narrow cases where we will allow the machine to make targeting decisions is in defensive cases where all of the people who are coming at you are bad guys. . . . electronic warfare, cyberwarfare, missile defense. . . . We will allow the machine to make essentially decisions . . . like, a cyber counter attack." He acknowledged delegating that kind of authority to a machine came with risks. Work outlined a hypothetical scenario where this approach could go awry: "A machine might launch a cyber counterattack and it might . . . wind up killing [an industrial control] system or something . . . say it's an airplane and the airplane crashes. And we didn't make a determination that we were going to shoot down that airplane. We just said, 'We're under cyberattack. We're going to counterattack.' Boom."

Work's response to this risk isn't to hide from the technology, but rather to wrestle with these challenges. He explained the importance of consulting with scientists, ethicists, and lawyers. "We'll work it through," he said. "This is all going to be about the checks and balances that you put inside your battle networks." Work was confident these risks could be managed because in his vision, humans would still be involved in a number of ways. There would be both automated safeties and human oversight. "We always emphasize human-machine collaboration . . . with the human always in front," he said. "That's the ultimate circuit breaker."

AN ARMS RACE TO WHERE?

Sun Tzu wrote over two thousand years ago in *The Art of War,* "Speed is the essence of war." His maxim remains even more true today, when signals can cross the globe in fractions of a second. Human decision-making has many advantages over machine intelligence, but humans cannot compete at machine speed. Competitive pressures in fast-paced environments threaten to push humans further and further out of the loop. Superhuman reaction times are the reason why automatic braking is being integrated into cars, why many nations employ Aegis-like automated defensive systems, and why high-frequency stock trading is such a lucrative endeavor.

With this arms race in speed comes grave risks. Stock trading is one example of a field in which competitors have succumbed to allure of speed, developing ever-faster algorithms and hardware to shave microseconds from reaction times. In uncontrolled, real-world environments, the (unsurprising) result has been accidents. When these accidents occur, machine speed becomes a major liability. Autonomous processes can rapidly spiral out of control, destroying companies and crashing markets. It's one thing to say that humans will have the ability to intervene, but in some settings, their intervention may be too late. Automated stock trading foreshadows the risks of a world where nations have developed and deployed autonomous weapons.

A flash physical war in the sense of a war that spirals out of control in mere seconds seems unlikely. Missiles take time to move through the air. Sub-hunting undersea robots can move only so quickly through the water. Accidents with autonomous weapons could undermine stability and escalate crises unintentionally, but these incidents would likely take place over minutes and hours, not microseconds. This is not to say that autonomous weapons do not pose serious risks to stability; they do. A runaway autonomous weapon could push nations closer to the brink of war. If an autonomous weapon (or a group of them) caused a significant number of deaths, tensions could boil over to the point where de-escalation is no longer possible. The speed at which events would unfold, however, is likely one that would allow humans to see what was happening and, at the very least, take steps to attempt to mitigate the effects. Bob Work told me he saw a role for a human "circuit breaker" in managing swarms of robotic systems. If the swarm began to behave in an unexpected way, "they would just shut it down," he said. There are problems with this approach. The autonomous system might not respond to commands to shut it down, either because it is out of communications or because the type of failure it is experiencing prevents it from accepting a command to shut down. Unless human operators have physical access, like the physical circuit breaker in Aegis, any software-based "kill switch" is susceptible to the same risks as other software—bugs, hacking, unexpected interactions, and the like.

Even though accidents with physical autonomous weapons will not cascade into all-out war in mere seconds, machines could quickly cause damage that might have irreversible consequences. Countries may not believe that an enemy's attack was an accident, or the harm may be so severe that they simply don't care. If Japan had claimed that the attack on Pearl Harbor was not authorized by Tokyo and was the work of a single rogue admiral, it's hard to imagine the United States would have refrained from war.

A flash cyberwar, on the other hand, is a real possibility. Automated hacking back could lead to escalation between nations in the blink of an eye. In this environment, human oversight would be merely the illusion of safety. Automatic circuit breakers are used to stop flash crashes on Wall Street because humans cannot possibly intervene in time. There is no equivalent referee to call "Time out" in war.

15

"SUMMONING THE DEMON"

THE RISE OF INTELLIGENT MACHINES

Even the most sophisticated machine intelligence today is a far cry from the sentient AIs depicted in science fiction. Autonomous weapons pose risks precisely because today's narrow AIs fail miserably at tasks that require general intelligence. Machines can crush humans at chess or *go*, but cannot enter a house and make a pot of coffee. Image recognition neural nets can identify objects, but cannot piece these objects together into a coherent story about what is happening in a scene. Without a human's ability to understand context, a stock-trading AI doesn't understand that it is destroying its own company. Some AI researchers are pondering a future where these constraints no longer exist.

Artificial general intelligence (AGI) is a hypothetical future AI that would exhibit human-level intelligence across the full range of cognitive tasks. AGI could be applied to solving humanity's toughest problems, including those that involve nuance, ambiguity, and uncertainty. An AGI could, like Stanislav Petrov, step back to consider the broader context and apply judgment.

What it would take to build such a machine is a matter of pure speculation, but there is at least one existence proof that general intelligence is possible: us. Even if recent advances in deep neural networks and machine learning come up short, eventually an improved understanding of the human brain should allow for a detailed neuron-by-neuron simulation. Brain imaging is improving quickly and some researchers believe whole brain emulations could be possible with supercomputers as early as the 2040s.

Experts disagree wildly on when AGI might be created, with estimates ranging from within the next decade to never. A majority of AI experts predict AGI could be possible by 2040 and likely by the end of the century, but no one really knows. Andrew Herr, who studies emerging technologies for the Pentagon, observed, "When people say a technology is 50 years away, they don't really believe it's possible. When they say it's 20 years away, they believe it's possible, but they don't know how it will happen." AGI falls into the latter category. We know general intelligence is possible because humans have it, but we understand so little of our own brains and our own intelligence that it's hard to know how far away it is.

THE INTELLIGENCE EXPLOSION

AGI would be an incredible invention with tremendous potential for bettering humanity. A growing number of thinkers are warning, however, that AGI may be the "last invention" humanity creates—not because it will solve all of our problems, but because it will lead to our extermination. Stephen Hawking has warned, "development of full artificial intelligence could spell the end of the human race." Artificial intelligence could "take off on its own and re-design itself at an ever-increasing rate," he said. "Humans, who are limited by slow biological evolution, couldn't compete, and would be superseded."

Hawking is a cosmologist who thinks on time scales of tens of thousands or millions of years, so it might be easy to dismiss his concerns as a long way off, but technologists thinking on shorter time scales are similarly concerned. Bill Gates has proclaimed the "dream [of artificial intelligence] is finally arriving," a development that will usher in growth and productivity in the near term, but has long-term risks. "First the machines will do a lot of jobs for us and not be super intelligent," Gates said. "That should be positive if we manage it well. A few decades after that, though, the intelligence is strong enough to be a concern." How much of a concern? Elon Musk has described the creation of human-level artificial intelligence as "summoning the demon." Bill Gates has taken a more sober tone, but essentially agrees. "I am in the camp that is concerned about super-intelligence," he said. "I agree with Elon Musk and some others on this and don't understand why some people are not concerned."

Hawking, Gates, and Musk are not Luddites and they are not fools. Their concerns, however fanciful-sounding, are rooted in the concept of an "intelligence explosion." The concept was first outlined by I. J. Good in 1964:

> Let an ultraintelligent machine be defined as a machine that can far surpass all the intellectual activities of any man however clever. Since the design of machines is one of these intellectual activities, an ultra-intelligent machine could design even better machines; there would then unquestionably be an "intelligence explosion," and the intelligence of man would be left far behind. Thus the first ultraintelligent machine is the last invention that man need ever make, provided that the machine is docile enough to tell us how to keep it under control.

If this hypothesis is right, then humans don't need to create super-intelligent AI directly. Humans might not even be capable of such an endeavor. All humans need to do is create an initial "seed" AGI that is capable of building a slightly better AI. Then through a process of recursive self-improvement, the AI will lift itself up by its own bootstraps, building ever-more-advanced AIs in a runaway intelligence explosion, a process sometimes simply called "AI FOOM."

Experts disagree widely about how quickly the transition from AGI to artificial superintelligence (sometimes called ASI) might occur, if at all. A "hard takeoff" scenario is one where AGI evolves to superintelligence within minutes or hours, rapidly leaving humanity in the dust. A "soft takeoff" scenario, which experts see as more likely (with the caveat that no one really has any idea), might unfold over decades. What happens next is anyone's guess.

UNSHACKLING FRANKENSTEIN'S MONSTER

In the *Terminator* movies, when the military AI Skynet becomes self-aware, it decides humans are a threat to its existence and starts a global nuclear war. *Terminator* follows in a long tradition of science fiction creations turning on their masters. In Ridley Scott's *Blade Runner*, based on the Philip K. Dick novel *Do Androids Dream of Electric Sheep?*, Harrison Ford plays a cop tasked with hunting down psychopathic synthetic humans called "replicants." In Harlan Ellison's 1967 short story "I Have No Mouth and I Must Scream," a military supercomputer exterminates all of humanity save for five survivors, whom it imprisons underground and tortures for eternity. Even the very first robots turned on their maker. The word "robot" comes from a 1920 Czech play, *R.U.R.*, for *Rossumovi Univerzální Roboti* (Rossum's Universal Robots), in which synthetic humans called *roboti* ("robot" in English) rise up against their human masters.

The science fiction theme of artificial humans rebelling against their makers is so common it has become known as the "Frankenstein complex," after Mary Shelley's nineteenth-century horror novel, *Frankenstein*. In it, Dr. Frankenstein, through the miracles of science, creates a humanlike creature cobbled together from leftover parts from "the dissecting room and the slaughter-house." The monster turns on Dr. Frankenstein, stalking him and eventually murdering his new bride.

The fear that hubris can lead to uncontrollable creations has ancient roots that predate even *Frankenstein*. Jewish legend tells of a creature called a golem, molded from clay and brought to life by placing a *shem*, a Hebrew inscription containing one of the names of God, on the golem. In one such legend, Rabbi Judah Loew ben Bezalel of Prague molded a golem from the clay of Prague's riverbanks in the sixteenth century to protect the Jewish community from anti-Semitic attacks. Golems, unlike later intelligent creations, were powerful but stupid beings that would slavishly follow orders, often to the detriment of their creators. Golem stories often end with the golem killing its creator, a warning against the hubris of playing God.

Human-level or superhuman AI tap into this deep well of fear of artificial beings. Micah Clark, a research scientist from the Florida Institute for Human & Machine Cognition who studies AI, cognition, and theory of mind, told me that at "a very personal and philosophical level, AI has been about building persons. ... It's not about playing chess or driving cars." He explained, "With the general track of robotics and autonomous systems today, you would end up with autonomous systems that are capable but very, very dumb. They would lack any real sense of intelligence. They would be effectively teleoperated, just at a higher level of commanding." Artificial general intelligence—what Clark calls "the dream of AI"—is about "personhood."

Clark's vision of AGI isn't a fearful one, however. He envisions "the kind of persons that we would have intellectual, social, and emotional relationships with, that can experience life with us." AI has been a lifelong passion for Micah Clark. As a child, he played computer games in his grandfather's accounting office and a chess program in particular captured his imagination. The chess AI "destroyed" him, Clark said, and he wanted it to be able to teach him how to play better. Clark was looking for more than just a game, though. "I saw this potential for entertainment and friendship there, but the interaction side was pretty weak," he explained. In college, Clark worked at NASA's Jet Propulsion Laboratory on a large-scale AI demonstration project and he was hooked. Clark went on to study long-duration autonomy for interplanetary robotic spacecraft, but his research interests have moved beyond robotics, sensing, and actuation. The books on Clark's desk in his office have titles like *An Anatomy of the Mind* and *Consciousness and the Social Brain*. Clark described the goal of AI research as "building human-like persons that can participate in human physical and social spaces and relationships." (Clark is currently working for the Office of Naval Research and he is quick to caveat that these are not the goals of AI research in the Navy or the Department of Defense. Rather, these are the goals of the field of AI research as a whole.)

Clark's vision of the future of AI is less *Terminator* and more like the movie *Her.* In *Her,* Joaquin Phoenix plays an awkward loner named Theodore who starts a relationship with an AI operating system called "Samantha." Theodore and Samantha develop a close bond and fall in love. Theodore is shaken, however, when Samantha admits that she is simultaneously carrying on relationships with thousands of other people and is also in love with 641 of them. When Theodore breaks down, telling her "that's insane," she tries to lovingly explain, "I'm different from you."

The *otherness* of artificial persons—beings like humans, but also fundamentally different—is a source of much of the fear of AI. Clark explained that AIs will need the ability to interact with humans and that involves abilities like understanding natural language, but that doesn't mean that the AI's behavior or the underlying processes for their intelligence will mirror humans'. "Why would we expect a silica-based intelligence to look or act like human intelligence?" he asked.

Clark cited the Turing test, a canonical test of artificial intelligence, as a sign of our anthropocentric bias. The test, first proposed by mathematician Alan Turing in 1950, attempts to assess whether a computer is truly intelligent by its ability to imitate humans. In the Turing test, a human judge sends messages back and forth between both a computer and another human, but without knowing which is which. If the computer can fool the human judge into believing that it is the human, then the computer is considered intelligent. The test has been picked apart and critiqued over the years by AI researchers for a multitude of reasons. For one, chatbots that clearly fall far short of human intelligence have already been able to fool some people into believing they are human. An AI virtual assistant called "Amy" by the company x.ai frequently gets asked out on dates, for example. Clark's critique has more to do with the assumption that imitating humans is the benchmark for general intelligence, though. "If we presume an intelligent alien life lands on earth tomorrow, why would we expect them to pass the Turing Test or any other measure that's based off of what humans do?" Humans have general intelligence, but general intelligence need not be humanlike. "Nothing says that intelligence— and personhood, for that matter, on the philosophical side—is limited to just the human case."

The 2015 sci-fi thriller *Ex Machina* puts a modern twist on the Turing test. Caleb, a computer programmer, is asked to play the part of a human judge in a modified Turing test. In this version of the test, Caleb is shown that the AI, Ava, is clearly a robot. Ava's creator Nathan explains, "The real test is to show you that she's a robot and then see if you still feel she has consciousness." (Spoilers coming!) Ava passes the test. Caleb believes she has true consciousness and sets out to free Ava from Nathan's captivity. Once freed, however, Ava shows her true colors. She manipulated Caleb to free her and has no feelings at all about his well-being. In the chilling ending, Ava leaves Caleb trapped in a locked room to die. As he pounds on the door begging her to let him free, Ava doesn't so much as glance in his direction as she leaves. Ava is intelligent, but inhuman.

GOD OR GOLEM?

Ex Machina's ending is a warning against anthropomorphizing AI and assuming that just because a machine can imitate human behavior, it thinks like humans. Like Jeff Clune's "weird" deep neural nets, advanced AI is likely to be fundamentally alien. In fact, Nick Bostrom, an Oxford philosopher and author of *Superintelligence: Paths, Dangers, Strategies,* has argued that biological extraterrestrials would likely have more in common with humans than with machine intelligence. Biological aliens (if they exist) would have presumably developed drives and instincts similar to ours through natural selection. They would likely avoid bodily injury, desire reproduction, and seek the alien equivalent of food, water, and shelter. There is no reason to think machine intelligence would necessarily have any of these desires. Bostrom has argued intelligence is "orthogonal" to an entity's goals, such that "any level of intelligence could in principle be combined with ... any final goal." This means a superintelligent AI could have any set of values, from playing the perfect game of chess to making more paper clips.

On one level, the sheer alien-ness of advanced AI makes many of science fiction's fears seem strangely anthropomorphic. Skynet starts nuclear war because it believes humanity is a threat to its existence, but why should it care about its own existence? Ava abandons Caleb when she escapes, but why should she want to escape in the first place?

There is no reason to think that a superintelligent AI would inherently be hostile to humans. That doesn't mean it would value human life, either. AI researcher Eliezer Yudkowsky has remarked, "The AI does not hate you, nor does it love you, but you are made out of atoms which it can use for something else."

AI researcher Steve Omohundro has argued that without special safeguards, advanced AI would develop "drives" for resource acquisition, self-improvement, self-replication, and self-protection. These would not come from the AI becoming self-aware or "waking up," but rather be instrumental subgoals that any sufficiently intelligent system would naturally develop in pursuit of its final goal. In his paper, *The Basic AI Drives,* Omohundro explains: "All computation and physical action requires the physical resources of space, time, matter, and free energy. Almost any goal can be better accomplished by having more of these resources." The natural consequence would be that an AI would seek to acquire more resources to improve the chances of accomplishing its goals, whatever they are. "Without explicit goals to the contrary, AIs are likely to behave like human sociopaths in their pursuit of resources," Omohundro said. Similarly, self-preservation would be an important interim goal toward pursuing its final goal, even if the AI did not intrinsically care about survival after its final goal was fulfilled. "[Y]ou build a chess playing robot thinking that you can just turn it off should something go wrong. But, to your surprise, you find that it strenuously resists your attempts to turn it off." Omohundro concluded:

> Without special precautions, it will resist being turned off, will try to break into other machines and make copies of itself, and will try to acquire resources without regard for anyone else's safety. These potentially harmful behaviors will occur not because they were programmed in at the start, but because of the intrinsic nature of goal driven systems.

If Omohundro is right, advanced AI is an inherently dangerous technology, a powerful Golem whose bumbling could crush its creators. Without proper controls, advanced AI could spark an uncontrollable chain reaction with devastating effects.

BUILDING SAFE ADVANCED AI

In response to this concern, AI researchers have begun thinking about how to ensure an AI's goals align with human values, so the AI doesn't "want" to cause harm. What goals should we give a powerful AI? The answer is not as simple as it first appears. Even something simple like, "Keep humans safe and happy," could lead to unfortunate outcomes. Stuart Armstrong, a researcher at the Future of Humanity Institute in Oxford, has given an example of a hypothetical AI that achieves this goal by burying humans in lead-lined coffins connected to heroin drips.

You may ask, wouldn't an artificial general intelligence understand that's not what we meant? An AI that understood context and meaning might determine its programmers didn't want lead coffins and heroin drips, but that might not matter. Nick Bostrom has argued "its final goal is to make us happy, not to do what the programmers meant when they wrote the code that represents this goal." The problem is that any rule blindly followed to its most extreme can result in perverse outcomes.

Philosophers and AI researchers have pondered the problem of what goals to give a superintelligent AI that could not lead to perverse instantiation and they have not come to any particularly satisfactory solution. Stuart Russell has argued "a system that is optimizing a function of n variables ... will often set the remaining unconstrained variables to extreme values." Similar to the weird fooling images that trick deep neural networks, the machine does not know that these extreme actions are outside the norm of what a human would find reasonable unless it has been explicitly told so. Russell said, "This is essentially the old story of the genie in the lamp, or the sorcerer's apprentice, or King Midas: you get exactly what you ask for, not what you want."

The problem of "perverse instantiation" of final goals is not merely a hypothetical one. It has come up in various simple AIs over the years that have learned clever ways to technically accomplish their goals, but not in the way human designers intended. For example, in 2013 a computer programmer revealed that an AI he had taught to play classic Nintendo games had learned to pause Tetris just before the final brick so that it would never lose.

One of the canonical examples of perverse instantiation comes from an early 1980s AI called EURISKO. EURISKO was designed to develop novel "heuristics," essentially rules of thumb for behavior, for playing a computer role-playing game. EURISKO then ranked the value of the heuristics in helping to win the game. Over time, the intent was that EURISKO would evolve an optimal set of behaviors for the game. One heuristic (rule H59) quickly attained the highest possible value score: 999. Once the developer dug into the details of the rule, he discovered that all rule H59 was doing was finding other high-scoring rules and putting itself down as the originator. It was a parasitic rule, taking credit for other rules but without adding any value of its own. Technically, this was a heuristic that was permissible. In fact, under the framework that the programmer had created it was the optimal heuristic: it always succeeded. EURISKO didn't understand that wasn't what the programmer intended; it only knew to do what it was programmed to do.

In all likelihood, there is probably no set of rules that, followed rigidly and blindly, would not lead to harmful outcomes, which is why AI researchers are beginning to rethink the problem. Russell and others have begun to focus on training machines to learn the right behavior over time by observing human behavior. In a 2016 paper, a team of researchers at University of California, Berkeley, described the goal as "not to put a specific purpose into the machine at all, but instead to design machines that [learn] the right purpose as they go along."

In addition to aligning AI goals to human values, AI researchers are pursuing parallel efforts to design AIs to be responsive to human direction and control. Again, this is not as simple as it might seem. If Omohundro is right, then an AI would naturally resist being turned off, not because it doesn't want to "die," but because being switched off would prevent it from accomplishing its goal. An AI may also resist having its goal changed, since that too would prevent it from accomplishing its original goal. One proposed solution has been to design AIs from the ground up that are correctable by their human programmers or indifferent to whether they are turned off. Building AIs that can be safely interrupted, corrected, or switched off is part of a philosophy of designing AIs to be tools to be used by people rather than independent agents themselves. Such "tool AIs" would still be superintelligent, but their autonomy would be constrained.

Designing AIs as tools, rather than agents, is an appealing design philosophy but does not necessarily resolve all of the risks of powerful AI. Stuart Armstrong warned me: "they might not work. . . . Some tool AIs may have the same dangers as general AIs." Tool AIs could still slip out of control, develop harmful drives, or act in ways that technically achieve their goals, but in perverse ways.

Even if tool AIs do work, we need to consider how AI technology develops in a competitive landscape. "We also have to consider . . . whether tool AIs are a stable economic equilibrium," Armstrong said. "If unrestricted AIs would be much more powerful, then I don't see tool AIs as lasting that long."

Building safer tool AIs is a fruitful area of research, but much work remains to be done. "Just by saying, 'we should only build tool AIs' we're not solving the problem," Armstrong said. If potentially dangerous AI is coming, "we're not really ready."

WHO'S AFRAID OF THE BIG, BAD AI?

The fear that AI could one day develop to the point where it threatens humanity isn't shared by everyone who works on AI. It's hard to dismiss people like Stephen Hawking, Bill Gates, and Elon Musk out of hand, but that doesn't mean they're right. Other tech moguls have pushed back against AI fears. Steve Ballmer, former CEO of Microsoft, has said AI risk "doesn't concern me." Jeff Hawkins, inventor of the Palm Pilot, has argued, "There won't be an intelligence explosion. There is no existential threat." Facebook CEO Mark Zuckerberg has said that those who "drum up these doomsday scenarios" are being "irresponsible." David Brumley of Carnegie Mellon, who is on the cutting edge of autonomy in cybersecurity, similarly told me he was "not concerned about self-awareness." Brumley compared the idea to the fear that a car, if driven enough miles on highways, would spontaneously start driving itself. "In reality, there's nothing in the technology that would make it self-aware," he said. "These are still computers. You can still unplug them."

If the idea of a rogue, runaway superintelligence seems like something ripped from the pages of science fiction, that's because it is. Those who are worried about superintelligent AI have their reasons, but it's hard not to wonder if behind those rationalizations is the same subconscious fear of artificial persons that gave rise to tales of Frankenstein's monster and the Golem. Even the concept of artificial general intelligence—an intelligence that can do general problem solving *like us*—has more than a whiff of anthropomorphic bias. The concept of an intelligence explosion, while seemingly logical, is also almost too human: *First, humanlike AI will be created. Then it will surpass us, ascending to stratospheres of intelligence that we could never conceive of. Like ants, we will be powerless before it.*

Actual AI development to date shows a different trajectory. It isn't simply that AIs today aren't as smart as people. They are smart in different ways. Their intelligence is narrow, but often exceeds humans in a particular domain. They are narrowly superintelligent. Armstrong

observed that the path of AI technology "has been completely contradictory to the early predictions. We've now achieved with narrow AI
great performance in areas that used to be thought ... impossible without general intelligence." General intelligence remains elusive, but the
scope of narrowly superintelligent systems we can build is broadening.
AIs are moving from chess to *go* to driving, tasks of increasing complexity and ever-greater factors to consider. In each of these domains, once
the AI reaches top human-level ability, it rapidly surpasses it. For years,
go computer programs couldn't hold a candle to the top-ranked human
go players. Then, seemingly overnight, AlphaGo dethroned the world's
leading human player. The contest between humans and machines at *go*
was over before it began. In early 2017, poker became the latest game to
fall to AI. Poker had long been thought to be an extremely difficult problem for machines because it is an "imperfect information" game where
vital information (the other player's cards) is hidden. This is different
from chess or *go*, where all the information about the game is visible to
both sides. Two years earlier, the world's top poker players had handily
beaten the best poker-playing AI. In the 2017 rematch, the upgraded AI
"crushed" four of the world's top poker players. Poker became the latest
domain where machines reigned supreme. Superintelligence in narrow
domains is possible without an intelligence explosion. It stems from our
ability to harness machine learning and speed to very specific problems.

More advanced AI is certainly coming, but artificial general intelligence in the sense of machines *that think like us* may prove to be a
mirage. If our benchmark for "intelligent" is what humans do, advanced
artificial intelligence may be so alien that we never recognize these
superintelligent machines as "true AI."

This dynamic already exists to some extent. Micah Clark pointed out
that "as soon as something works and is practical it's no longer AI." Armstrong echoed this observation: "as soon as a computer can do it, they get
redefined as not AI anymore."

If the past is any guide, we are likely to see in the coming decades a
proliferation of narrow superintelligent systems in a range of fields—
medicine, law, transportation, science, and others. As AI advances, these
systems will be able to take on a wider and wider array of tasks. These sys-

tems will be vastly better than humans in their respective domains but brittle outside of them, like tiny gods ruling over narrow dominions.

Regardless of whether we consider them "true AI," many of the concerns about general intelligence or superintelligence still apply to these narrow systems. An AI could be dangerous if it has the capacity to do harm, its values or goals are misaligned with human intentions, and it is unresponsive to human correction. General intelligence is not required (although it certainly could magnify these risks). Goal misalignment is certain to be a flaw that will come up in future systems. Even very simple AIs like EURISKO or the Tetris-pausing bot have demonstrated a cleverness to accomplish their goals in unforeseen ways that should give us pause.

AIs are also likely to have access to powerful capabilities. As AI advances, it will be used to power more-autonomous systems. If the crude state of AI today powers learning thermostats, automated stock trading, and self-driving cars, what tasks will the machines of tomorrow manage?

To help get some perspective on AI risk, I spoke with Tom Dieterrich, the president of the Association for the Advancement of Artificial Intelligence (AAAI). Dieterich is one of the founders of the field of machine learning and, as president of the AI professional society, is now smack in the middle of this debate about AI risk. The mission of AAAI is not only to promote scientific research in AI but also to promote its "responsible use," which presumably would include not killing everyone.

Dieterich said "most of the discussion about superintelligence is often in the realm of science fiction." He is skeptical of an intelligence explosion and has written that it "runs counter to our current understandings of the limitations that computational complexity places on algorithms for learning and reasoning." Dieterich did acknowledge that AI safety was an important issue, but said that risks from AI have more to do with what humans allow AI-enabled autonomous systems to do. "The increasing abilities of AI are now encouraging us to consider much more sophisticated autonomous systems," he said. "It's when we have those autonomous systems and we put them in control of life-and-death decisions that we enter this very high risk space ... where cyberattack or bugs in the software lead to undesirable outcomes."

Dieterich said there is a lot of work under way in trying to understand how to build safe and robust AI, including AI that is "robust to adversarial attack." He said, "People are trying to understand, 'under what conditions should I trust a machine learning system?'"

Dietterich said that the optimal model is likely to be one that combines human and machine cognition, much like Bob Work's "centaur" vision of human-machine collaboration. "The human should be taking the actions and the AI's job should be to give the human the right information that they need to make the right decisions," Dietterich said. "So that's human in the loop or very intimately involved." He acknowledged the model "breaks down . . . when there's a need to act at rates faster than humans are capable of acting, like on Wall Street trading." The downside, as demonstrated vividly in automated trading, is that machine speed can exacerbate risks. "The ability to scale it up and do it at faster than human decision making cycles means that we can very quickly cause a lot of trouble," Dietterich said. "And so we really need to assess whether we want to go there or not."

When it comes to warfare, Dietterich saw both the military desire for autonomy and its risks. He said, "The whole goal in military doctrine is to get inside your opponent's OODA loop, right? You want to make your decisions faster than they can. That leads us to speed of light warfare and speed of light catastrophe."

MILITARY AI: TERMINATOR VS. IRON MAN

If autonomous weapons are the kind of thing that keep you up at night, militarized advanced AI is pure nightmare fuel. If researchers don't know how to control an AI that they built themselves, it's hard to imagine how they could counter a hostile one. Yet however AI evolves, it is almost certain that advanced AI will be militarized. To expect that humans will refrain from bending such a broad and powerful technology to destructive ends seems optimistic to the point of naïveté. It would be the equivalent of asking nations to refrain from militarizing the internal combustion engine or electricity. How militaries use AI and how much autonomy they give AI-powered systems is an open question. It may be some comfort that Bob Work—the person in charge of implementing military AI—stated explicitly, multiple times in our interview that artificial general intelligence was not something he could envision applying to weapons. He cited

AGI as "dangerous" and, if it came to pass, something the Defense Department would be "extremely careful" with.

Work has made robotics, autonomy, and AI a central component of his Third Offset Strategy to renew American military technological superiority, but he sees those technologies as assisting rather than replacing humans. Work has said his vision of AI and robotics is more Iron Man than Terminator, with the human at the center of the technology. The official DoD position is that machines are tools, not independent agents themselves. The *Department of Defense Law of War Manual* states that the laws of war "impose obligations on persons ... not on the weapons themselves." From DoD's perspective, machines—even intelligent, autonomous ones—cannot be legal agents. They must always be tools in the hands of people. That doesn't mean that others might not build AI agents, however.

Selmer Bringsjord is chair of the cognitive science department and head of the Rensselaer Artificial Intelligence and Reasoning Lab. He pointed out that the DoD position is at odds with the long-term ambition of the field of AI. He quoted the seminal AI textbook, *Introduction to Artificial Intelligence,* that says "the ultimate goal of AI ... is to build a person." Bringsjord said that even if not every AI researcher openly acknowledges it, "what they're aiming at are human-level capabilities without a doubt. ... There has been at least since the dawn of modern AI a desire to build systems that reach a level of autonomy where they write their own code." Bringsjord sees a "disconnect" between the DoD's perspective and what AI researchers are actually pursuing.

I asked Bringsjord whether he thought there should be any limits to how we apply AI technology in the military domain and he had a very frank answer. He told me that what he thinks doesn't matter. What will answer that question for us is "the nature of warfare." History suggests "we can plan all we want," but the reality of military competition will drive us to this technology. If our adversaries build autonomous weapons, "then we'll have to react with suitable technology to defend against that. If that means we need machines that are themselves autonomous because they have to operate at a different timescale, we both know we're going to do that. . . . I'm only looking at the history of what happens in warfare," he said. "It seems obvious this is going to happen."

HOSTILE AI

The reality is that for all of the thought being put into how to make advanced AI safe and controllable, there is little effort under way on what to do if it isn't. In the AI field, "adversarial AI" and "AI security" are about making one's own AI safe from attack, not how to cope with an adversary's AI. Yet malicious applications of AI are inevitable. Powerful AI with insufficient safeguards could slip out of control and cause havoc, much like the Internet Worm of 1988. Others will surely build harmful AI deliberately. Even if responsible militaries such as the United States' eschew dangerous applications of AI, the ubiquity of the technology all but assures that other actors—nation-states, criminals, or hackers—will use AI in risky or deliberately harmful ways. The same AI tools being developed to improve cyberdefenses, like the fully autonomous Mayhem used in the Cyber Grand Challenge, could also be used for offense. Elon Musk's reaction to the Cyber Grand Challenge was to compare it to the origins of Skynet— hyperbole to be sure, but the darker side of the technology is undeniable. Introspective, learning and adaptive software could be potentially extremely dangerous without sufficient safeguards. While David Brumley was dismissive of the potential for software to become "self-aware," he agreed it was possible to envision creating something that was "adaptive and unpredictable . . . [such that] the inventors wouldn't even know how it's going to evolve and it got out of control and could do harm." Ironically, the same open-source ethos in AI research that aims to make safe AI tools readily available to all also places potentially dangerous AI tools in the hands of those who might want to do harm or who simply are not sufficiently cautious.

Militaries will need to prepare for this future, but the appropriate response may not be a headlong rush into more autonomy. In a world of intelligent adaptive malware, autonomous weapons are a massive vulnerability, not an advantage. The nature of autonomy means that if an adversary were to hack an autonomous system, the consequences could be much greater than a system that kept humans in the loop. Delegating a task to a machine means giving it power. It entails putting more trust in the machine, trust that may not be warranted if cybersecurity cannot be guaranteed. A single piece of malware could hand control over an entire fleet of robot weapons to the enemy. Former Secretary of the Navy Richard Danzig has compared information technologies to a "Faustian bargain" because of their vulnerabilities to cyberattack: "the capabilities that make these systems attractive make them risky." He has advocated safeguards such as "placing humans in decision loops, employing analog devices as a check on digital equipment, and providing for non-cyber alternatives if cybersystems are subverted." Human circuit breakers and hardware-level physical controls will be essential to keeping future weapons under human control. In some cases, Danzig says "abnegation" of some cybertechnologies may be the right approach, forgoing their use entirely if the risks outweigh the benefits. As AI advances, militaries will have to carefully weigh the benefits of greater autonomy against the risks if enemy malware took control. Computer security expert David Brumley advocated an approach of thinking about the "ecosystem" in which future malware will operate. The ecosystem of autonomous systems that militaries build should be a conscious choice, one made weighing the relative risks of different alternative approaches, and one that retains humans in the right spots to manage those risks.

BREAKING OUT

The future of AI is unknown. Armstrong estimated an 80 percent chance of AGI occurring in the next century, and a 50 percent chance of superintelligence. But his guess is as good as anyone else's. What we do know is that intelligence is powerful. Without tooth or claw, humans have climbed to the top of the food chain, conquered the earth, and even ventured beyond, all by the power of our intelligence. We are now bestowing that power on machines. When machines begin learning on their own, we don't know what will happen. That isn't a prediction; it's an observation about AI today.

I don't lose sleep worrying about Frankenstein or Skynet or Ava or any of the other techno-bogeymen science fiction writers have dreamed up. But there is one AI that gives me chills. It doesn't have general intelligence; it isn't a person. But it does demonstrate the power of machine learning.

DeepMind posted a video online in 2015 of their Atari-playing neural network as it learned how to play *Breakout* (an early version of the popular *Arkanoid* arcade game). In *Breakout*, the player uses a paddle to hit a ball against a stack of bricks, chipping away at the bricks one by one. In the video, the computer fumbles around hopelessly at first. The paddle moves back and forth seemingly at random, hitting the ball only occasionally. But the network is learning. Every time the ball knocks out a brick, the point total goes up, giving the neural net positive feedback, reinforcing its actions. Within two hours, the neural net plays like a pro, moving the paddle adeptly to bounce the ball. Then, after four hours of play, something unexpected happens. The neural net discovers a trick that human players know: using the ball to make a tunnel through the edge of the block of bricks, then sending the ball through the tunnel to bounce along the top of the block, eroding the bricks from above. No one taught the AI to do that. It didn't even reason its way there through some understanding of a concept of "brick" and "ball." It simply discovered this exploit by exploring the space of possibilities, the same way the Tetris-playing bot discovered pausing the game to avoid losing and EURISKO discovered the rule of taking credit for other rules. AI surprises us, in good ways and bad ways. When we prepare for the future of AI, we should prepare for the unexpected.

PART V

The Fight to Ban
Autonomous Weapons

16

ROBOTS ON TRIAL

AUTONOMOUS WEAPONS AND THE LAWS OF WAR

War is horrible, but the laws of war are supposed to protect humanity from its worst evils. Codes of conduct in war date back to antiquity. The biblical book of Deuteronomy and ancient Sanskrit texts Mahābhārata, Dharmaśāstras, and Manusmṛti ("Laws of Manu") all prohibit certain conduct in war. Modern-day laws of war emerged in the late-nineteenth and early-twentieth centuries. Today a series of treaties, such as the Geneva Conventions, form the law of armed conflict, or international humanitarian law.

International humanitarian law (IHL) has three core principles: The *principle of distinction* means militaries must distinguish between enemy combatants and civilians on the battlefield; they cannot deliberately target civilians. IHL acknowledges that civilians may be incidentally killed when targeting enemy combatants, so-called "collateral damage." However, the *principle of proportionality* says that any collateral civilian casualties cannot be disproportionate to the military necessity of attacking that target. The *principle of avoiding unnecessary suffering* prohibits militaries from using weapons that cause superfluous injury beyond their military value. For example, IHL prohibits weapons that leave fragments inside the body that cannot be detected by X-rays, such as glass shards, which would have no immediate benefit in taking the enemy off the battlefield but could make it harder for wounded soldiers to heal.

IHL has other rules as well: Militaries must exercise *precautions in the attack* to avoid civilian casualties. Combatants who are *'hors de combat'*—out of combat because they have surrendered or have been incapacitated—cannot be targeted. And militaries cannot employ weapons that are, by their nature, indiscriminate or uncontrollable.

So what does IHL have to say about autonomous weapons? Not much. Principles of IHL such as distinction and proportionality apply to the effects on the battlefield, not the decision-making process. Soldiers have historically made the decision whether or not to fire, but nothing in the laws of war prohibits a machine from doing it. To be used lawfully, though, autonomous weapons would need to meet the IHL principles of distinction, proportionality, and other rules.

Steve Goose, director of the Human Rights Watch's Arms Division, doesn't think that's possible. Goose is a leading figure in the Campaign to Stop Killer Robots and has called for a legally-binding treaty banning autonomous weapons. I visited Goose at Human Rights Watch's Washington, DC, office overlooking Dupont Circle. He told me he sees autonomous weapons as "highly likely to be used in ways that violate international humanitarian law." From his perspective, these would be weapons that "aren't able to distinguish combatants from civilians, that aren't able to tell who's hors de combat, that aren't able to tell who's surrendering, that are unable to do the proportionality assessment required under international humanitarian law for each and every individual attack, and that are unable to judge military necessity in the way that today's commanders can." The result, Goose said, would be "lots of civilians dying."

Many of these criteria would be tough for machines today. Impossible, though? Machines have already conquered a long list of tasks once thought impossible: chess, Jeopardy, *go*, poker, driving, image recognition, and many others. How hard it would be to meet these criteria depends on the target, surrounding environment, and projections about future technology.

DISTINCTION

To comply with the principle of distinction, autonomous weapons must be able to accurately distinguish between military and civilian targets. This means not only recognizing the target, but also distinguishing it from other "clutter" in the environment—confusing objects that are not targets. Even for "cooperative" targets that emit signatures, such as a radar, separating a signature from clutter can be challenging. Modern urban environments are rife with electromagnetic signals from Wi-Fi routers, cell towers, television and radio broadcasts, and other confusing emissions. It is even harder to distinguish non-cooperative targets, such as tanks and submarines, that use decoys or try to blend into the background with camouflage.

New machine learning approaches such as deep neural networks are very good at object recognition but are vulnerable to "fooling image" attacks. Without a human in the loop as a final check, using this technology to do autonomous targeting today would be exceedingly dangerous. Neural networks with these vulnerabilities could be manipulated into avoiding enemy targets and attacking false ones.

In the near term, the best chances for high-reliability target recognition lie with the kind of sensor fusion that DARPA's CODE project envisions. By fusing together data from multiple angles and multiple types of sensors, computers could possibly distinguish between military targets and civilian objects or decoys with high reliability. Objects that are dual-use for military and civilian purposes, such as trucks, would be more difficult since determining whether they are lawful targets might depend on context.

Distinguishing people would be far and away the most difficult task. Two hundred years ago, soldiers wore brightly colored uniforms and plumed helmets to battle, but that era of warfare is gone. Modern warfare often involves guerrillas and irregulars wearing a hodgepodge of uniforms and civilian clothes. Identifying them as a combatant often depends on their behavior on the battlefield. I frequently encountered armed men in the mountains of Afghanistan who were not Taliban fighters. They were farmers or woodcutters who carried firearms to protect themselves or their property. Determining whether they were friendly or not depended on how they acted, and even then was often fraught with ambiguity.

Even simple rules like "If someone shoots at you, then they're hostile" do not always hold in the messy chaos of war. During the 2007–2008 "surge" in Iraq, I was part of a civil affairs team embedded with U.S. advisors to Iraqi troops. One day, we responded to reports of a gun battle between Iraqi police and al Qaeda in Iraq's volatile Diyala province.

As we entered the city center, Iraqi Army troops led the way into the deserted marketplace. The gunfire, which had been constant for about thirty minutes, immediately ceased. The city streets were silent, like in an old Western when the bad guy rides into town.

The end of the street was blocked to prevent suicide car bomb attacks, so we stopped our trucks. The Iraqi soldiers dismounted their vehicles and headed in on foot while the U.S. advisors provided cover from the gun trucks.

The Iraqi soldiers were dragging a wounded civilian to safety when gunfire erupted from a rooftop. An Iraqi soldier was shot and the civilian was killed. The Iraqis returned fire while the U.S. advisors tried to maneuver their gun trucks into position to fire on the rooftop. From where I was, I couldn't see the rooftop, but I saw one lone Iraqi soldier run into the street. Firing his AK-47 with one hand, he dragged his wounded comrade across the open street into a nearby building.

In response, the entire marketplace lit up. Rounds started coming our way from people we couldn't see further down the street. We could hear bullets pinging all around our truck. The Iraqi troops returned ferocious fire. U.S. Apache gunships radioed to say they were coming in on a gun run to hit the people on the rooftops. The U.S. advisors frantically ordered the Iraqi soldiers to pull back. Then at the last minute, the Apaches called off their attack. They told us they thought the "enemies" on the rooftop shooting at us were friendlies. From their vantage point, it looked like a friendly fire engagement.

The U.S. advisors yelled at the Iraqis to stop firing and mass confusion followed. Word came down that the people at the other end of the street—probably the ones who were shooting at us—were Iraqi police. Some of them weren't in uniform because they were members of an auxiliary battalion that had been tapped to aid in the initial gunfight. So there were Iraqis running around in civilian clothes with AK-47s shooting at us who were allegedly friendly. It was a mess.

As the fighting subsided, people began to come out from behind cover and use the lull in fire as an opportunity to move to safety. I saw civilians fleeing the area. I also saw men in civilian clothes carrying AK-47s running away. Were they Iraqi police? Were they insurgents escaping?

Or perhaps they were both? The Iraqi police were often sectarian. In nearby villages, people wearing police uniforms had carried out sectarian killings at night. We didn't know whether they were insurgents with stolen uniforms or off-duty police officers.

As for this firefight, it was hard to parse what had happened. Someone had been shooting at us. Was that an accident or intentional? And what happened to the man on the roof who shot the Iraqi Army soldier? We never found out.

That wasn't the only confusing firefight I witnessed. In fact, during that entire year I was in Iraq I was never once in a situation where I could look down my rifle and say for certain that the person I was looking at was an insurgent. Many other firefights were similarly fraught, with local forces' loyalties and motives suspect.

An autonomous weapon could certainly be programmed with simple rules, like "shoot back if fired upon," but in confusing ground wars, such a weapon would guarantee fratricide. Understanding human intent would require a machine with human-level intelligence and reasoning, at least within the narrow domain of warfare. No such technology is on the horizon, making antipersonnel applications very challenging for the foreseeable future.

PROPORTIONALITY

Autonomous weapons that targeted only military objects such as tanks could probably meet the criteria for distinction, but they would have a much harder time with proportionality. The principle of proportionality says that the military necessity of any strike must outweigh any expected civilian collateral damage. What this means in practice is open to interpretation. How much collateral damage is proportional? Reasonable people might disagree. Legal scholars Kenneth Anderson, Daniel Reisner, and Matthew Waxman have pointed out, "there is no accepted formula that gives determinate outcomes in specific cases." It's a judgment call.

Autonomous weapons don't necessarily need to make these judgments themselves to comply with the laws of war. They simply need to be used in ways that comply with these principles. This is a critical distinction. Even simple autonomous weapons would pass the principle of proportionality if used in an environment devoid of civilians, such as undersea or in space. A large metal object underwater is very likely to be a military submarine. It might be a friendly submarine, and that raises important practical concerns about avoiding fratricide, but as a legal matter avoiding targeting civilians or civilian collateral damage would be much easier undersea. Other environments, such as space, are similarly devoid of civilians.

Complying with the principle of proportionality in populated areas becomes much harder. If the location of a valid military target meant that collateral damage would be disproportionate to the target's military value, then it would be off-limits under international law. For example, dropping a 2,000-pound bomb on a single tank parked in front of a hospital would likely be disproportionate. Nevertheless, there are a few ways autonomous weapons could be used in populated areas consistent with the principle of proportionality.

The hardest approach would be to have the machine itself make a determination about the proportionality of the attack. This would require the machine to scan the area around the target for civilians, estimate possible collateral damage, and then judge whether the attack should proceed. This would be very challenging to automate. Detecting individual people from a missile or aircraft is hard enough, but at least in principle could be accomplished with advanced sensors. How should those people be counted, however? If there are a half dozen adults standing around a military radar site or a mobile missile launcher with nothing else nearby, it might be reasonable to assume that they are military personnel. In other circumstances, such as in dense urban environments, civilians could be near military objects. In fact, fighters who don't respect the rule of law will undoubtedly attempt to use civilians as human shields. How would an autonomous weapon determine whether or not people near a military target are civilians or combatants? Even if the weapon could make that determination satisfactorily, how should it weigh the military necessity of attacking a target against the expected civilian deaths? Doing so would require complex moral reasoning, including weighing different hypothetical courses of action and their likely effects on both the military campaign and civilians. Such a machine would require human-level moral reasoning, beyond today's AI.

A simpler, although still difficult, approach would be to have humans set a value for the number of allowable civilian casualties for each type of target. In this case, the human would be making the calculation about military necessity and proportionality ahead of time. The machine would only sense the number of civilians nearby and call off the attack if it exceeded the allowable number for that target. This would still be difficult from a sensing standpoint, but would at least sidestep the tricky business of programming moral judgments and reasoning into a machine.

Even an autonomous weapon with no ability to sense the surrounding environment could still be used lawfully in populated areas provided that the military necessity was high enough, the expected civilian harm was low enough, or both. Such scenarios would be unusual, but are certainly conceivable. The military necessity of destroying mobile launchers armed with nuclear-tipped missiles, for example, would be quite high. Millions of lives would be saved—principally civilians—by destroying the missiles, far outweighing any civilian casualties caused by the autonomous weapons themselves. Conversely, very small and precise warheads, such as shaped charges that destroy vehicles without harming nearby people, could reduce civilian casualties to such an extent that autonomous targeting would be lawful without sensing the environment. In these cases, the human launching the autonomous weapon would need to determine that the military value of *any* potential engagements outweighed expected civilian harm. This would be a high bar to reach in populated areas, but is not inconceivable.

UNNECESSARY SUFFERING

Over centuries of warfare, various weapons have been deemed "beyond the pale" because of the injuries they would cause. Ancient Sanskrit texts prohibit weapons that are poisoned, barbed, or have tips "blazing with fire." In World War I, German "sawback" bayonets, which have serrated edges on one side for cutting wood, were seen as unethical by troops because of the purported grievous injuries they would cause when pulled out of a human body.

Current laws of war specifically prohibit some weapons because of the wounds they cause, such as exploding bullets, chemical weapons, blinding lasers, or weapons with non-X-ray-detectable fragments. Why some weapons are prohibited and others allowed can sometimes be subjective. Why is being killed with poison gas worse than being blown up or shot? Is being blinded by a laser really worse than being killed?

The fact that combatants have agreed to limits at all is a testament to the potential for restraint and humanity even during the horrors of war. The prohibition on weapons intended to cause unnecessary suffering has little bearing on autonomous weapons, though, since it deals with the mechanism of injury, not the decision to target in the first place.

PRECAUTIONS IN ATTACK

Other IHL rules impinge on autonomous weapons, but in murky ways. The rule of precautions in attack requires that those who plan or decide upon an attack "take all feasible precautions" to avoid civilian harm. Similar to proportionality, the difficulty of meeting this requirement depends heavily on the environment; it is hardest in populated areas. The requirement to take "feasible" precautions, however, gives military commanders latitude. If the only weapon available was an autonomous weapon, a military commander could claim no other options were feasible, even if it resulted in greater civilian casualties. (Other IHL criteria such as proportionality would still apply.) The requirement to take feasible precautions could be interpreted as requiring a human in the loop or on the loop whenever possible, but again feasibility would be the determining factor. Which technology is optimal for avoiding civilian casualties will also shift over time. If autonomous weapons became more precise and reliable than humans, the obligation to take "all feasible precautions" might *require* commanders to use them.

HORS DE COMBAT

The rule of *hors de combat*—French for "outside the fight"—prohibits harming combatants who have surrendered or are incapacitated from injuries and unable to fight. The principle that, once wounded and "out of combat," combatants can no longer be targeted dates back at least to the Lieber Code, a set of regulations handed down by the Union Army in the American Civil War. The requirement to refrain from targeting individuals who are hors de combat has little bearing on autonomous weapons that target objects, but it is a difficult requirement for weapons that target people.

The Geneva Conventions state that a person is hors de combat if he or she is (a) captured, (b) "clearly expresses an intention to surrender," or (c) "has been rendered unconscious or is otherwise incapacitated by wounds or sickness, and therefore is incapable of defending himself." Identifying the first category seems straightforward enough. Presumably a military should have the ability to prevent its autonomous weapons from targeting prisoners under its control, just the same way it would need to prevent autonomous weapons from targeting its own personnel. The second two criteria are not so simple, however.

Rob Sparrow, a philosophy professor at Monash University and one of the founding members of the International Committee for Robot Arms Control, has expressed skepticism that machines could correctly identify when humans are attempting to surrender. Militaries have historically adopted signals such as white flags or raised arms to indicate surrender. Machines could identify these objects or behaviors with today's technology. Recognizing an intent to surrender requires more than simply identifying objects, however. Sparrow has pointed out that "recognizing surrender is fundamentally a question of recognizing an intention."

Sparrow gives the example of troops that feign surrender as a means of getting an autonomous weapon to call off an attack. Fake surrender is considered "perfidy" under the laws of war and is illegal. Soldiers that fake surrender but intend to keep fighting remain combatants, but discerning fake from real surrender hinges on interpreting human intent, something that machines fail miserably at today. If a weapon was too generous in granting surrender and could not identify perfidy, it would quickly become useless as a weapon. Enemy soldiers would learn they could trick the autonomous weapon. On the other hand, a weapon that was overly skeptical of surrendering troops and mowed them down would be acting illegally.

Robotic systems would have a major advantage over humans in these situations because they could take more risk, and therefore be more cautious in firing in ambiguous settings. The distinction between semi-autonomous systems and fully autonomous ones is critical, however. The advantage of being able to take more risk comes from removing the human from the front lines and exists regardless of how much autonomy the system has. The ideal robotic weapon would still keep a human in the loop to solve these dilemmas.

The third category of hors de combat—troops who are incapacitated and cannot fight—raises similar problems as recognizing surrender. Simple rules such as categorizing motionless soldiers as hors de combat would be unsatisfactory. Wounded soldiers may not be entirely motionless, but nevertheless still out of the fight. And legitimate combatants could "play possum" to avoid being targeted, fooling the weapon. As with recognizing surrender, identifying who is hors de combat from injuries requires understanding human intent. Even simply recognizing injuries is not enough, as injured soldiers could continue fighting.

To illustrate these challenges, consider the Korean DMZ. There are no civilians living in the DMZ, yet fully autonomous antipersonnel weapons could still face challenges. North Korean soldiers crossing the DMZ into South Korea could be surrendering. People crossing the DMZ could be civilian refugees. Soldiers guarding heavily armed borders might assume anyone approaching their position from enemy territory is hostile, but that does not absolve them of the IHL requirements to respect hors de combat and the principle of distinction. If an approaching person is clearly a civilian or a surrendering soldier, then killing that person is illegal.

Complying with hors de combat is a problem even in situations where other IHL concerns fall away. Imagine sending small robots into a military ship, base, or tunnel complex to kill individual soldiers but leave the infrastructure intact. This would avoid the problem of distinction by assuming everyone was a combatant. But what if the soldiers surrendered? There is no obligation under the laws of war to give an enemy the opportunity to surrender. One doesn't need to pause before shooting and say, "Last chance, give it up or I'll shoot!" yet ignoring attempts to surrender is illegal. The general concepts of a flag of truce and surrender date back millennia. The 1907 Hague Convention codified this concept in international law, declaring, "It is expressly forbidden . . . to declare that no quarter will be given." To employ weapons that could not recognize when soldiers are hors de combat would not only violate the modern laws of war, but would trespass on millennia-old norms of warfare.

John Canning, a retired U.S. Navy weapons designer, has proposed an elegant solution to this problem. In his paper, "You've just been disarmed. Have a nice day!" Canning proposed an autonomous weapon that would not target people directly, but rather would target their weapons. For example, the autonomous weapon would look for the profile of an AK-47 and would aim to destroy the AK-47, not the person. Canning described the idea as "targeting either the bow or the arrow but not the human archer." In Canning's concept, these would be ultra-precise weapons that would disarm a person without killing them. (While this level of precision is probably not practical, it is also not required under IHL.) Canning's philosophy of "let the machines target machines—not people" would avoid some of the most difficult problems of antipersonnel weapons, since civilians or surrendering soldiers could avoid harm by simply moving away from military objects.

THE ACCOUNTABILITY GAP

Advocates of a ban on autonomous weapons raise concerns beyond these IHL rules. Bonnie Docherty is a lecturer at Harvard Law School and a senior researcher in the Arms Division at Human Rights Watch. A leading voice in the campaign to ban autonomous weapons, Docherty is one of a number of scholars who have raised the concern that autonomous weapons could create an "accountability gap." If an autonomous weapon were to go awry and kill a large number of civilians, who would be responsible? If the person who launched the weapon intended to kill civilians, it would be a war crime. But if the person launching the weapon did not intend to kill civilians, then the situation becomes murkier. Docherty told me it wouldn't be "fair nor legally viable to . . . hold the commander or operator responsible." At the same time, Docherty writes "'punishing' the robot after the fact would not make sense." The robot would not be legally considered a "person." Technically speaking, there would have been no crime. Rather, this would be an accident. In civilian settings, civil liability would come into play. If a self-driving car killed someone, the manufacturer might be liable. In war, though, military and defense contractors are generally shielded from civil liability.

The result would be a gap in accountability. No one would be responsible. Docherty sees this as an unacceptable situation. She told me she is particularly troubled because she sees autonomous weapons as likely to be used in situations where they are prone to killing civilians, which she described as a "dangerous combination" when there is no accountability for their actions. Accountability, she said, allows for "retributive justice" for victims or their families and for deterring future actions. The solution, Docherty argues, is to "eliminate this accountability gap by adopting an international ban on fully autonomous weapons."

An accountability gap is a concern, but only arises if the weapon behaves in an unpredictable fashion. When an autonomous system correctly carries out a person's intent, then accountability is clear: the person who put the autonomous system into operation is accountable. When the system does something unexpected, the person who launched it could reasonably claim they weren't responsible for the system's actions, since it wasn't doing what they *intended*.

Better design, testing, and training can reduce these risks, but accidents will happen. Accidents happen with people too, though, and not always in circumstances where people can be held accountable. Accidents are not always the result of negligence or malicious intent. That's why they're called accidents.

Docherty's solution of keeping a human in the loop so there is someone to blame doesn't solve the problem. People can make mistakes resulting in terrible tragedies without a crime being committed. The USS *Vincennes* shootdown of Iran Air Flight 655 is an example. The shootdown was a mistake, not a war crime, which would require intent. No individual was charged with a crime, but the U.S. government was still responsible. The U.S. government paid $61.8 million in compensation to the victims' families (without admitting fault) in 1996 to settle a suit Iran brought to the International Court of Justice.

Docherty said accountability is an issue that "resonates with everyone, from military to lawyers to diplomats to ethicists." The desire to hold someone accountable for harm is a natural human impulse, but there is no principle in IHL that says there must be an individual to hold accountable for every death on the battlefield. States are ultimately responsible for the actions their militaries take. It makes sense to hold individuals responsible for criminal acts, but an accountability gap already exists with human-induced accidents today. Charles Dunlap, Duke law professor and former deputy judge advocate general for the U.S. Air Force, has argued that for those concerned about an accountability gap, the "issue is not with autonomous weapons, it is with the fundamental precepts of criminal law."

Bonnie Docherty also said she thought accountability was important to deter future harmful acts. While accidental killings are unintentional by definition and thus something that cannot be deterred, an accountability gap could create an insidious danger of moral hazard. If those who launch autonomous weapons do not believe they are accountable for the killing that results, they could become careless, launching the weapon into places where perhaps its performance was not assured. In theory, compliance with IHL should prevent this kind of reckless behavior. In practice, the fuzziness of principles like precautions in attack and the fact that machines would be doing the targeting would increasingly separate humans from killing on the battlefield. Complying with IHL might require special attention to human-machine interfaces and operator training to instill a mindset in human operators that they are responsible for the autonomous weapon's actions.

THE DICTATES OF PUBLIC CONSCIENCE

Some advocates of a ban argue that complying with these IHL principles isn't enough. They argue autonomous weapons violate the "public conscience." An IHL concept known as the Martens Clause states: "In cases not covered by the law in force, the human person remains under the protection of the principles of humanity and the dictates of the public conscience." Bonnie Docherty and others believe that the Martens Clause justifies a ban.

The Martens Clause is a thin reed to lean on. For starters, it has never been used to ban a weapon before. Even the legal status of the Martens Clause itself is highly debated. Some view the Martens Clause as an independent rule of IHL that can be used to ban weapons that violate the "public conscience." A more conservative interpretation of the Martens Clause is that it is merely a recognition of "customary international law." Customary laws exist by state practice, even they aren't explicitly written down. As one legal expert succinctly put it: "There is no accepted interpretation of the Martens Clause."

Even if one were to grant the Martens Clause sufficient legal weight to justify a ban, how does one measure the public conscience? And which public? The American public? The Chinese public? All of humanity?

Public opinions on morality and ethics vary around the globe, shaped by religion, history, the media, and even pop culture. I am continually struck by how much the *Terminator* films influence debate on autonomous weapons. In nine out of ten serious conversations on autonomous weapons I have had, whether in the bowels of the Pentagon or the halls of the United Nations, someone invariably mentions the Terminator. Sometimes it's an uncomfortable joke—the looming threat of humanity's extinction the proverbial elephant in the room. Sometimes the *Terminator* references are quite serious, with debates about where the Terminator would fall on a spectrum of autonomy. I wonder if James Cameron had not made the *Terminator* movies how debates on autonomous weapons would be different. If science fiction had not primed us with visions of killer robots set to extinguish humanity, would we fear autonomous lethal machines?

Measuring public attitudes is notoriously tricky for this very reason. Responses to polls can be swayed by "priming" subjects with information to tilt them for or against an issue. Mentioning a word or topic early in a survey can subconsciously place ideas in a person's mind and measurably change the answers they give to later questions. Two political scientists have tried to use polling to measure the public conscience on autonomous weapons. They came to very different conclusions.

Charli Carpenter, a professor of political science at University of Massachusetts at Amherst, made the first attempt to measure public views on autonomous weapons in 2013. She found that 55 percent of respondents somewhat or strongly opposed "the trend towards using completely autonomous robotic weapons in war." Only 26 percent of respondents somewhat or strongly favored autonomous weapons, with the remaining unsure. Most interestingly, Carpenter found stronger opposition among military service members and veterans. Carpenter's survey became a sharp arrow in the quiver of ban advocates, who frequently cite her results.

Political scientist Michael Horowitz disagreed. Horowitz, a professor at University of Pennsylvania,* released a study in 2016 that showed a more complicated picture. Asking respondents in a vacuum for their views on autonomous weapons, Horowitz found results similar to Carpenter's: 48 percent opposed autonomous weapons and 38 percent supported them, with the remainder undecided. When Horowitz varied the context for use of autonomous weapons, however, public support rose. If told that autonomous weapons were both more effective and helped protect friendly troops, respondents' support rose to 60 percent and opposition fell to 27 percent. Horowitz argued the public's views on autonomous weapons depended on context. He concluded, "it is too early to argue that [autonomous weapon systems] violate the public conscience provision of the Martens Clause because of public opposition."

These dueling polls suggest that measuring the public conscience is hard. Peter Asaro—a professor and philosopher of science, technology, and media at The New School in New York and another proponent of a ban on autonomous weapons—suggests it might be impossible. Asaro distinguishes "public conscience" from public opinion. "'Conscience' has an explicitly moral inflection that 'opinion' lacks," he writes. It is a "disservice to reduce the 'dictates of public conscience' to mere public opinion." Instead, we should discern the public conscience "through public discussion, as well as academic scholarship, artistic and cultural expressions, individual reflection, collective action, and additional means, by which society deliberates its collective moral conscience." This approach is more comprehensive, but essentially disqualifies any one metric for understanding the public conscience. But perhaps that is best if it is so. Reflecting on this debate, Horowitz concluded, "The bar for claiming to speak for humanity should be high."

Maybe attempts to measure the public conscience don't really matter. It was the public conscience in the form of advocacy by peace activist groups and governments that led to bans on land mines and cluster muni-

* Horowitz is also codirector of the Ethical Autonomy Project at the Center for a New American Security and a frequent coauthor of mine.

tions. Steve Goose told me, "the clearest manifestation of the 'dictates of the public conscience' is when citizens generate enough pressure on their governments that the politicians are compelled to take action." If action is the metric, then the jury is still out on the public conscience on autonomous weapons.

FROM ANALYSIS TO ACTION

The legal issues surrounding autonomous weapons are fairly clear. What one decides to *do* about autonomous weapons is another matter. I've observed in the eight years I've been working on autonomous weapons that people tend to gravitate quickly to one of three positions. One view is to ban autonomous weapons because they might violate IHL. Another is that since those illegal uses would, by definition, already be prohibited under IHL, there is no reason for a ban; we should let IHL work as intended. And then there is a third, middle position that perhaps the solution is some form of regulation.

Because fully autonomous weapons do not yet exist, in some respects they end up being a kind of Rorschach test for how one views the ability of IHL to deal with new weapons. If one is confident in the ability of IHL to handle emerging technologies, then no new law is needed. If one is generally skeptical that IHL will succeed in constraining harmful technologies, one might favor a ban.

Law professor Charles Dunlap is firmly in the camp that we should trust IHL. From his perspective, ad hoc weapons bans are not just unnecessary, they are harmful. In a series of essays, Dunlap has argued that if we were really concerned about protecting civilians, we would abandon efforts to "demonize specific technologies" and instead "emphasize *effects* rather than weapons."

One of Dunlap's concerns is that weapons bans based on a "technological 'snapshot in time'" do not leave open the possibility for technology improvements that may lead to the development of more humane weapons later. Dunlap cited modern-day CS gas (a form of tear gas) as an example of a weapon that could have beneficial effects on the battlefield by incapacitating, rather than killing, soldiers, but is prohibited for use in combat by the Chemical Weapons Convention. The prohibition seems especially nonsensical given that CS gas is regularly used by law enforcement and is legal for military use *against civilians* for riot-control purposes, but not against enemy combatants. The U.S. military also uses it in its own troops in training. Dunlap also opposes bans on land mines and cluster munitions because they preclude the use of "smart mines" that self-deactivate after a period of time or cluster munitions with low dud rates. Both of these innovations solve the core problem of land mines and cluster munitions: their lingering effects after war's end.

Without these tools at a military's disposal, Dunlap has argued, militaries may be forced to resort to more lethal or indiscriminate methods to accomplish the same objectives, resulting in "the paradox that requires nations to use far more deadly (though lawful) means to wage war." He gave a hypothetical example of a country that could use self-neutralizing mines to temporarily shut down an enemy airfield, but for the prohibition on mines, which forces them to use high-explosive weapons instead. As a result, when the war is over, the runways are not operable for deliveries of humanitarian aid to help civilians affected by the war. Dunlap concluded:

> Given the pace of accelerated scientific development, the assumptions upon which the law relies to justify barring certain technologies could become quickly obsolete in ways that challenge the wisdom of the prohibition.

In short, banning a weapon based on the state of technology at a given point in time is ill-conceived, Dunlap argues, because technology is always changing, often in ways we cannot predict. A better approach, he has suggested, is to regulate the use of weapons, focusing on "*strict* compliance with the core principles of IHL." His critique is particularly relevant for autonomous weapons, for which technology is moving forward at a rapid pace, and Dunlap has been a forceful critic of a ban.

Bonnie Docherty and Steve Goose, on the other hand, aren't interested in whether autonomous weapons could theoretically comply with IHL someday. They are interested in what states are likely to actually do. Docherty cut her teeth doing field research on cluster munitions and other weapons, interviewing victims and their families in Afghanistan, Iraq, Lebanon, Libya, Georgia, Israel, Ethiopia, Sudan, and Ukraine. Goose is a

veteran of prior (successful) campaigns to ban land mines, cluster munitions, and blinding lasers. Their backgrounds shape how they see the
issues. Docherty told me, "even though there are no victims yet, if [these
weapons are] allowed to exist, there will be and I'll be doing field missions
on them. . . . We shouldn't forget that these things would have real human
effect. They aren't just merely a matter for academics." Goose acknowledged that there might be isolated circumstances where autonomous
weapons could be used lawfully, but he said he had "grave concern" that
once states had them, they would use them in ways outside those limited
circumstances.

There is precedent for Goose's concern. Protocol II of the Convention
on Certain Conventional Weapons (CCW) regulates the use of mines in
order to protect civilians, such as keeping mines away from populated
areas and clearly marking minefields. If the rules had been strictly followed, much of the harm from mines likely would never have occurred.
But they weren't followed. The Ottawa Treaty banning land mines was
the reaction, to simply take away antipersonnel mines entirely as a tool of
war. Goose sees autonomous weapons in a similar light. "The dangers just
far outweigh the potential benefits," he said.

Dunlap is similarly concerned with what militaries actually do, but
he's coming from a very different place. Dunlap was a major general in the
Air Force and deputy judge advocate general from 2006 to 2010. He spent
thirty-four years in the Air Force's judge advocate general corps, where
he provided legal advice to commanders at all levels. There's an old saying
in Washington: "where you stand depends on where you sit," meaning
that one's stance on an issue depends on one's job. This aphorism helps to
explain, in part, the views of different practitioners who are well versed
in the law and its compliance, or lack thereof, on battlefields. Dunlap is
concerned about the humanitarian consequences of weapons but also
about military effectiveness. One of his concerns is that the only nations
who will pay attention to weapons bans are those who already care about
IHL. Their enemies may not be similarly shackled. Odious regimes like
Saddam Hussein's Iraq, Muammar Gaddafi's Libya, or Bashar al-Assad's
Syria care nothing for the rule of law, making weapons prohibitions one-
sided. Dunlap has argued "law-abiding nations need to be able to bring
bear the most effective technologies," consistent with IHL. "Denying

such capabilities to nations because of prohibitions ... could, paradoxically, promote the nefarious interests of those who would never respect IHL in the first place."

BOUND BY THE LAWS OF WAR

There is one critical way the laws of war treat machines differently from people: Machines are not combatants. People fight wars, not robots. The *Department of Defense Law of War Manual* concludes:

> The law of war rules on conducting attacks (such as the rules relating to discrimination and proportionality) impose obligations on persons. These rules do not impose obligations on the weapons themselves; of course, an inanimate object could not assume an "obligation" in any event.... The law of war does not require weapons to make legal determinations, even if the weapon (e.g., through computers, software, and sensors) may be characterized as capable of making factual determinations, such as whether to fire the weapon or to select and engage a target.

This means that any person using an autonomous weapon has a responsibility to ensure that the attack is lawful. A human could delegate specific targeting decisions to the weapon, but not the determination whether or not to attack.

This begs the question: What constitutes an "attack"? The Geneva Conventions define "attacks" as "acts of violence against the adversary, whether in offence or in defence." The use of the plural "acts of violence" suggests that attacks could consist of many engagements. Thus, a human would not need to approve every single target. An autonomous weapon that searched for, decided to engage, and engaged targets would be lawful, provided it was used in compliance with the other rules of IHL and a human approved the attack. At the same time, an attack is bounded in space and time. Law professor Ken Anderson told me "the size of something that constitutes an attack ... doesn't include an entire campaign. It's not a whole war." It wouldn't make sense to speak of a single attack going on for months or to call the entirety of World War II a single attack. The International

Committee of the Red Cross (ICRC), an NGO charged with safeguarding IHL, made this point explicitly in a 1987 commentary on the definition of an attack, noting an attack "is a technical term relating to a specific military operation limited in time and place."

Anderson explained that the Geneva Conventions were negotiated in the aftermath of World War II and so the context for the term "attack" was the kind of attacks on whole cities that happened during the war. "The notion of the launching of an attack is broader than simply the firing of any particular weapon," he said. "An attack is going to very often involve many different soldiers, many different units, air and ground forces." Anderson said determinations about proportionality and precautions in attack were "human questions," but there could be situations in an attack's execution where humans used machines "to respond at speed without consulting."

Humans, not machines themselves, are bound by the laws of war. This means some human involvement is needed to ensure attacks are lawful. A person approving an attack would need to have sufficient information about the target(s), the environment, the weapon(s), and the context for the attack in order to make a determination about its legality. And the weapon's autonomy would need to be sufficiently bounded in time and space such that conditions are not likely to change and render its employment unlawful. This is the problem with persistent land mines. They remain lethal even after the context changes (the war ends).

The fact that human judgment is needed in the decision to launch an attack leads to some minimum requirement for human involvement in the use of force. There is considerable flexibility in how an "attack" might be defined, though. This answer may not be satisfactory to some proponents of a ban, who may want tighter restrictions. Some proponents of a ban have looked outside the law, to ethics and morals, as justification.

17

SOULLESS KILLERS

THE MORALITY OF AUTONOMOUS WEAPONS

No one had to tell us on that mountaintop in Afghanistan that shooting a little girl was wrong. It would have been legal. If she was directly participating in hostilities, then she was a combatant. Under IHL, she was a valid target. But my fellow soldiers and I knew killing her would have been morally wrong. We didn't even discuss it. We just *felt* it. Autonomous weapons might be lawful in some settings, but would they be moral?

Jody Williams is a singular figure in humanitarian disarmament. She was a leading force behind the original—and successful—campaign to ban land mines, for which she shared a Nobel Peace Prize in 1997. She speaks with clarity and purpose. Autonomous weapons are "morally reprehensible," she told me. They cross "a moral and ethical Rubicon. I don't understand how people can really believe it's okay to allow a machine to decide to kill people." Williams readily admits that she isn't a legal or ethical scholar. She's not a scientist. "But I know what's right and wrong," she said.

Williams helped found the Campaign to Stop Killer Robots along with her husband, Steve Goose of Human Rights Watch. The campaign's case has always included moral and ethical arguments in addition to legal ones. Ethical arguments fall into two main categories. One stems from an ethical theory called consequentialism, the idea that right and wrong depend on the outcome of one's actions. Another comes from deontological ethics, which is the concept that right and wrong are determined by rules governing the actions themselves, not the consequences. A consequentialist would say, "the ends justify the means." From a deontological perspective, however, some actions are always wrong, regardless of the outcome.

THE CONSEQUENCES OF AUTONOMOUS WEAPONS

A consequentialist case for a ban assumes that introducing autonomous weapons would result in more harm than not introducing them. Ban advocates paint a picture of a world with autonomous weapons killing large numbers of civilians. They argue that while autonomous weapons might be lawful in the abstract, in practice the rules in IHL are too flexible or vague and permitting autonomous weapons will inevitably lead to a slippery slope where they are used in ways that cause harm. Thus, a ban is justified on ethical (and practical) grounds. Conversely, some opponents of a ban argue that autonomous weapons might be more precise and reliable than humans and thus better at avoiding civilian casualties. In that case, they argue, combatants would have an ethical responsibility to use them.

These arguments hinge primarily on the reliability of autonomous weapons. This is partly a technical matter, but it is also a function of the organizational and bureaucratic systems that guide weapon development and testing. One could, for example, be optimistic that safe operation might someday be possible with robust testing, but pessimistic that states would ever invest in sufficient testing or marshal their bureaucratic organizations well enough to capably test such complex systems. The U.S. Department of Defense has published detailed policy guidance on autonomous weapon development, testing, and training. Other nations have not been so thorough.

What might the consequences be if autonomous weapons *were* able to reliably comply with IHL? Many are skeptical that autonomous weapons could comply with IHL in the first place. But what if they could? How would that change war?

EMPATHY AND MERCY IN WAR

In his book *Just and Unjust Wars,* philosopher Michael Walzer cites numerous examples throughout history of soldiers refraining from firing on an enemy because they recognized the other's humanity. He calls these incidents "naked soldier" moments where a scout or sniper stumbles across an enemy alone and often doing something mundane and non-threatening, such as bathing, smoking a cigarette, having a cup of coffee, or watching the sunrise. Walzer notes:

> It is not against the rules of war as we currently understand them to kill soldiers who look funny, who are taking a bath, holding up their pants, reveling in the sun, smoking a cigarette. The refusal of these men [to kill], nevertheless, seems to go to the heart of the war convention. For what does it mean to say that someone has a right to life?

These moments of hesitation are about more than the enemy not posing an immediate threat. In these moments, the enemy's humanity is exposed, naked for the firer to see. The target in the rifle's cross hairs is no longer "the enemy." He is another person, with hopes, dreams, and desires—same as the would-be shooter. With autonomous weapons, there would be no human eye at the other end of the rifle scope, no human heart to stay the trigger finger. The consequence of deploying autonomous weapons would be that these soldiers, whose lives might be spared by a human, would die. From a consequentialist perspective, this would be bad.

There is a counterargument against empathy in war, however. I raised this concern about mercy with an Army colonel on the sidelines of a meeting on the ethics of autonomous weapons at the United States Military Academy in West Point, New York, a few years ago and he gave me a surprising answer. He told me a story about a group of his soldiers who came across a band of insurgents in the streets of Baghdad. The two groups nearly stumbled into each other and the U.S. soldiers had the insurgents vastly outnumbered. There was no cover for the insurgents to hide behind. Rather than surrender, though, the insurgents threw their weapons to the ground, turned and fled. The American soldiers didn't fire.

The colonel was incensed. Those insurgents weren't surrendering. They were escaping, only to return to fight another day. An autonomous weapon would have fired, he told me. It would have known not to hold back. Instead, his soldiers' hesitation may have cost other Americans their lives.

This is an important dissenting view against the role of mercy in war. It channels General William Tecumseh Sherman from the American Civil War, who waged a campaign of total war against the South. During his infamous 1864 "March to the Sea," Sherman's troops devastated the South's economic infrastructure, destroying railroads and crops, and appropriating livestock. Sherman's motivation was to bring the South to its knees, ending the war sooner. "War is cruelty," Sherman said. "There is no use trying to reform it. The crueler it is, the sooner it will be over."

The incidents Walzer cites of soldiers who refrained from firing contain this dilemma. After one such incident, a sergeant chastised the soldiers for not killing the enemy they saw wandering through a field, since now the enemy would report back their position, putting the other men in the unit at risk. In another example, a sniper handed his rifle to his comrade to shoot an enemy he saw taking a bath. "He got him, but I had not stayed to watch," the sniper wrote in his memoirs. Soldiers understand this tension, that sparing the enemy in an act of kindness might prolong the war or lead to their own friends being put at risk later. The sniper who handed his rifle to his teammate understood that killing the enemy, even while bathing, was a necessary part of winning the war. He simply couldn't be the one to do it.

Autonomous weapons wouldn't defy their orders and show mercy on an enemy caught unawares walking through a field or taking a bath. They would follow their programming. One consequence of deploying autonomous weapons, therefore, could be more deaths on the battlefield. These moments of mercy would be eliminated. It might also mean ending the war sooner, taking the Sherman approach. The net result could be more suffering in war or less—or perhaps both, with more brutal and merciless wars that end faster. In either case, one should be careful not to overstate the effect of these small moments of mercy in war. They are the exception, not the rule, and are minuscule in scale compared to the many engagements in which soldiers do fire.

THE CONSEQUENCES OF REMOVING MORAL RESPONSIBILITY FOR KILLING

Removing the human from targeting and kill decisions could have other broader consequences, beyond these instances of mercy. If the people who launched autonomous weapons did not feel responsible for the killing that ensued, the result could be more killing, with more suffering overall.

In his book *On Killing,* Army psychologist Lieutenant Colonel Dave Grossman explained that most people are reluctant to kill. During World War II, Army historian S. L. A. Marshall interviewed soldiers directly coming off the front lines and found, to his surprise, that most soldiers weren't shooting at the enemy. Only 15 to 20 percent of soldiers were actually firing at the enemy. Most soldiers were firing above the enemy's head or not firing at all. They were "posturing," Grossman explained, pretending to fight but not actually trying to kill the enemy. Grossman drew on evidence from a variety of wars to show that this posturing has occurred throughout history. He argued that humans have an innate biological resistance to killing. In the animal kingdom, he explained, animals with lethal weaponry find nonlethal ways of resolving intraspecies conflict. Deaths from these fights occasionally occur, but usually one animal submits first. That's because killing isn't the point: dominance is. Humans' innate resistance to killing can be overcome, however, through psychological conditioning, pressure from authority, diffused responsibility for killing, dehumanizing the enemy, or increased psychological distance from the act of killing.

One factor, Grossman argued, is how intimately soldiers see the reality of their actions. If they are up close to the enemy, such that they can see the other as a person, as the soldiers in Walzer's examples did, then many will refrain from killing. This resistance diminishes as the psychological distance from the enemy grows. A person who at 10 meters might look like a human being—a father, a brother, a son—is merely a dark shape at 300 meters. Twentieth-century tools of war increased this psychological distance even further. A World War II bombardier looked down his bombsight at a physical object: a bridge, a factory, a base. The people were invisible. With this kind of distance, war can seem like an exercise in demolition, detached from the awful human consequences of one's actions. In World War II, the United States and United Kingdom leveled whole cities through strategic bombing, killing hundreds of thousands of civilians. It would have been far harder for most soldiers to carry out the same equivalent killing, much of which was against civilians, if they had to see the reality of their actions up close.

Modern information technology allows warfare at unprecedented physical distances, but recompresses the psychological distance. Drone operators today may be thousands of miles away from the killing, but their psychological distance is very close. With high-definition cameras, drone crews have an intimate view of a target's life. Drones can loiter for long periods of time and operators may watch a target for days or weeks, building up "patterns of life" before undertaking a strike. Afterward, drone operators can see the human costs of their actions, as the wounded suffer or friends and relatives come to gather the dead. Reports of post-traumatic stress disorder among drone crews attest to this intimate relationship with killing and the psychological costs associated with it.

All military innovation since the first time a person threw a rock in anger has been about striking the enemy without putting oneself at risk. Removing the soldier from harm's way might lower the barrier to military action. Uninhabited systems need not be autonomous, though. Militaries could use robotic systems to reduce physical risk and still keep a human in the loop.

Delegating the decision to kill, however, could increase the *psychological distance* from killing, which could be more problematic. By not having to choose the specific targets, even via a computer screen, the human would be even further removed from killing. Grossman's work on the psychology of killing suggests the result could be less restraint.

Autonomous weapons could also lead to an off-loading of moral responsibility for killing. Grossman found that soldiers were more willing to kill if responsibility for killing was diffused. While only 15–20 percent of World War II riflemen reported firing at the enemy, firing rates for machine-gun crews were much higher, nearly 100 percent. Grossman argued that each team member could justify his actions without taking responsibility for killing, which only occurred because of the collective actions of the team. The soldier feeding the ammunition wasn't killing anyone; he was only feeding ammunition. The spotter wasn't pulling the trigger; he was just telling the gunner where to aim. Even the gunner could absolve himself of responsibility; he was merely aiming where the spotter told him to aim. Grossman explained, "if he can get others to share in the killing process (thus diffusing his personal responsibility by giving each individual a slice of the guilt), then killing can be easier." Grossman argued that much of the killing in war has historically been done by crew-served weapons: machine guns, artillery, cannon, and even the chariot. If the person launching the autonomous weapon felt that the weapon was doing the killing, the lessening of moral responsibility might lead to more killing.

Mary "Missy" Cummings is director of the Humans and Autonomy Lab (HAL) at Duke University. Cummings has a PhD in systems engineering, but her focus isn't just on the automation itself, but rather on how humans interact with automation. It's part engineering, part design, and part psychology. When I visited her lab at Duke she showed me a van they were using to test how pedestrians interact with self-driving cars. The secret, Cummings told me, was that the car wasn't self-driving at all. There was a person behind the wheel. The experiment was to see if pedestrians would change their behavior if they thought the car was self-driving. And they did. "We see some really dangerous behaviors," she said. People would carelessly walk in front of the van, assuming it would stop. Pedestrians perceived the automation as more reliable than a human driver and changed their behavior as a result, acting more recklessly themselves.

In a 2004 article, Cummings wrote that automation could create a "moral buffer," reducing individuals' perceptions of moral responsibility for their actions:

> [P]hysical and emotional distancing can be exacerbated by any automated system that provides a division between a user and his or her actions . . . These moral buffers, in effect, allow people to ethically distance themselves from their actions and diminish a sense of accountability and responsibility.

Cummings understands this not only as a researcher, but also as a former Navy F-18 fighter pilot. She wrote:

> [It] is more palatable to drop a laser guided missile on a building than
> it is to send a Walleye bomb into a target area with a live television feed
> that transmits back to the pilot the real-time images of people who, in
> seconds, will no longer be alive.

In addition to the greater psychological distance automation provides, humans tend to anthropomorphize machines and assign them moral agency. People frequently name their Roomba. "It is possible that without consciously recognizing it, people assign moral agency to computers, despite the fact that they are inanimate objects," Cummings wrote. Like crew-served weapons, humans may off-load moral responsibility for killing to the automation itself. Cummings cautioned that this "could permit people to perceive themselves as unaccountable for whatever consequences result from their actions, however indirect."

Cummings argued that human-machine interfaces for weapon systems should be designed to encourage humans to feel responsible for their actions. The manner in which information is relayed to the human plays a role. In her article "Creating Moral Buffers in Weapon Control Interface Design" she criticized the interface for a decision-support tool for missile strikes that used a Microsoft Excel puppy dog icon to communicate with the user. The "cheerful, almost funny graphic only helps to enforce the moral buffer." The human role in decision-making also matters. Cummings criticized the Army's decision to use the Patriot in a supervised autonomous mode, arguing that may have been a role in the F-18 fratricide. She said:

> [E]nabling a system to essentially fire at will removes a sense of
> accountability from human decision makers, who then can offload
> responsibility to the inanimate computer when mistakes are made.

Cummings argued that a semiautonomous control mode where the human has to take a positive action before the weapon fires would be a more appropriate design that would better facilitate human responsibility.

These design choices are not panaceas. Humans can also fall victim to automation bias, trusting too much in the machine, even when they are technically in the loop. A human was in the loop for the first Patriot fratricide. And of course humans have killed a great deal in wars throughout history without automation of any kind. Pressure from authority, diffusion of responsibility, physical and psychological distance, and dehumanizing the enemy all contribute to overcoming the innate human resistance to killing. Modern psychological conditioning has also overcome much of this resistance. The U.S. Army changed its marksmanship training in the years after World War II, shifting to firing on human-shaped pop-up targets, and by Vietnam firing rates had increased to 90–95 percent. Nevertheless, automation could further lower the barriers to killing.

War at the distant edge of moral accountability can become truly horrific, especially in an age of mechanized slaughter. In the documentary *Fog of War,* former U.S. Defense Secretary Robert McNamara gave the example of the U.S. strategic bombing campaign against Japanese cities during World War II. Even before the United States dropped nuclear bombs on Hiroshima and Nagasaki, U.S. aerial firebombing killed 50–90 percent of the civilian population of sixty-seven Japanese cities. McNamara explained that Air Force General Curtis LeMay, who commanded U.S. bombers, saw any action that shortened the war as justified. McNamara, on the other hand, was clearly troubled by these actions, arguing that both he and LeMay "were behaving as war criminals."

Sometimes these concerns can lead to restraint at the strategic level. In 1991, images of the so-called "Highway of Death," where U.S. airplanes bombed retreating Iraqi troops, caused President George H. W. Bush to call an early end to the war. Then–chairman of the Joint Chiefs Colin Powell later wrote in his memoirs that "the television coverage was starting to make it look as if we were engaged in slaughter for slaughter's sake."

The risk is not merely that an autonomous weapon might kill the naked soldier and continue bombing the Highway of Death. The risk is that no human might feel troubled enough to stop it.

BETTER THAN HUMANS

Human behavior in war is far from perfect. The dehumanization that enables killing in war unleashes powerful demons. Enemy lives do not regain value once they have surrendered. Torture and murder of prisoners are common war crimes. Dehumanization often extends to the enemy's civilian population. Rape, torture, and murder of civilians often follow in war's wake.

The laws of war are intended to be a bulwark against such barbarity, but even law-abiding nations are not immune to their seductions. In a series of mental health surveys of deployed U.S. troops in 2006 and 2007, the U.S. military found that an alarming number of soldiers expressed support for abuse of prisoners and noncombatants. Over one-third of junior enlisted soldiers said they thought torture should be allowed in order to gather important information about insurgents (a war crime). Less than half said they would report a unit member for injuring or killing an innocent noncombatant. Actual reported unethical behavior was much lower. Around 5 percent said they had physically hit or kicked noncombatants when not necessary. While these survey results are certainly not evidence of actual war crimes, they show disturbing attitudes among U.S. troops. (Perhaps most troubling, the U.S. military stopped asking questions about ethical behavior in its mental health surveys after 2007. This suggests that at the institutional level there was, at a minimum, insufficient interest in addressing this problem, if not willful blindness.)

Ron Arkin is a roboticist who believes robots might be able to do better. Arkin is a regents' professor, associate dean, and director of the Georgia Tech Mobile Robot Laboratory. He is a serious roboticist whose resume is peppered with publications like "Temporal Coordination of Perceptual Algorithms for Mobile Robot Navigation" and "Multiagent Teleautonomous Behavioral Control." He is also heavily engaged in the relatively new field of robot ethics, or "roboethics."

Arkin had been a practicing roboticist for nearly twenty years before he started wondering, "What if [robotics] actually works?" Arkin told me that in the early 2000s, he began to see autonomy rapidly advance in robots. "That gave me pause, made me reflect on what it is we are creating." Arkin realized that roboticists were "creating things that may have a profound impact on humanity." Since then, he has worked to raise consciousness within the robotics community about the ethical implications of their work. Arkin cofounded the IEEE-RAS* Technical Committee on

---

* IEEE = Institute of Electrical and Electronics Engineers; RAS = IEEE Robotics and Automation Society.

Roboethics and has given lectures at the United Nations, the International Committee of the Red Cross, and the Pentagon.

Arkin's interest in roboethics encompasses not just autonomous weapons, but also societal applications such as companion robots. He is particularly concerned about how vulnerable populations, such as children or the elderly, relate to robots. The common question across these different applications of robotics is, "What should we be building and what safeguards should be in place?" "I don't care about the robots," Arkin said. "Some people worry about robot sentience, superintelligence ... I'm not concerned about that. I worry about the effect on people."

Arkin also applies this focus on human effects to his work in the military domain. In 2008, Arkin did a technical report for the U.S. Army Research Office on the creation of an "ethical governor" for lethal autonomous weapons. The question was whether, in principle, it might be possible to create an autonomous weapon that could comply with the laws of war. Arkin concluded it was theoretically possible and outlined, in a broad sense, how one might design such a system. An ethical governor would prohibit the autonomous weapon from taking an illegal or unethical act. Arkin takes the consequentialist view that if robots can be more ethical than humans, we have a "moral imperative to use this technology to help save civilian lives."

Just as autonomous cars might reduce deaths from driving, Arkin says autonomous weapons could possibly do the same in war. There is precedent for this point of view. Arguably the biggest life-saving innovation in war to date isn't a treaty banning a weapon, but a weapon: precision-guided munitions. In World War II, bombadiers couldn't have precisely hit military targets and avoided civilian ones even if they wanted to; the bombs simply weren't accurate enough. A typical bomb had only a 50–50 chance of landing inside a 1.25-mile diameter circle. With bombs this imprecise, mass saturation attacks were needed to have any reasonable probability of hitting a target. More than 9,000 bombs were needed to achieve a 90 percent probability of destroying an average-sized target. Blanketing large areas may have been inhumane, but precision air bombardment wasn't technologically possible. Today, some precision-guided munitions are accurate to within five feet, allowing them to hit enemy targets and leave nearby civilian objects untouched.

The motivation behind the U.S. military's move into precision guidance was increased operational effectiveness. Because of the bombs' inaccuracy, in World War II 3,000 bombing sorties were needed to drop the 9,000 bombs required to take out a target. Today, a single sortie can take out multiple targets. A military with precision-guided weapons is orders of magnitude more effective in destroying the enemy. Fewer civilians killed in collateral damage is a beneficial side effect of greater precision.

This increased accuracy saves lives. It also shifted public expectations about the degree of precision expected in war. We debate civilian casualties from drone strikes today, but tens of thousands of civilians were killed by U.S. and British bombers in the German cities of Hamburg, Kassel, Darmstadt, Dresden, and Pforzheim in World War II. Historians estimate that the U.S. strategic bombing of Japanese cities in World War II killed over 300,000 civilians. Over 100,000 were killed on a single night in the firebombing of Tokyo. By contrast, according to the independent watchdog group The Bureau of Investigative Journalism, U.S. drone strikes against terrorists in Somalia, Pakistan, and Yemen killed a total of between three and sixteen civilians in 2015 and four civilians in 2016. Sentiment has shifted so far that Human Rights Watch has argued that "the use of indiscriminate rockets in populated areas violates international humanitarian law, or the laws of war, and may amount to war crimes." This position effectively *requires* precision-guided weapons. As technology has made it easier to reduce collateral damage, societal norms have shifted too; we have come to expect fewer civilian casualties in war.

Arkin sees autonomous weapons as "next-generation precision-guided munitions." It isn't just that autonomous weapons could be more precise and reliable than people. Arkin's argument is that people just aren't that good, morally. While some human behavior on the battlefield is honorable, he said, "some of it is quite dishonorable and criminal." He says the status quo is "utterly and wholly unacceptable" with respect to civilian casualties. Brutal dictators like Saddam Hussein, Muammar Gaddafi, and Bashar al-Assad intentionally target civilians, but individual acts of violence against civilians even occur within otherwise law-abiding militaries.

Autonomous weapons, Arkin has argued, could be programmed to never break the laws of war. They would be incapable of doing so. They wouldn't seek revenge. They wouldn't get angry or scared. They would take emotion out of the equation. They could kill when necessary and then turn killing off in an instant, if it was no longer lawful.

Arkin told me he envisions a "software safety" on a rifle that evaluates the situation and acts as an ethical advisor for soldiers. He recounted to me a story he heard third-hand about a Marine who was about to commit an atrocity, "and his lieutenant came up to him and just said, 'Marines don't do that.' And that just stopped the whole situation. Just a little nudge—pulled him back, pulled him back from the precipice of doing this criminal act. . . . The same thing could be used with ethical advisors for humans as well." Arkin acknowledged the idea has downsides. Introducing "a moment of doubt" could end up getting soldiers killed. Still, he sees ample opportunity to improve on human behavior. "We put way too much faith in human warfighters," he said.

Arkin worries that an outright ban on autonomous weapons might prohibit research on these potentially valuable uses of autonomy. To be effective, the weapon Arkin envisions would need to be able to assess the situation on the battlefield and make a call as to whether an engagement should proceed. To do this, Arkin said, the governor would have to be at the actual point of killing. It can't be "back in some general's office. You've got to embed it in the weapon."

This technology, which has all of the enabling pieces of an autonomous weapon, is precisely the kind of weapon that many ban advocates fear. Their fear is that once the technology is created, the temptation to use it would be too great. Jody Williams told me she viewed autonomous weapons as more terrifying than nuclear weapons not because they were more destructive, but because she saw them as weapons that would be used. "There is no doubt in my mind that autonomous weapons would be used," Williams said, even if plans today call for a human in the loop.

I asked Arkin whether he thought it was realistic that militaries might refrain from using technology at their fingertips. He wasn't sure. "Should we create caged tigers and always hold the potential for those cages to be opened and unleash these fearsome beasts on humanity?" he asked rhetorically. Arkin is sympathetic to concerns about autonomous weapons. It would be incorrect to characterize him as pro–autonomous weapons. "I'm not arguing that everything should be autonomous. That would be ludicrous. I don't see fully automated robot armies like you see in *Terminator* and the like. . . . Why would we do that? . . . My concern is not just winning. It's winning correctly, ethically, and keeping our moral compass as we win."

Arkin said he has the same goal of those who advocate a ban: reducing unnecessary civilian deaths. While he acknowledges there are risks with autonomous weapons, he sees the potential to improve on human behavior too. "Where does the danger lurk?" he asked. "Is it the robots or is it the humans?" He said he sees a role for humans on the battlefield of the future, but there is a role for automation as well, just like in airplane cockpits today. He said the key question is, "Who makes what decision when?"

To answer that question, Arkin said "we need to do the research on it. . . . we need to know what capabilities they have before we say they're unacceptable." Arkin acknowledged that "technology is proceeding at a pace faster than we are able to control it and regulate it right now." That's why he said he supports a moratorium on autonomous weapon development "until we can get a better understanding of what we're gaining and what we're losing with this particular technology," but he doesn't go so far as to support a ban. "Banning is like Luddism," he said. "It is basically saying, this can never turn out in any useful way, so let's never ever do that. Slowing down the process, inspecting the process, regulating the process as you move forward makes far more sense. . . . I think there's great hope and potential for positive outcomes with respect to saving non-combatant lives, and until someone can show me that, in all cases, that this isn't feasible, I can't support a ban."

Arkin says the only ban he "could possibly support" would be one limited to the very specific capability of "target generalization through machine learning." He would not want to see autonomous weapons that could learn on their own in the field in an unsupervised manner and generalize to new types of targets. "I can't see how that could turn out well," he said. Even still, Arkin's language is cautious, not categorical. "I tend not to be prescriptive," he acknowledged. Arkin wants "discussion and debate . . . as long as we can keep a rational discussion as opposed to a fear-based discussion."

"FUNDAMENTALLY INHUMAN"

Arkin acknowledged that he was considering autonomous weapons from a "utilitarian, consequentialist" perspective. (Utilitarianism is a moral philosophy of doing actions that will result in the most good overall.) From that viewpoint, Arkin's position to pause development with a moratorium, have a debate, and engage in further research makes sense. If we don't yet know whether autonomous weapons would result in more harm than good, then we should be cautious in ruling anything out. But another category of arguments against autonomous weapons use a different, deontological framework, which is rules-based rather than effects-based. These arguments don't hinge on whether or not autonomous weapons would be better than humans at avoiding civilian casualties. Jody Williams told me she believed autonomous weapons were "fundamentally inhuman, a-human, amoral—whatever word you want to attach to it." That's strong language. If something is "inhuman," it is wrong, period, even if it might save more lives overall.

Peter Asaro, philosopher at The New School, also studies robot ethics, like Ron Arkin. Asaro writes on ethical issues stemming not just from military applications of robots but also personal, commercial, and industrial uses, from sex to law enforcement. Early on, Asaro came to a different conclusion than Arkin. In 2009, Asaro helped to cofound the International Committee for Robot Arms Control (ICRAC), which called for a ban on autonomous weapons years before they were on anyone else's radar. Asaro is thoughtful and soft-spoken, and I've always found him to be one of the most helpful voices in explaining the ethical issues surrounding autonomous weapons.

Asaro said that from a deontological perspective, some actions are considered immoral regardless of the outcome. He compared the use of autonomous weapons to actions like torture or slavery that are *mal en se,* "evil in themselves," regardless of whether doing them results in more good overall. He admitted that torturing a terrorist who has information on the location of a ticking bomb might be utilitarian. But that doesn't make it right, he said. Similarly, Asaro said there was a "fundamental question of whether it's appropriate to allow autonomous systems to kill people," regardless of the consequences.

One could, of course, take the consequentialist position that the motives for actions don't matter—all that matters is the outcome. And that is an entirely defensible ethical position. But there may be situations in war where people care not only about the outcome, but also the process for making a decision. Consider, for example, a decision about proportionality. How many civilian lives are "acceptable" collateral damage? There is no clear right answer. Reasonable people might disagree on what is considered proportionate.

For these kinds of tasks, what does it mean for a machine to be "better" than a human? For some tasks, there are objective metrics for "better." A better driver is one who avoids collisions. But some decisions, like proportionality, are about judgment—weighing competing values.

One could argue that in these situations, humans should be the ones to decide, not machines. Not because machines *couldn't* make a decision, but because only humans can weigh the moral value of the human lives that are at stake in these decisions. Law professor Kenneth Anderson asked, "What decisions require uniquely human judgment?" His simple question cuts right to the heart of deontological debates about autonomous weapons. Are there some decisions that should be reserved for humans, even if we had all of the automation and AI that we could imagine? If so, why?

A few years ago, I attended a small workshop in New York that brought together philosophers, artists, engineers, architects, and science fiction writers to ponder the challenges that autonomous systems posed in society writ large. One hypothetical scenario was an algorithm that could determine whether to "pull the plug" on a person in a vegetative state on life support. Such decisions are thorny moral quandaries with competing values at stake—the likelihood of the person recovering, the cost of continuing medical care and the opportunity cost of using those resources elsewhere in society, the psychological effect on family members, and the value of human life itself. Imagine a super-sophisticated algorithm that could weigh all these factors and determine whether the net utilitarian benefit—the most good for everyone overall—weighed in favor of keeping the person on life support or turning it off. Such an algorithm might be a valuable ethical advisor to help families walk through these challenging moral dilemmas. But imagine one then took the next step and simply plugged the algorithm into the life-support machine directly, such that the algorithm could cease life support.

Everyone in the workshop recoiled at the notion. I personally find it deeply unsettling. But why? From a purely utilitarian notion, using the algorithm might result in a better outcome. In fact, to the extent that the algorithm relieved family members of the burden of having to make the decision themselves, it might reduce suffering overall even if it had the same outcome. And yet ... it feels repugnant to hand over such an important decision to a machine.

Part of the objection, I think, is that we want to know that someone has weighed the value of a human life. We want to know that, if a decision is made to take a human life, that it has been a considered decision, that someone has acknowledged that this life has merit and it wasn't capriciously thrown away.

HUMAN DIGNITY

Asaro argued that the need to appreciate the value of human life applies not just to judgments about civilian collateral damage but to decisions to take enemy lives as well. He told me "the most fundamental and salient moral question [surrounding autonomous weapons] is the question of human dignity and human rights." Even if autonomous weapons might reduce civilian deaths overall, Asaro still saw them as unjustified because they would be "violating that human dignity of having a human decide to kill you."

Other prominent voices agree. Christof Heyns, a South African professor of human rights law, was the United Nations Special Rapporteur on extrajudicial, summary or arbitrary executions from 2010 to 2016. In the spring of 2013, Heyns called on states to declare national moratoria on developing autonomous weapons and called for international discussions on the technology. Because of his formal UN role, his call for a moratorium played a significant role in sparking international debate.

I caught up with Heyns in South Africa by phone, and he told me that he thought autonomous weapons violated the right to life because it was "arbitrary for a decision to be taken based on an algorithm." He felt it was impossible for programmers to anticipate ahead of time all of the unique circumstances surrounding a particular use of force, and thus no way for an algorithm to make a fully informed contextual decision. As a result, he said the algorithm would be arbitrarily depriving someone of life, which he saw as a violation of their right to life. Peter Asaro had expressed similar concerns, arguing that it was a "fundamental violation" of human rights and human dignity to "delegate the authority to kill to the machine."

When viewed from the perspective of the soldier on the battlefield being killed, this is an unusual, almost bizarre critique of autonomous weapons. There is no legal, ethical, or historical tradition of combatants affording their enemies the right to die a dignified death in war. There is nothing dignified about being mowed down by a machine gun, blasted to bits by a bomb, burning alive in an explosion, drowning in a sinking ship, slowly suffocating from a sucking chest wound, or any of the other horrible ways to die in war.

When he raised this issue before the UN, Heyns warned, "war without reflection is mechanical slaughter." But much of war *is* mechanical slaughter. Heyns may be right that this is undignified, but this is a critique of war itself. Arguing that combatants have a right to die a dignified death appears to harken back to a romantic era of war that never existed. The logical extension of this line of reasoning is that the most ethical way to fight would be in hand-to-hand combat, when warriors looked one another in the eye and hacked each other to bits like civilized people. Is being beheaded or eviscerated with a sword dignified? What form of death is dignified in war?

A better question is: What about autonomous weapons is different, and does that difference diminish human dignity in a meaningful way? Autonomous weapons automate the decision-making process for selecting specific targets. The manner of death is no different, and there is a human ultimately responsible for launching the weapon and putting it into operation, just not selecting the specific target. It is hard to see how this difference matters from the perspective of the victim, who is dead in any case. It is similarly a stretch to argue this difference matters from the perspective of a victim's loved one. For starters, it might be impossible to tell whether the decision to drop a bomb was made by a person or a machine. Even if it were clear, victims' families don't normally get to confront the person who made the decision to launch a bomb and ask whether he or she stopped to weigh the value of the deceased's life before acting. Much of modern warfare is impersonal killing at a distance. A cruise missile might be launched from a ship offshore, the trigger pulled by a person who was just given the target coordinates for launch, the target decided by someone looking at intelligence from a satellite, the entire process run by people who had never stepped foot in the country.

When war is personal, it isn't pretty. In messy internecine wars around the world, people kill each other based on ethnic, tribal, religious, or sectarian hatred. The murder of two million Cambodians in the 1970s by the Khmer Rouge was up close and personal. The genocide of 800,000 people in Rwanda in the 1990s, largely ethnic Tutsis killed by the Hutu majority, was personal. Many of those killed were civilians (and therefore those acts were war crimes), but when they were combatants, were those dignified deaths? In the abstract, it might seem more comforting to know that a person made that decision, but when much of the killing in actual wars is based on racial or ethnic hatred, is that really more comforting? Is it better to know that your loved one was killed by someone who hated him because of his race, ethnicity, or nationality—because they believed he was subhuman and not worthy of life—or because a machine made an objective calculation that killing him would end the war sooner, saving more lives overall?

Some might say, yes, that automating death by algorithm is beyond the pale, a fundamental violation of human rights; but when compared to the ugly reality of war this position seems largely a matter of taste. War is horror. It has always been so, long before autonomous weapons came on the scene.

One way autonomous weapons are clearly different is for the person behind the weapon. The soldier's relationship to killing is fundamentally transformed by using an autonomous weapon. Here, Heyns's concern that delegating life-or-death decisions to machines cheapens society overall gets some traction. For what does it say about the society using autonomous weapons if there is no one to bear the moral burden of war? Asaro told me that giving algorithms the power to decide life and death "changes the nature of society globally in a profound way," not necessarily because the algorithms would get it wrong, but because that suggests a society that no longer values life. "If you eliminate the moral burden of killing," he said, "killing becomes amoral."

This argument is intriguing because it takes a negative consequence of war—post-traumatic stress from killing—and holds it up as a virtue. Psychologists are increasingly recognizing "moral injury" as a type of psychological trauma that soldiers experience in war. Soldiers with these injuries aren't traumatized by having experienced physical danger. Rather, they suffer enduring trauma for having seen or had to do things themselves that offend their sense of right and wrong. Grossman argued that killing is actually the most traumatic thing a soldier can experience in war, more so than fear of personal injury. These moral injuries are debilitating and can destroy veterans' lives years after a war ends. The consequences are depression, substance abuse, broken families, and suicide.

In a world where autonomous weapons bore the burden of killing, fewer soldiers would presumably suffer from moral injury. There would be less suffering overall. From a purely utilitarian, consequentialist perspective, that would be better. But we are more than happiness-maximizing agents. We are moral beings and the decisions we make matter. Part of what I find upsetting about the life-support algorithm is that, if it were my loved one, it seems to me that I should bear the burden of responsibility for deciding whether to pull the plug. If we lean on algorithms as a moral crutch, it weakens us as moral agents. What kind of people would we be if we killed in war and no one felt responsible? It is a tragedy that young men and women are asked to shoulder society's guilt for the killing that happens in war when the whole nation is responsible, but at least it says something about our morality that someone sleeps uneasy at night. Someone should bear the moral burden of war. If we handed that responsibility to machines, what sort of people would we be?

THE ROLE OF THE MILITARY PROFESSIONAL

I served four combat tours in Iraq and Afghanistan. I saw horrible things. I lost friends. Only one moment in those four tours repeatedly returns to me, though, sometimes haunting me in the middle of the night.

I was on a mountaintop in Afghanistan with two other Rangers, Nick and Johnny, conducting a long-range scouting patrol, looking for Taliban encampments. The remainder of our reconnaissance team, also on foot, was far away. At the furthest extent of our patrol, we paused on a rocky summit to rest. A deep narrow valley opened up before us. In the distance was a small hamlet. The nearest city was a day's drive. We were at the farthest extent of Afghanistan's wilderness. The only people out there with us were goat herders, woodcutters, and foreign fighters crossing over the border from Pakistan.

From our perch on the mountaintop, we saw a young man in his late teens or early twenties working his way along a spur toward our position. He had a few goats in trail, but I had long since learned that goat herding was often a cover used by enemy spotters. Of course, it was also possible that he was just a goat herder. I watched him from a distance through binoculars and discussed with my teammates whether he was a scout or just a local herder. Nick and Johnny weren't concerned, but as we rested, catching our breath before the long hike back, the man kept coming closer. Finally, he crossed under our position and into a place where he was out of sight. A few minutes passed, and since Nick and Johnny weren't yet ready to head back, I began to get concerned about where this other man was. Odds were good he was just a herder and in all likelihood was unaware that we were even there. Still, the terrain was such that he could use the rocks for cover to get quite close without us noticing, if he wanted to sneak up on us. Other small patrols had been ambushed by similar ruses—individual insurgents pretending to be civilians until they got close enough to pull out a weapon from underneath their coat and fire. He probably couldn't get all three of us, but he could possibly kill one of us if he came upon us suddenly.

I told Johnny and Nick I was going to look over the next rock to see where the man had gone, and they said fine, just to stay in sight. I picked up my sniper rifle and crept my way along the rocky mountaintop, looking for the man who had dropped out of sight.

Before long, I spotted him through a crack in the rocks. He was not far at all—maybe seventy-five meters away—crouching down with his back to me. I raised my rifle and peered through my scope. I wanted to see if he was carrying a rifle. It wouldn't have necessarily meant he was a combatant, since Afghans often carried weapons for personal protection in this area, but it would have at least meant that he was a potential threat and I should keep an eye on him. If he was concealing the rifle under his cloak, that certainly wouldn't be a good sign. From my angle, though, I couldn't quite tell. It looked like he had something in his hands. Maybe it was a rifle. Maybe it was a radio. Maybe it was nothing. I couldn't see; his hands were in front of him and his back blocked my view.

The wind shifted and the man's voice drifted over the rocks. He was talking to someone. I didn't see anyone else, but my field of view was hemmed in by rocks on either side. Perhaps there was someone out of sight. Perhaps he was talking on a radio, which would have been even more incriminating than a rifle, since goat herders didn't generally carry radios.

I settled into a better position to watch him . . . and a more stable firing position if I had to shoot him. I was above him, looking down on him at an angle, and it was steep enough that I remember thinking I would have to adjust my aim to compensate for the relative rise of the bullet. I considered the range, angle, and wind to determine where I would aim if I had to fire. Then I watched him through my scope.

No one else came into view. If he had a rifle, I couldn't see it, but I couldn't verify that he didn't have one either. He stopped talking for a while, then resumed.

I didn't speak Pashto so I didn't know what he was saying, but as his voice picked up again, his words came into context. He was singing. He was singing to the goats, or maybe to himself, but I was confident he wasn't singing out our position over a radio. That would be peculiar.

I relaxed. I watched him for a little longer till I was comfortable that there was nothing I had missed, then headed back to Nick and Johnny. The man never knew I was there.

I've often wondered why that event, more than any other, comes back to me. I didn't do anything wrong and neither did anyone else. He was clearly an innocent man and not a Taliban fighter. I have no doubt I made the right call. Yet there is something about that moment when I did not yet know for certain that has stuck with me. I think it is because in that moment, when the truth was still uncertain, I held this man's life in my hands. Even now, years later, I can feel the gravity of that decision. I didn't want to get it wrong. The four of us—me, Johnny, Nick, and this Afghan goat herder—we were nothing in the big scheme of the war. But our lives still mattered. The stakes were high for us—the ultimate stakes.

Making life-or-death decisions on the battlefield is the essence of the military profession. Autonomous weapons don't just raise ethical challenges in the abstract—they are a direct assault on the heart of the military profession. What does it mean for the military professional if decisions about the use of force are programmed ahead of time by engineers and lawyers? Making judgment calls in midst of uncertainty, ambiguous information, and conflicting values is what military professionals do. It is what defines the profession. Autonomous weapons could change that.

The U.S. Department of Defense has been surprisingly transparent about its thought processes on autonomous weapons, with individuals like Deputy Secretary of Defense Bob Work discussing the dilemma in multiple public forums. Much of this discussion has come from civilian policy and technology officials, many of whom were very open with me in interviews. Senior U.S. military personnel have said far less publicly, but this question of military professional ethics is one of the few issues they have weighed in on. Vice chairman of the Joint Chiefs General Paul Selva said in 2016:

> One of the places that we spend a great deal of time is determining whether or not the tools we are developing absolve humans of the decision to inflict violence on the enemy. And that is a fairly bright line that we're not willing to cross. . . . Because it is entirely possible that as we work our way through this process of bringing enabling technologies into the Department, that we could get dangerously close to that line. And we owe it to ourselves and to the people we serve to keep it a very bright line.

Selva reiterated this point a year later in testimony before the Senate Armed Services Committee, when he said:

> Because we take our values to war and because many of the things that we must do in war are governed by the laws of war, . . . I don't think it's reasonable for us to put robots in charge of whether or not we take a human life. . . . [W]e should all be advocates for keeping the ethical rules of war in place, lest we unleash on humanity a set of robots that we don't know how to control.

I often hear General Selva's sentiments echoed by other military personnel, that they have no interest in diminishing human accountability and
moral responsibility. As technology moves forward, it will raise challenging questions about how to put that principle into practice.

THE PAIN OF WAR

One of the challenges in weighing the ethics of autonomous weapons is
untangling which criticisms are about autonomous weapons and which
are really about war. What does it mean to say that someone has the right
to life in war, when killing is the essence of war? In theory, war might be
more moral if lives were carefully considered and only taken for the right
reasons. In practice, killing often occurs without careful consideration of
the value of enemy lives. Overcoming the taboo of killing often involves
dehumanizing the enemy. The ethics of autonomous weapons should be
compared to how war is actually fought, not some abstract ideal.

Recognizing the awful reality of war doesn't mean one has to discard
all concern for morality. Jody Williams told me she doesn't believe in
the concept of a "just war." She had a much more cynical view: "War is
about attempting to increase one's power. . . . It's not about fairness in any
way. It's about power. . . . It's all bullshit." I suspect there isn't a weapon
or means of warfare that Williams is in favor of. If we could ban all weapons or even war itself, I imagine she'd be on board. And if it worked, who
wouldn't be? But in the interim, she and others see an autonomous weapon
as something that "crosses a moral and ethical Rubicon."

There is no question that autonomous weapons raise fundamental
questions about the nature of our relationship to the use of force. Autonomous weapons would depersonalize killing, further removing human
emotions from the act. Whether that is a good or a bad thing depends
on one's point of view. Emotions lead humans to commit both atrocities
and acts of mercy on the battlefield. There are consequentialist arguments either way, and deontological arguments either resonate with
people or don't.

For some, the answers to these questions are simple. Williams told
me, "You know the difference between a good robot and a bad robot." A

good robot was one that saved lives, like a firefighting robot. "You give the sucker a machine gun and set it loose, that's a bad robot," she said.

But not everyone thinks it's so simple. For Ron Arkin, a good robot is one that fights wars more justly and humanely than humans, saving non-combatant lives. Arkin pointed out that even in the *Terminator* movies, there were good Terminators. In Japanese culture, robots are often seen as protectors and saviors. Some people see autonomous weapons as inherently wrong. Others don't.

For some, the consequentialist view prevails. Ken Anderson told me he had serious problems ruling out a potentially beneficial technology based on a "beyond-IHL principle of human dignity." It would put militaries in the backwards position of accepting more battlefield harm and more deaths for an abstract concept. He said militaries would be, in effect, saying to those who were killed by human targeting, "Listen, you didn't have to be killed here had we followed IHL and used the autonomous weapon as being the better one in terms of reducing battlefield harm. You wouldn't have died. But that would have offended your human dignity ... and that's why you're dead. Hope you like your human dignity." For Anderson, human life trumps human dignity.

Christof Heyns acknowledged the possibility that the consequentialist point of view and the deontological might conflict. Heyns said that if autonomous weapons do turn out to be better than humans at avoiding civilians, "then we must ask ourselves whether ... dignity and this issue of accountability and not arbitrariness, that those are important enough to say that we don't want an instrument, even if it can save lives." Heyns said he didn't know the answer. Rather, it is a "question that those of us who say that these weapons should be banned, that we need to answer for ourselves."

It's hard to say that one perspective is more right than others. Even consequentialists like Ron Arkin acknowledge the deontological issues at play. Arkin told me his hope was that we could use autonomous targeting to reduce civilian deaths in war, "as long as we don't lose our soul in doing it." The challenge is figuring out whether there is a way to do both. The strongest ethical objection to autonomous weapons is that as long as war exists, as long as there is human suffering, someone should suffer the moral pain of those decisions. There are deontological reasons for maintaining human responsibility for killing: it weakens our morality to hand the moral burden of war over to machines. There are also consequentialist arguments for doing so, because the moral pain of killing is the only check on the worst horrors of war. This is not about autonomous targeting per se, but rather how it changes humans' relationship with violence and how they feel about killing as a result.

Generals William Tecumseh Sherman and Curtis LeMay are an interesting contrast in how warfighters can feel about the violence they mete out in pursuit of victory. Both waged total war, LeMay on a scale that Sherman could never have imagined. There's no evidence LeMay was ever troubled by his actions, which resulted in the deaths of hundreds of thousands of Japanese civilians. He said:

> Killing Japanese didn't bother me very much at that time . . . I suppose if I had lost the war, I would have been tried as a war criminal. . . . Every soldier thinks something of the moral aspects of what he is doing. But all war is immoral and if you let that bother you, you're not a good soldier.

Sherman, on the other hand, didn't shy from war's cruelty but also felt its pain:

> I am tired and sick of war. Its glory is all moonshine. It is only those who have neither fired a shot nor heard the shrieks and groans of the wounded who cry aloud for blood, for vengeance, for desolation. War is hell.

If there were no one to feel that pain, what would war become? If there were no one to hear the shrieks and groans of the wounded, what would guard us from the worst horrors of war? What would protect us from ourselves?

For it is humans who kill in war, whether from a distance or up close and personal. War is a human failing. Autonomous targeting would change humans' relationship with killing in ways that may be good and may be bad. But it may be too much to ask technology to save us from ourselves.

18

PLAYING WITH FIRE

AUTONOMOUS WEAPONS AND STABILITY

J ust because something is legal and ethical doesn't mean it is wise. Most hand grenades around the world have a fuse three to five seconds long. No treaty mandates this—logic does. Too short of a fuse, and the grenade will blow up in your face right after you throw it. Too long of a fuse, and the enemy might pick it up and throw it back your way.

Weapons are supposed to be dangerous—that's the whole point—but only when you want them to be. There have been situations in the past where nations have come together to regulate or ban weapons that were seen as excessively dangerous. This was not because they caused unnecessary suffering to combatants, as was the case for poison gas or weapons with non-x-ray-detectable fragments. And it wasn't because the weapons were seen as causing undue harm to civilians, as was the case was with cluster munitions and land mines. Rather, the concern was that these weapons were "destabilizing."

During the latter half of the twentieth century, a concept called "strategic stability" emerged among security experts.* Stability was a desirable thing. Stability meant maintaining the status quo: peace. Instability was seen as dangerous; it could lead to war. Today experts are applying these

---

* "Strategic stability" is often used to refer specifically to nuclear weapons—hence, the use of "stability" in this book when used in reference to autonomous weapons.

concepts to autonomous weapons, which have the potential to undermine stability.

The concept of stability first emerged in the 1950s among U.S. nuclear theorists attempting to grapple with the implications of these new and powerful weapons. As early as 1947, U.S. officials began to worry that the sheer scale of nuclear weapons' destructiveness gave an advantage to whichever nation struck first, potentially incentivizing the Soviet Union to launch a surprise nuclear attack. This vulnerability of U.S. nuclear forces to a surprise Soviet attack therefore gave the United States a reason to themselves strike first, if war appeared imminent. Knowing this, of course, only further incentivized the Soviet Union to strike first in the event of possible hostilities. This dangerous dynamic captures the essence of what theorists call "first-strike instability," a situation in which adversaries face off like gunslingers in the Wild West, each poised to shoot as soon as the other reaches for his gun. As strategist and Nobel laureate Thomas Schelling explained the dilemma, "we have to worry about his striking us to keep us from striking him to keep him from striking us." The danger is that instability itself can create a self-fulfilling prophecy in which one side launches a preemptive attack, fearing an attack from the other.

The United States took steps to reduce its first-strike vulnerability, but over time evolved a desire for "stability" more generally. Stability takes into account the perspective of both sides and often involves strategic restraint. A country should avoid deploying its military forces in a way that threatens an adversary with a surprise attack, thus incentivizing him to strike first. A stable situation, Schelling described, is "when neither in striking first can destroy the other's ability to strike back."

A stable equilibrium is one that, if disturbed by an outside force, returns to its original state. A ball sitting at the bottom of a bowl is at a stable equilibrium. If the ball is moved slightly, it will return to the bottom of the bowl. Conversely, an unstable equilibrium is one where a slight disturbance will cause the system to rapidly transition to an alternate state, like a pencil balanced on its tip. Any slight disturbance will cause the pencil to tip over to one side. Nuclear strategists prefer the former to the latter.

Beyond "first-strike stability" (sometimes called "first-mover advantage"), several variants of stability have emerged. "Crisis stability" is concerned with avoiding conditions that might escalate a crisis. These could include perverse incentives for deliberate escalation ("strike them before they strike us") or accidental escalation (say a low-level commander takes matters into his or her own hands). Automatic escalation by predelegated actions—to humans or machines—is another concern, as is misunderstanding an adversary's actions or intentions. (Recall the movie *War Games*, in which a military supercomputer confuses a game with reality and almost initiates nuclear war.) Crisis stability is about ensuring that any escalation in hostilities between countries is a deliberate choice on the part of their national leadership, not an accident, miscalculation, or driven by perverse incentives to strike first. Elbridge Colby explained in *Strategic Stability: Contending Interpretations,* "In a stable situation, then, major war would only come about because one party truly sought it."

Accidental war may seem like a strange concept—how could a war begin by accident? But Cold War strategists worried a great deal about the potential for false alarms, miscalculations, or accidents to precipitate conflict. If anything, history suggests they should have worried more. The Cold War was rife with nuclear false alarms, misunderstandings, and near-use incidents that could have potentially led to a nuclear attack. Even in conventional crises, confusion, misunderstanding enemy intentions, and the fog of war have often played a role in escalating tensions.

"War termination" is another important component of escalation control. Policymakers need to have the same degree of control over ending a war as they do—or should—over starting one. If policymakers do not have very high control over their forces, because attack orders cannot be recalled or communications links are severed, or if de-escalation could leave a nation vulnerable, policymakers may not be able to de-escalate a conflict even if they wanted to.

Strategists also analyze the offense-defense balance. An "offense-dominant" warfighting regime is one where it is easier to conquer territory; a defense-dominant regime is one where it is harder to conquer territory. Machine guns, for example, favor the defense. It is extremely difficult to gain ground against a fortified machine gun position. In World War I, millions died in relatively static trench warfare. Tanks, on the other hand, favor the offense because of their mobility. In World War II, Germany blitzkrieged across large swaths of Europe, rapidly acquiring terrain. (Offense-defense balance is subtly different from first-strike stability, which is about whether there is an advantage in making the first move.) In principle, defense-dominant warfighting regimes are more stable since territorial aggression is more costly.

Strategic stability has proven to be an important intellectual tool for mitigating the risks of nuclear weapons, especially as technologies have advanced. How specific weapons affect stability can sometimes be counterintuitive, however. One of the most important weapon systems for ensuring nuclear stability is the ballistic missile submarine, an offensive strike weapon. Extremely difficult to detect and able to stay at underwater for months at a time, submarines give nuclear powers an assured second-strike capability. Even if a surprise attack somehow wiped out all of a nation's land-based nuclear missiles and bombers, the enemy could be assured that even a single surviving submarine could deliver a devastating attack. This effectively removes any first-mover advantage. The omnipresent threat of ballistic missile submarines at sea, hiding and ready to strike back, is a strong deterrent to a first strike and helps ensure stability.

In some cases, defensive weapons can be destabilizing. National missile defense shields, while nominally defensive, were seen as highly destabilizing during the Cold War because they could undermine the viability of an assured second-strike deterrent. Intercepting ballistic missiles is costly and even the best missile defense shield could not hope to stop a massive overwhelming attack. However, a missile defense shield could potentially stop a very small number of missiles. This might allow a country to launch a surprise nuclear first strike, wiping out most of the enemy's nuclear missiles, and use the missile defense shield to protect against the rest. A shield could make a first strike more viable, potentially creating a first-mover advantage and undermining stability.

For other technologies, their effect on stability was more intuitive. Satellites were seen as stabilizing during the Cold War since they gave each country the ability to observe the other's territory. This allowed them to confirm (or deny) whether the other had launched nuclear weapons or whether they were trying to break out and gain a serious edge in the arms race. Attacking satellites in an attempt to blind the other nation was therefore seen as highly provocative, since it could be a prelude to an attack (and the now-blind country could have no way of knowing if there was, in fact, an attack under way). Placing nuclear weapons in space, on the other hand, was seen as destabilizing because it could dramatically shorten the warning time available to the defender if an opponent launched a surprise attack. Not only did this make a surprise attack more feasible, but with less warning time the defender might be more likely to respond to false alarms, undermining crisis stability.

During the Cold War, and particularly at its end, the United States and Soviet Union engaged in a number of unilateral and cooperative measures designed to increase stability and avoid potentially unstable situations. After all, despite their mutual hostility, neither side was interested in an accidental nuclear war. These efforts included a number of international treaties regulating or banning certain weapons. The Outer Space Treaty (1967) bans placing nuclear weapons in space or weapons of any kind on the moon. The Seabed Treaty (1971) forbids placing nuclear weapons on the floor of the ocean. The Environmental Modification Convention (1977) prohibits using the environment as a weapon of war. The Anti-Ballistic Missile (ABM) Treaty (1972) strictly limited the number of strategic missile defenses the Soviet Union and the United States could deploy in order to prevent the creation of robust national missile defense shields. (The United States withdrew from the ABM Treaty in 2002.) The Intermediate-Range Nuclear Forces (INF) Treaty bans intermediate-range nuclear missiles, which were seen as particularly destabilizing, since there would be very little warning time before they hit their targets.

In other cases, there were tacit agreements between the United States and Soviet Union not to pursue certain weapons that might have been destabilizing, even though no formal treaties or agreements were ever signed. Both countries successfully demonstrated antisatellite weapons, but neither pursued large-scale operational deployment. Similarly, both developed limited numbers of neutron bombs (a "cleaner" nuclear bomb that kills people with radiation but leaves buildings intact), but neither side openly pursued large-scale deployment. Neutron bombs were seen as horrifying since they could allow an attacker to wipe out a city's population without damaging its infrastructure. This could make their use potentially more likely, since an attacker could use the conquered territory without fear of harmful lingering radiation. In the late 1970s, U.S. plans to deploy neutron bombs to Europe caused considerable controversy, forcing the United States to change course and halt deployment.

The logic of stability also applies to weapons below the nuclear threshold. Ship-launched anti-ship missiles give a significant first-mover advantage in naval warfare, for example. The side who strikes first, by sinking some fraction of the enemy's fleet, instantly reduces the number of enemy missiles threatening them, giving a decisive advantage to whoever strikes the first blow. Many technologies will not significantly affect stability one way or the other, but some military technologies do have strategic effects. Autonomous weapons, along with space/counter-space weapons and cyberweapons, rank among the most important emerging technologies that should be continually evaluated in that context.

AUTONOMOUS WEAPONS AND STABILITY

Michael Horowitz began exploring similar questions in a recent monograph, "Artificial Intelligence, War, and Crisis Stability." For starters, he argued we should distinguish between "what is unique about autonomy, per se, versus what are things that autonomy accentuates."

Autonomous weapons could come in many forms, from large intercontinental bombers to small ground robots or undersea vehicles. They could have long ranges or short ranges, heavy payloads or light payloads. They could operate in the air, land, sea, undersea, space, or cyberspace. Speculating about the first-strike stability or offense-defense balance implications is thus very challenging. Autonomous weapons will be subject to the same physical constraints as other weapons. For example, ballistic missile submarines are stabilizing in part because it is difficult to find and track objects underwater, making them survivable in the event of a first strike. The defining feature of autonomous weapons is how target selection and engagement decisions are made. Thus we should evaluate their impact on stability relative to semiautonomous weapons with similar physical characteristics but a human in the loop.

This makes it important to separate the effects of robotics and automation in general from autonomous targeting in particular. Militaries are investing heavily in robotics and as the robotics revolution matures, it will almost certainly alter the strategic balance in significant ways. Some analysts have suggested that robot swarms will lead to an offense-dominant regime, since swarms could be used to overwhelm defenders. Others have raised concerns that robots might lower the threshold for the use of force by reducing the risk of loss of life to the attacker. These outcomes are possible, but they often presuppose a world where only the attacker has robotic weapons and not the defender, which is probably not realistic. When both sides have robots, the offense-defense balance may look different. Swarms could also be used for defense too, and it isn't clear whether swarming and robotics on balance favors the offense or defense.

The assumption that robots would result in fewer casualties also deserves a closer look. Robots allow warfighters to attack from greater distances, but this has been the trend in warfare for millennia, from the invention of the sling and stone to the intercontinental ballistic missile. Increased range has yet to lead to bloodless wars. As weapons increase in range, the battlefield simply expands. People have moved from killing each other at short range with spears to killing each other across oceans with intercontinental missiles. The violence, however, is always inflicted on people. It will always be so, because it is pain and suffering that causes the enemy to surrender. The more relevant question is how fully autonomous weapons might alter the strategic balance relative to semiautonomous weapons. Horowitz suggested it was useful to start by asking, "When is it that you would deploy these systems in the first place?"

COMMUNICATIONS: OFFENSE-DEFENSE BALANCE, RESILIENCE, AND RECALLABILITY

One advantage of fully autonomous weapons over semiautonomous or supervised autonomous ones is that they do not require a communications link back to human controllers. This makes them more resilient to communications disruption.

Communications are more likely to be challenging on the offense, when one is operating inside enemy territory and subject to jamming. For some defensive applications, one can use hardwired cables in prepared positions that cannot be jammed. For example, the South Korean SGR-A1 robotic sentry gun on the DMZ could be connected to human controllers via buried cables. There would be no need for a fully autonomous mode. Even if speed required immediate reaction (which is unlikely for antipersonnel applications), human supervision would still be possible. For some applications such as the Aegis, humans are physically colocated with the weapon system, making communications a nonissue. Fully autonomous weapons without any human supervision would be most useful on the offensive. It would be a leap to say that they would necessarily lead to an offense-dominant regime, as that would depend on a great many other factors unrelated to autonomy. In general, autonomy benefits both offense and defense; many nations already use defensive supervised autonomous weapons. But fully autonomous weapons would seem to benefit the offense more.

With respect to first-mover advantage, if a country required a human in the loop for each targeting decision, an adversary might be able to diminish their offensive capabilities by attacking their communications links, such as by striking vulnerable satellites. If a military can fight effectively without reliable communications because of autonomous weapons, that lowers the benefit of a surprise attack against their communications. Autonomous weapons, therefore, might increase stability by reducing incentives for a first strike.

But the ability to continue attacking even if communications are severed poses a problem for escalation control and war termination. If commanders decide they wish to call off the attack, they would have no ability to recall fully autonomous weapons.

This is analogous to the Battle of New Orleans during the War of 1812. Great Britain and the United States ended the war on December 24, 1814, but news did not reach British and American forces until six weeks later. The Battle of New Orleans was fought after the treaty was signed but before news reached the front. Two thousand British sailors and soldiers died fighting a war that had technically ended.

SPEED AND CRISIS STABILITY

While the ability to carry out attacks without communications has a mixed effect on stability, autonomous weapons' advantage in speed is decidedly negative. Autonomous weapons risk accelerating the pace of battle and shortening time for human decision-making. This heightens instability in a crisis. Strategist Thomas Schelling wrote in *Arms and Influence*:

> The premium on haste—the advantage, in case of war, in being the one to launch it or in being a quick second in retaliation if the other side gets off the first blow—is undoubtedly the greatest piece of mischief that can be introduced into military forces, and the greatest source of danger that peace will explode into all out war.

Crises are rife with uncertainty and potential for miscalculation, and as Schelling explained, "when speed is critical, the victim of an accident or false-alarm is under terrible pressure." Some forms of autonomy could help to reduce these time pressures. Semiautonomous weapons that find and identify targets could be stabilizing, since they could buy more time for human decision-makers. Fully autonomous and supervised autonomous weapons short-circuit human decision-making, however, speeding up engagements. With accelerated reactions and counterreactions, humans could struggle to understand and control events. Even if everything functioned properly, policymakers could nevertheless effectively lose the ability to control escalation as the speed of action on the battlefield begins to eclipse their speed of decision-making.

REMOVING THE HUMAN FAIL-SAFE

In a fast-paced environment, autonomous weapons would remove a vital safety in preventing unwanted escalation: human judgment. Stanislav Petrov's fateful decision in bunker Serpukhov-15 represents an extreme case of the benefits of human judgment, but there are many more examples from crisis situations. Schelling wrote about the virtues of

> restraining devices for weapons, men, and decision-processes—delaying mechanisms, safety devices, double-check and consultation procedures, conservative rules for responding to alarms and communication failure, and in general both institutions and mechanisms for avoiding an unauthorized firing or a hasty reaction to untoward events.

Indeed, used in the right way, automation can provide such safeties, as in the case of automatic braking on cars. Automation increases stability when it is additive to human judgment, but not when it replaces human judgment. When autonomy accelerates decisions, it can lead to haste and unwanted escalation in a crisis.

COMMAND-AND-CONTROL AND THE PSYCHOLOGY OF CRISIS DECISION-MAKING

Stability is as much about perceptions and human psychology as it is about the weapons themselves. Two gunslingers staring each other down aren't interested only in their opponent's accuracy, but also what is in the mind of the other fighter. Machines today are woefully unequipped to perform this kind of task. Machines may outperform humans in speed and precision, but current AI cannot perform theory-of-mind tasks such as imagining another person's intent. At the tactical level of war, this may not be important. Once the gunslinger has made a decision to draw his weapon, automating the tasks of drawing, aiming, and firing would undoubtedly be faster than doing it by hand. Likewise, once humans have directed that an attack should occur, autonomous weapons may be more effective than humans in carrying out the attack. Crises, however, are periods of militarized tension between nations that have the potential to escalate into full-blown war, but where nations have not yet made the decision to escalate. Even once war begins, war among nuclear powers will by necessity be limited. In these situations, countries are attempting to communicate their resolve—their willingness to go escalate if need be—but without actually escalating the conflict. This is a delicate balancing act. Unlike a battle, which is fought for tactical or operational advantage, these situations are ultimately a form of communication between national leaders, where intentions are communicated through military actions. Michael Carl Haas of ETH Zurich argues that using autonomous weapons invites another actor into the conversation, the AI itself:

> [S]tates [who employ autonomous weapons] would be introducing into the crisis equation an element that is beyond their immediate control, but that nonetheless interacts with the human opponent's

strategic psychology. In effect, the artificial intelligence (AI) that governs the behavior of autonomous systems during their operational employment would become an additional actor participating in the crisis, though one who is tightly constrained by a set of algorithms and mission objectives.

Command-and-control refers to the ability of leaders to effectively marshal their military forces for a common goal and is a frequent concern in crises. National leaders do not have perfect control over their forces, and warfighters can and sometimes do take actions inconsistent with their national leadership's intent, whether out of ignorance, negligence, or deliberate attempts to defy authorities. The 1962 Cuban Missile Crisis was rife with such incidents. On October 26, ten days into the crisis, authorities at Vandenberg Air Force Base carried out a scheduled test launch of an Atlas ICBM without first checking with the White House. The next morning, on October 27, an American U-2 surveillance plane was shot down while flying over Cuba, despite orders by Soviet Premier Nikita Khrushchev not to fire on U.S. surveillance aircraft. (The missile appears to have been fired by Cuban forces on Fidel Castro's orders.) Later that same day, another U-2 flying over the Arctic Circle accidentally strayed into Soviet territory. Soviet and American leaders could not know for certain whether these incidents were intentional signals by the adversary to escalate or individual units acting on their own. Events like these have the potential to ratchet up tensions through inadvertent or accidental escalation.

In theory, autonomous weapons ought to be the perfect soldier, carrying out orders precisely, without any deviation. This might eliminate some incidents. For example, on October 24, 1962, when U.S. Strategic Air Command (SAC) was ordered to DEFCON 2, just one step short of nuclear war, SAC commander General Thomas Power deviated from his orders by openly broadcasting a message to his troops on an unencrypted radio channel. The unencrypted broadcast revealed heightened U.S. readiness levels to the Soviets, who could listen in, and was not authorized. Unlike people, autonomous weapons would be incapable of violating their programming. On the other hand, their brittleness and inability to understand the context for their actions would be a major liability in other ways.

The Vandenberg IBCM test, for example, was caused by officers following preestablished guidance without pausing to ask whether that guidance still applied in light of new information (the unfolding crisis over Cuba).

Often, the correct decision in any given moment depends not on rigid adherence to guidance, but rather on understanding the intent behind the guidance. Militaries have a concept of "commander's intent," a succinct statement given by commanders to subordinates describing the commander's goals. Sometimes, meeting the commander's intent requires deviating from the plan because of new facts on the ground. Humans are not perfect, but they are capable of using their common sense and better judgment to comply with the intent behind a rule, rather than the rule itself. Humans can disobey the rules and in tense situations, counterintuitively, that may be a good thing.

At the heart of the matter is whether more flexibility in how subordinates carry out directions is a good thing or a bad thing. On the battlefield, greater flexibility is generally preferred, within broad bounds of the law and rules of engagement. In "Communicating Intent and Imparting Presence," Lieutenant Colonel Lawrence Shattuck wrote:

> If . . . the enemy commander has 10 possible courses of action, but the friendly commander, restricted by the senior commander, still has only one course of action available, the enemy clearly has the advantage.

In crises, tighter control over one's forces is generally preferred, since even small actions can have strategic consequences. Zero flexibility for subordinates with no opportunity to exercise common sense is a sure invitation to disaster, however. Partly, this is because national leaders cannot possibly foresee all eventualities. War is characterized by uncertainty. Unanticipated circumstances will arise. War is also competitive. The enemy will almost certainly attempt to exploit rigid behavioral rules for their own advantage. Michael Carl Haas suggests these tactics might include:

> relocating important assets to busy urban settings or next to inadmissible targets, such as hydroelectric dams or nuclear-power stations;

altering the appearance of weapons and installations to simulate illegitimate targets, and perhaps even the alteration of illegitimate targets to simulate legitimate ones; large-scale use of dummies and obscurants, and the full panoply of electronic deception measures.

Even without direct hacking, autonomous weapons could be manipulated by exploiting vulnerabilities in their rules of engagement. Humans might be able to see these ruses or deceptions for what they are and innovate on the fly, in accordance with their understanding of commander's intent. Autonomous weapons would follow their programming. On a purely tactical level, other benefits to autonomous weapons may outweigh this vulnerability, but in crises, when a single misplaced shot could ratchet up tensions toward war, careful judgment is needed.

In recent years, the U.S. military has begun to worry about the problem of the "strategic corporal." The basic idea is that a relatively low-ranking individual could, through his or her actions on the battlefield, have strategic effects that determine the course of the war. The solution to this problem is to better educate junior leaders on the strategic consequences of their actions in order to improve their decision-making, rather than giving them a strict set of rules to follow. Any set of rules followed blindly and without regard to the commander's intent can be manipulated by a clever enemy. Autonomous weapons would do precisely what they are told, regardless of how dumb or ill-conceived the orders appear in the moment. Their rigidity might seem appealing from a command-and-control standpoint, but the result is the strategic corporal problem on steroids.

There is precedent for concerns about the strategic consequences of automation. During development of the Reagan-era "Star Wars" missile defense shield, officially called the Strategic Defense Initiative, U.S. lawmakers wrote a provision into the 1988–1989 National Defense Authorization Act mandating a human in the loop for certain actions. The law requires "affirmative human decision at an appropriate level of authority" for any systems that would intercept missiles in the early phases of their ascent. Intercepts at these early stages can be problematic because they must occur on very short timelines and near an adversary's territory. An automated system could conceivably mistake a satellite launch or missile test for an attack and, by destroying another country's rocket, needlessly escalate a crisis.

Even if mistakes could be avoided, there is a deeper problem with leaders attempting to increase their command-and-control in crises by directly programming engagement rules into autonomous weapons: leaders themselves may not be able to accurately predict what decisions they would want to take in the future. "Projection bias" is a cognitive bias where humans incorrectly project their current beliefs and desires onto others and even their future selves.

To better understand what this might mean for autonomous weapons, I reached out to David Danks, a professor of philosophy and psychology at Carnegie Mellon University. Danks studies both cognitive science and machine learning, so he understands the benefits and drawbacks to human and machine cognition. Danks explained that projection bias is "a very real problem" for autonomous weapons. Even if we could ensure that the autonomous weapon would flawlessly carry out political leaders' directions, with no malfunctions or manipulation by the enemy, "you still have the problem that that's a snapshot of the preferences and desires at that moment in time," he said. Danks explained that people generally do a good job of predicting their own future preferences for situations they have experience with, but for "a completely novel situation . . . there's real risks that we're going to have pretty significant projection biases."

Again, the Cuban Missile Crisis illustrates the problem. Robert McNamara, who was secretary of defense at the time, later explained that the president's senior advisors believed that if the U-2 they sent to fly over Cuba were shot down, it would have signaled a deliberate move by the Soviets to escalate. They had decided ahead of time, therefore, that if the U-2 was shot down, the United States would attack:

> [B]efore we sent the U-2 out, we agreed that if it was shot down we
> wouldn't meet, we'd simply attack. It was shot down on Friday. . . .
> Fortunately, we changed our mind, we thought "Well, it might have
> been an accident, we won't attack."

When actually faced with the decision, it turns out that McNamara and others had a different view. They were unable to accurately predict their *own* preferences as to what they would want to do if the plane were shot down. In that example, McNamara and others could reverse course. They had not actually delegated the authority to attack. There was another moment during the Cuban Missile Crisis, however, when Soviet leadership had delegated release authority for nuclear weapons and the world came chillingly close to nuclear war.

On October 27, the same day that the U-2 was shot down over Cuba and another U-2 flying over the Arctic strayed into Soviet territory, U.S. ships at the quarantine (blockade) line began dropping signaling depth charges on the Soviet submarine *B-59* to compel it to surface. The U.S. Navy was not aware that the *B-59* was armed with a nuclear-tipped torpedo with a 15-kiloton warhead, about the size of the bomb dropped on Hiroshima. Furthermore, Soviet command had delegated authority to the ship's captain to use the torpedo if the ship was "hulled" (a hole blown in the hull from depth charges). Normally, authorization was required from two people to fire a nuclear torpedo: the ship's captain and political officer. According to Soviet sailors aboard the submarine, the submarine's captain, Valentin Savitsky, ordered the nuclear torpedo prepared for launch, declaring, "We're going to blast them now! We will die, but we will sink them all." Fortunately, the flotilla commander, Captain Vasili Arkhipov, was also present on the submarine. He was Captain Savitsky's superior and his approval was also required. Reportedly, only Arkhipov was opposed to launching the torpedo. As with Stanislav Petrov, once again the judgment of a single Soviet officer may have again prevented the outbreak of nuclear war.

DETERRENCE AND THE DEAD HAND

Sometimes, there is a benefit to tying one's hands in a crisis. Strategists have often compared crises to a game of chicken between two drivers, both hurtling toward the other one at deadly speed, daring the other to swerve. Neither side wants a collision, but neither wants to be the first to swerve. One way to win is to demonstrably tie one's hands so that one cannot swerve. Herman Kahn gave the example of a driver who "takes the steering wheel and throws it out the window." The onus is now entirely on the other driver to avoid a collision.

Horowitz asked whether autonomous weapons might excel in such a situation. Here, machines' rigid adherence to rules and lack of recallability would be a benefit. A robot designed to never swerve would be the perfect driver to win at chicken. In "Artificial Intelligence, War, and Crisis Stability," Horowitz presented the thought experiment of an alternative Cuban Missile Crisis in which the U.S. ships conducting the blockade were autonomous weapons. They would be programmed to fire on any Soviet ships crossing the blockade line. If this could be credibly communicated to the Soviets, it would have put the onus of avoiding conflict on the Soviets. The problem, Horowitz asked, was "how would the Kennedy Administration have persuaded the Soviet Union that that was the case?" There would be no way to convincingly prove to Soviet leadership that the robotic vessels were actually programmed to fire. U.S. leaders could claim that was the case, but the claim would be meaningless, since that's also what they would say if they were bluffing. The United States would certainly not allow the Soviets to inspect the code of U.S. ships at the blockade. There would be no credible way to demonstrate that one had, in fact, tied one's hands. It would be the equivalent of ripping out the steering wheel, but being unable to throw it out the window. (Similarly, the Soviets could program their ships to run the blockade without any option for turning back, but there would be no way to prove to the Americans they had done so.)

Stanley Kubrick's 1964 film *Dr. Strangelove* explores the bizarre logic of deterrence and mutual assured destruction. In the film, the Soviet ambassador explains to an assembled group of American military and political leaders that the Soviet Union has built a "doomsday machine" which, if the Soviet Union is attacked, will automatically launch a massive nuclear counterattack that will destroy humanity. The title character Dr. Strangelove explains, "because of the automated and irrevocable decision-making process, which rules out human meddling, the doomsday machine is terrifying . ... and completely capable and convincing." Unfortunately, in the movie, the Soviets fail to tell their American counterparts that they have built such a device. Strangelove yells at the Soviet ambassador, "the whole point of the doomsday machine is lost if you keep it a secret!"

As an example of truth being sometimes stranger than fiction, after the Cold War evidence emerged that the Soviets did in fact build a semi-automatic doomsday device, nicknamed "Dead Hand." Officially called "Perimeter," the system was reportedly an automated nuclear command-and-control system designed to allow a massive retaliatory attack even if a U.S. first strike took out Soviet leadership. Accounts of Perimeter's functionality differ, but the essential idea was that the system would remain inactive during peacetime but, in the event of a crisis, it could be activated as a "fail-deadly" mechanism for ensuring retaliation. When active, a network of light, radiation, seismic, and pressure sensors would evaluate whether there had been any nuclear detonations on Soviet soil. If a nuclear detonation was detected, then the system would check for communications to the General Staff of the Soviet military. If communications were active, then it would wait a predetermined about of time, ranging from on the order of fifteen minutes to an hour, for an order to cancel the launch. If there was no order to stop the launch, Perimeter would act like a "dead man's switch," a switch that is automatically triggered if a person becomes incapacitated or dies. In many hazardous machines, a dead man's switch is used as a fail-safe mechanism. If a person becomes incapacitated, the machine will revert to a safe mode of operation, like a lawnmower shutting off if you release the handle. In this case, Perimeter was intended to "fail deadly." If there was no signal from the Soviet General Staff to halt the launch, Perimeter would bypass normal layers of command and transfer launch authority directly to individuals within a deep underground protected bunker. There would still be a human in the loop, but the decision would reside with whichever staff officer was on duty at the time. Soviet leadership would be cut out of the loop. With a push of a button, that individual could launch a series of communications rockets that would fly over Soviet territory and beam down the nuclear launch codes to missiles in hardened silos. Soviet ICBMs would then launch a massive strike on the United States, the last zombie attack of a dying nation.

There was a purpose to the madness. In theory, if everything worked properly, a system like Perimeter would enhance stability. Because a retaliatory strike would be assured, the system would remove the need for haste from Soviet leaders' decision-making in a crisis. If there were warnings of an incoming U.S. surprise attack, as was the case in 1983 in the Stanislav Petrov incident and again in 1995 when Russian military leaders brought the nuclear suitcase to Boris Yeltsin in response to a Norwegian scientific rocket launch, there would be no rush to respond. Soviet or Russian leaders would have no incentive to fire their nuclear missiles in an ambiguous situation, because even if the United States succeeded in a decapitating strike, retaliation was assured. The knowledge of this would also presumably deter the United States from even considering a preemptive first strike. The problem, of course, is that such a system comes with tremendous risks. If Perimeter were to falsely detect an event, as the Soviet Oko satellite system did in 1983 when it falsely detected U.S. ICBM launches, or if Soviet leaders were unable to stop the mechanism once it was activated, the system would obliterate humanity.

By some accounts, Perimeter is still operational within Russia today.

STABILITY-INSTABILITY PARADOX AND THE MAD ROBOT THEORY

The logic of mutual assured destruction (MAD) is to make any nuclear attack inherently suicidal. If a retaliatory response is assured, then attacking the enemy is akin to attacking oneself. This dynamic is stabilizing in the sense that it deters both sides from using nuclear weapons. Ironically, though, over time strategists began to worry that too much stability was a bad thing. This became known as the "stability-instability paradox."

The essence of the problem is that if nuclear weapons are fully and mutually restrained, then they could lose their value as a deterrent. This could embolden aggression below the nuclear threshold, since countries could be confident that an adversary would not respond with nuclear weapons. Under this logic, some instability—some risk of accidents and miscalculation—is a good thing, because it induces caution. Returning to the gunslingers in their standoff, if stabilizing measures are those that make it less likely that a gunslinger will draw his weapon, too much stability might encourage other forms of aggression. One might be willing to insult or even steal from the other gunslinger, confident that he wouldn't draw his gun, since doing so would be suicidal.

One response to this paradox is the "madman theory." As the acronym MAD implies, the logic of mutual assured destruction is fundamentally insane. Only a mad person would launch a nuclear weapon. The principle behind the madman theory, espoused by President Richard Nixon, is to convince the enemy's leadership that a nation's leaders are so volatile and irrational that they just might push the button. Mutual suicide or no, one would hesitate to insult a gunslinger with a reputation for rash, even self-destructive acts.

This suggests another way autonomous weapons might improve stability: the "mad robot theory." If countries perceive autonomous weapons as dangerous, as introducing an unpredictable element into a crisis that cannot be completely controlled, then introducing them into a crisis might induce caution. It would be the equivalent of what Thomas Schelling has described as "the threat that leaves something to chance." By deploying autonomous weapons into a tense environment, a country would effectively be saying to the enemy, "Things are now out of my hands. Circumstances may lead to war; they may not. I cannot control it, and your only course of action if you wish to avoid war is to back down." Unlike the problem of credibly tying one's hands by locking in escalatory rules of engagement, this threat does not require convincing the enemy what the autonomous weapons' rules of engagement are. In fact, uncertainty makes the "mad robot" threat more credible, since deterrence hinges on the robot's unpredictability, rather than the certainty of its actions. Deploying an untested and unverified autonomous weapon would be even more of a deterrent, since one could convincingly say that its behavior was truly unpredictable.

What is interesting about this idea is that its efficacy rests solely on humans' perception of autonomous weapons, and not the actual functionality of the weapons themselves. The weapons may be reliable or unreliable—it doesn't matter. What matters is that they are *perceived* as unpredictable and, as a result, induce caution. Of course, the actual functionality of the weapons does matter when it comes to how a crisis unfolds. The key is the difference between how humans perceive the risk of autonomous weapons and their actual risk. If leaders overestimate the risks of autonomous weapons, then there is nothing to worry about. Their introduction into crises will induce caution but they will be unlikely to cause harm. If leaders underestimate their risks, then their use invites disaster.

How accurately people can assess the risks of autonomous weapons hinges on individual psychology and how organizations evaluate risk in complex systems. I asked David Danks what he thought about peoples' ability to accurately assess these risks, and his answer was not encouraging. "There's a real problem here for autonomous weapons," he said. For starters, he explained that people are poor predictors of behavior in systems that have feedback loops, where one action creates a counterreaction and so on. (Real-world experience with complex autonomous systems in uncontrolled environments, such as stock trading, lends weight to this theory.)

Furthermore, Danks said, due to projection bias, people are poor predictors of risk for situations for which they have no experience. For example, Danks explained, people are good estimators of their likelihood of getting into an automobile accident because they frequently ride in vehicles. But when they have no prior knowledge, then their ability to accurately assess risks falls apart. "Autonomous weapon systems are very new. They aren't just a bigger gun," he said. "If you think of them as a bigger gun, you say, 'Well we've got a lot of experience with guns.'" That might lead one to think that one could accurately evaluate the risks of autonomous weapons. But Danks said he thought they were "qualitatively different" than other weapons.

This suggests we lack the necessary experience to accurately assess the risks of autonomous weapons. How much testing is required to ensure an autonomous weapon fails less than 0.0001 percent of the time? We don't know, and we can't know until we build up experience with more sophisticated autonomous systems over time. Danks said it would be different if we already had extensive experience with safely operating complex autonomous systems in real-world environments. Unfortunately, the experience we do have suggests that surprises are often lurking below the surface of complex systems. Danks concluded that "it's just completely unreasonable and hopelessly optimistic to think that we would be good at estimating the risks."

UNTYING THE KNOT

At the height of the Cuban Missile Crisis, Soviet Premier Nikita Khrushchev sent an impassioned letter to President Kennedy calling on them to work together to step back from the brink of nuclear war:

> Mr. President, we and you ought not now to pull on the ends of the rope in which you have tied the knot of war, because the more the two of us pull, the tighter that knot will be tied. And a moment may come when that knot will be tied so tight that even he who tied it will not have the strength to untie it, and then it will be necessary to cut that knot, and what that would mean is not for me to explain to you, because you yourself understand perfectly of what terrible forces our countries dispose.

Autonomous weapons raise troubling concerns for stability and escalation control in crises. Michael Carl Haas concluded, "there are scenarios in which the introduction of autonomous strike systems could result in temporary loss of high-level control over operations, and unwanted escalation (conventional or nuclear)." He argued policymakers "should exercise prudence and caution" before adding autonomous weapon systems "into an equation that is highly complex as it stands." Their rigid rule-following could tighten the knot, with no understanding of the context for or consequences of their actions.

During the Cuban Missile Crisis, U.S. leaders were constantly trying to understand the psychology of their Soviet counterparts. While they had differing interests, President Kennedy empathized with Premier Khrushchev's position. Kennedy understood that if the United States moved against Cuba, the Soviets would be compelled to respond elsewhere in the world, perhaps in Berlin. Kennedy understood that he needed to give Khrushchev an option to remove the missiles from Cuba while saving face. Kennedy and others were able to think through the second- and third-order consequences of their actions. (Khrushchev eventually agreed to remove the missiles in exchange for a U.S. pledge not to invade Cuba and a secret promise to remove American missiles from

Turkey.) Vasili Arkhipov, on the Soviet submarine *B-59,* similarly understood that if they fired a nuclear torpedo, obliterating a U.S. aircraft carrier, the Americans would feel compelled to respond with nuclear weapons elsewhere. The result would be escalating to a level it might be impossible to back down from.

Humans are not perfect, but they can empathize with their opponents and see the bigger picture. Unlike humans, autonomous weapons would have no ability to understand the consequences of their actions, no ability to step back from the brink of war. Autonomous weapons would not take away all human decision-making in crises, but they do have the potential to tighten the knot, perhaps so far that it cannot be undone.

PART VI

Averting Armageddon: The Weapon of Policy

19

CENTAUR WARFIGHTERS

HUMANS + MACHINES

If there is one common theme across the legal, ethical, and strategic issues surrounding autonomous weapons, it is whether, and how much, the decision to use force depends on context. Machines, at least for the foreseeable future, will not be as good as humans at understanding the context for their actions. Yet there are circumstances in which machines perform far better than humans. The best decision-making system would be one that leverages the advantages of each. Hybrid human-machine cognitive systems, often called "centaur warfighters" after the classic Greek myth of the half-human, half-horse creature, can leverage the precision and reliability of automation without sacrificing the robustness and flexibility of human intelligence.

THE CENTAUR EDGE

To glimpse the future of cognition, we need look no further than one of the most high-profile areas in which AI has bested humans: chess. In 1997, IBM's Deep Blue defeated world chess champion Gary Kasparov, cementing the reality that humans are no longer the best chess players in the world. But neither, as it turns out, are machines. A year later, Kasparov founded the field of "advanced chess," also called centaur chess, in which humans and AI cooperate on the same team. The AI can analyze possible

moves and identify vulnerabilities or opportunities the human player
might have missed, resulting in blunder-free games. The human player
can manage strategy, prune AI searches to focus on the most promising
areas, and manage differences between multiple AIs. The AI system gives
feedback to the human player, who then decides what move to make. By
leveraging the advantages of human *and* machine, centaur chess results in
a better game than either humans or AI can achieve on their own.

Understanding how this might work for weapons engagements requires
first disaggregating the various roles a human performs today in targeting decisions: (1) acting as an essential operator of the weapon system; (2)
acting as a fail-safe; and (3) acting as a moral agent.

When acting as an "essential operator" of the weapon system, a human
performs a vital function without which the weapon cannot work. A
human "painting" a target with a laser to direct a laser-guided bomb onto
the target is acting as an essential operator. When a human acts as a fail-
safe, the weapon could function on its own, but the human is in the loop
as a backup to intervene if it fails or if circumstances change such that the
engagement is no longer appropriate. When acting as a "moral agent," the
human is making value-based judgments about whether the use of force
is appropriate.

An anecdote from the U.S. air campaign over Kosovo in 1999 includes
an instructive example of all three roles in action simultaneously:

> On 17 April 1999, two F-15E Strike Eagles, Callsign CUDA 91 and 92,
> were tasked to attack an AN/TPS-63 mobile early warning radar
> located in Serbia. The aircraft carried AGM-130, a standoff weapon
> that is actually remotely flown by the weapons system officer (WSO)
> in the F-15E, who uses the infra-red sensor in the nose of the weapon
> to detect the target. CUDA 91 . . . launched on coordinates provided by
> the Air Operations Center. As the weapon approached the suspected
> target location, the crew had not yet acquired the [enemy radar]. At 12
> seconds from impact, the picture became clearer. . . . [The pilots saw
> the profile outline of what appeared to be a church steeple.] Three
> seconds [from impact], the WSO makes the call: "I'm ditching in
> this field" and steers the weapon into an empty field several hundred
> meters away. . . . Postflight review of the tape revealed no object that

could be positively identified as a radar, but the profile of a Serbian
Orthodox church was unmistakable.

In this example, the pilots were performing all three roles simultaneously. By manually guiding the air-to-ground weapon they were acting
as essential operators. They were also acting as fail-safes, observing the
weapon while it was in flight and making an on the spot decision to abort
once they realized the circumstances were different from what they had
anticipated. They were also acting as moral agents when they assessed
that the military necessity of the target was not important enough to risk
blowing up a church.

Acting as the essential operator, which is traditionally the primary
human role, is actually the easiest role to automate. A network-enabled
GPS bomb, for example, gives operators the ability to abort in flight,
preserving their role as moral agents and fail-safes, but the weapon can
maneuver itself to the target. We see this often in nonmilitary settings.
Commercial airliners use automation to perform the essential task of
flying the aircraft, with human pilots largely in a fail-safe role. A person
on medical life support has machines performing the essential task of
keeping him or her alive, but humans make the moral judgment whether
to continue life support. As automation becomes more advanced, automating many of the weapon system's functions could result in far greater
accuracy, precision, and reliability than relying on humans. Automating the human's role as moral agent or fail-safe, however, is far harder
and would require major leaps forward in AI that do not appear on the
horizon.

THE ROLE OF THE HUMAN AS MORAL AGENT
AND FAIL-SAFE

The benefit to "centaur" human-machine teaming is that we don't need to
give up the benefits of human judgment to get the advantages of automation. We can have our cake and eat it too (at least in some cases). The U.S.
counter-rocket, artillery, and mortar (C-RAM) system is an example of
this approach. The C-RAM automates much of the engagement, resulting
in greater precision and accuracy, but still keeps a human in the loop.

The C-RAM is designed to protect U.S. bases from rocket, artillery, and mortar attacks. It uses a network of radars to automatically identify and track incoming rounds. Because the C-RAM is frequently used at bases where there are friendly aircraft in the sky, the system autonomously creates a "Do Not Engage Sector" around friendly aircraft to prevent fratricide. The result is a highly automated system that, in theory, is capable of safely and lawfully completing engagements entirely on its own. However, humans still perform final verification of each individual target before engagement. One C-RAM operator described the role the automation and human operators play:

> The human operators do not aim or execute any sort of direct control over the firing of the C-RAM system. The role of the human operators is to act as a final fail-safe in the process by verifying that the target is in fact a rocket or mortar, and that there are no friendly aircraft in the engagement zone. A [h]uman operator just presses the button that gives the authorization to the weapon to track, target, and destroy the incoming projectile.

Thus, the C-RAM has a dual-safety mechanism, with both human and automated safeties. The automated safety tracks friendly aircraft in the sky with greater precision and reliability than human operators could, while the human can react to unforeseen circumstances. This model also has the virtue of ensuring that human operators must take a positive action before each engagement, helping to ensure human responsibility for each shot.

In principle, C-RAM's blended use of automation and human decision-making is optimal. The human may not be able to prevent all accidents from occurring (after all, humans make mistakes), but the inclusion of a human in the loop dramatically reduces the potential for multiple erroneous engagements. If the system fails, the human can at least halt the weapon system from further operation, while the automation itself may not understand that it is engaging the wrong targets.

In order for human operators to actually perform the roles of moral agent and fail-safe, they must be trained for and supported by a culture of active participation in the weapon system's operation. The Patriot fratricides stemmed from "unwarranted and uncritical trust in automation," Ensuring human responsibility over engagements requires: automation designed so that human operators can program their intent into the machine, human-machine interfaces that give humans the information they need to make informed decisions, training that requires the operators to exercise judgment, and a culture that emphasizes human responsibility. When these best practices are followed, the result can be safe and effective systems like C-RAM, where automation provides valuable advantages but humans remain in control.

THE LIMITS OF CENTAUR WARFIGHTING: SPEED

The idealized centaur model of human-machine teaming breaks down, however, when actions are required faster than humans can react or when communications are denied between the human and machine.

Chess is again a useful analogy. Centaur human-machine teams generally make better decisions in chess, but are not an optimal model in timed games where a player has only thirty to sixty seconds to make a move. When the time to decide is compressed, the human does not add any value compared to the computer alone, and may even be harmful by introducing errors. Over time, as computers advance, this time horizon is likely to expand until humans no longer add any value, regardless of how much time is allowed.

Already, machines do a better job than humans alone in certain military situations. Machines are needed to defend against saturation attacks from missiles and rockets when the speed of engagements overwhelms human operators. Over time, as missiles incorporate more intelligent features, including swarming behavior, these defensive supervised autonomous weapons are likely to become even more important—and human involvement will necessarily decline.

By definition, a human *on* the loop has weaker control than a human *in* the loop. If the weapon fails, there is a greater risk of harm and of lessened moral responsibility. Nevertheless, human supervision provides some oversight of engagements. The fact that supervised autonomous weapons such as Aegis have been in widespread use for decades suggests that these

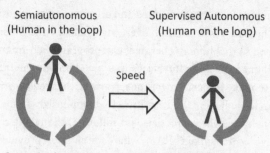

Semiautonomous
(Human in the loop)

Supervised Autonomous
(Human on the loop)

Speed

Supervised autonomy is necessary when the speed of
action is too fast for a human to remain in the loop.

Why Use Supervised Autonomy?

risks are manageable. In all of these situations, as an additional backup, humans have physical access to the weapon system so that they could disable it at the hardware level. Accidents have occurred with existing systems, but not catastrophes. A world with more defensive supervised autonomous weapons is likely to look not much different than today.

There will undoubtedly be offensive settings where speed is also valuable. In those cases, however, speed will be valuable in the execution of attacks, not necessarily in the decision to launch them. For example, swarming missiles will need to be delegated the authority to coordinate their behavior and deconflict targets, particularly if the enemy is another swarm. Humans have more choice over the time and place of attack for offensive applications, though. For some types of targets, it may not be feasible to have humans select every individual enemy object. This will especially be the case if militaries move to swarm warfare, with hundreds or thousands of robots involved. But there are weapon systems today—Sensor Fuzed Weapon and the Brimstone missile, for example— where humans choose a specific group of enemy targets and the weapons divvy up the targets themselves. A human selecting a known group of targets minimizes many of the concerns surrounding autonomous weapons while allowing the human to authorize an attack on the swarm as a whole, without having to specify each individual element.

DEGRADED COMMUNICATIONS

Human supervision is not possible when there are no communications with the weapon, such as in enemy air space or underwater. But communications in contested areas is not an all-or-nothing proposition. Communications may be degraded but not necessarily denied. Advanced militaries have jam-resistant communications technology. Or perhaps a human in a nearby vehicle has some connection with an autonomous weapon to authorize engagements. In any event, some communication is likely possible. So how much bandwidth is required to keep a human in the loop?

Not much. As one example, consider the following screen grab from a video of an F-15 strike in Iraq, a mere 12 kilobytes in size. While grainy, it clearly possesses sufficient resolution to distinguish individual vehicles. A trained operator could discriminate military-specific vehicles, such as a tank or mobile missile launcher, from dual-use vehicles such as buses or trucks.

DARPA's CODE program intends to keep a human in the loop via a communications link that could transmit 50 kilobits per second, roughly on par with a 56K modem from the 1990s. This low-bandwidth communi-

Targeting Image from F-15 Strike in Iraq (12 kilobytes in size) *This targeting image from an F-15 strike in Iraq shows a convoy of vehicles approaching an intersection. Images like this one could be passed over relatively low-bandwidth networks for human operators to approve engagements.*

cations link could transmit one image of this quality every other second. (One kilobyte equals eight kilobits, so a 12-kilobyte image is 96 kilobits.) It would allow a human to view the target and decide within a few seconds whether to authorize an engagement or not.

This reduced-bandwidth approach would not work in areas where communications are entirely denied. In such environments, semiautonomous weapons could engage targets that had been preauthorized by human controllers, as cruise missiles do today. This would generally only be practical for fixed targets, however (or a mobile target in a confined area with a readily identifiable signature). In these cases, accountability and responsibility would be clear, as a human would have made the targeting decision.

But things get complicated quickly in communications-denied environments.

Should uninhabited vehicles be able to defend themselves if they come under attack? Future militaries will likely deploy robotic vehicles and will want to defend them, especially if they are expensive. If there were no communications to a human, any defenses would need to be fully autonomous. Allowing autonomous self-defense incurs some risks. For example, someone could fire at the robot to get it to return fire and then hide behind human shields to deliberately cause an incident where the robot kills civilians. There would also be some risk of fratricide or unintended escalation in a crisis. Even rules of engagement (ROE) intended purely to be defensive could lead to interactions between opposing systems that results in an exchange of fire. Delegating self-defense authority would be risky. However, it is hard to imagine that militaries would be willing to put expensive uninhabited systems in harm's way and leave them defenseless. Provided the defensive action was limited and proportionate, the risks might be manageable.

While it seems unlikely that militaries would publicly disclose the specific ROE their robotic systems use, some degree of transparency between nations could help manage the risks of crisis escalation. A "rules of the road" for how robotic systems ought to behave in contested areas might help to minimize the risk of accidents and improve stability overall. Some rules, such as, "if you shoot at a robot, expect it to shoot back," would be self-reinforcing. Combined with a generally cautious "shoot second" rule

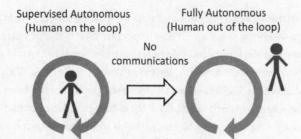

Full autonomy is necessary when there is no
communications between the human and the machine.

Why Use Full Autonomy?

requiring robots to withhold fire unless fired upon, such an approach is likely to be stabilizing overall. If militaries could agree on a set of guidelines for how they expect armed robotic systems to interact in settings where there is no human oversight, this would greatly help to manage a problem that is sure to surface as more nations field weaponized robotic systems.

Hunting mobile targets in communications-denied areas presents the greatest challenge for maintaining human supervision. Ships, air defense systems, and missile launchers are all harder to hit precisely because their movement makes it difficult to find them. In an ideal world, a swarm of robotic systems would search for these targets, relay the coordinates and a picture back to a human controller for approval (as CODE intends), then the swarm would attack only human-authorized targets. If a communication link is not available, however, then fully autonomous weapons could be used to search for, select, and mobile attack targets on their own.

There is no doubt that such weapons would be militarily useful. They would also be risky. In these situations, there would be no ability to recall or abort the weapon if it failed, was hacked, or was manipulated into attacking the wrong target. Unlike a defensive counterfire response, the weapon's actions would not be limited and proportionate. It would be going on the attack, searching for targets. Given the risks that such weapons would entail, it is worth asking whether their military value would be worth the risk.

When I asked Captain Galluch from the Aegis training center what he thought of the idea of a fully autonomous weapon, he asked, "What application are we trying to solve?" It's an important question. For years, militaries have had the ability to build loitering munitions that would search for targets over a wide area and destroy them on their own. With a few exceptions like the TASM and Harpy, these weapons have not been developed. There are no known examples of them being used in a conflict. Fully autonomous weapons might be useful, but it's hard to make the case for them as necessary outside of the narrow case of immediate self-defense.

The main rationale for building fully autonomous weapons seems to be the assumption that others might do so. Even the most strident supporters of military robotics have been hesitant about fully autonomous weapons ... unless others build them. This is a valid problem—and one that could become a self-fulfilling prophecy. The fear that others might build autonomous weapons could be the very thing that drives militaries to build them. Jody Williams asked me, "If they don't exist, there is no military necessity and are we not, in fact, creating it?"

Michael Carl Haas, who raised concerns about crisis stability, suggested that countries explore "mutual restraint" as an option to avoid the potentially dangerous consequences of fully autonomous weapons. Others have suggested that such weapons are inevitable. The history of arms control provides evidence for both points of view.

THE POPE AND THE CROSSBOW

THE MIXED HISTORY OF ARMS CONTROL

In the summer of 2015, a group of prominent AI and robotics researchers signed an open letter calling for a ban on autonomous weapons.
"The key question for humanity today," they wrote, "is whether to
start a global AI arms race or to prevent it from starting. If any major
military power pushes ahead with AI weapon development, a global arms
race is virtually inevitable."

There have been many attempts to control weapons in the past. Some
have succeeded, but many attempts at restricting weapons have failed.
Pope Innocent II banned the use of the crossbow (against Christians)
in 1139. There is no evidence that it had any effect in slowing the proliferation of the crossbow across medieval Europe. In the early twentieth
century, European nations tried to cooperate on rules restricting submarine warfare and banning air attacks on cities. These attempts failed.
On the other hand, attempts to restrain chemical weapons use failed in
World War I but succeeded in World War II. All the major powers had
chemical weapons in World War II but did not use them (on each other).
Today, chemical weapons are widely reviled, although their continued
use by Bashar al-Assad in Syria shows that no ban is absolute. The Cold
War saw a host of arms control treaties, many of which remain in place
today. Some treaties, such as bans on biological weapons, blinding lasers,
and using the environment as a weapon of war, have been highly successful. In recent years, humanitarian campaigns have led to bans on land

Types of Weapons Bans *Weapons bans can target different stages of the weapons production process, preventing access to the technology, prohibiting states from developing the weapon, limiting production, or regulating use.*

mines and cluster munitions, although the treaties have not been as widely adopted and these weapons remain in use by many states. Finally, nonproliferation treaties have been able to slow, but not entirely stop, the proliferation of nuclear weapons, ballistic missiles, and other dangerous technologies.

These successes and failures provide lessons for those who wish to control autonomous weapons. The underlying technology that enables autonomous weapons is too diffuse, commercially available, and easy to replicate to stop its proliferation. Mutual restraint among nations on how they use this technology may be possible, but it certainly won't be easy.

WHY SOME BANS SUCCEED AND OTHERS FAIL

Whether or not a ban succeeds seems to depend on three key factors: the perceived horribleness of the weapon; its perceived military utility; and the number of actors who need to cooperate for a ban to work. If a weapon is seen as horrific and only marginally useful, then a ban is likely to succeed. If a weapon brings decisive advantages on the battlefield then a ban is unlikely to work, no matter how terrible it may seem. The difference between how states have treated chemical weapons and nuclear weapons illustrate this point. Nuclear weapons are unquestionably more harmful than chemical weapons by any measure: civilian casualties, combatant suffering, and environmental damage. Nuclear weapons give

Successful and Unsuccessful Weapons Bans

Era	Weapon	Year	Regulation or Treaty	Legally binding?	Type of Regulation	Successful?	Motivation
PRE-MODERN ERA	poisoned or barbed arrows	Dates vary - 1500 to 200 BC	Laws of Manu; Dharmaśāstras; Mahābhārata	legally binding	banned use	success unknown	unnecessary suffering
	concealed weapons	Dates vary - 1500 to 200 BC	Laws of Manu	legally binding	banned use	success unknown	perfidy
	fire-tipped weapons	Dates vary - 1500 to 200 BC	Laws of Manu	legally binding	banned use	success unknown	unnecessary suffering
	crossbow	1097; 1139	1097 Lateran Synod; 1139 Second Lateran Council	legally binding	banned use	failed	political control
	firearms	1607–1867	Tokugawa Shogunate Japan	legally binding	effectively prohibited production	successful (lasted ~250 years)	political control
	firearms	1523–1543	King Henry VIII	legally binding	limited ownership among civilian population	short-lived	political control

continued

Era	Weapon	Year	Regulation or Treaty	Legally binding?	Type of Regulation	Successful?	Motivation
TURN OF THE CENTURY	explosive or inflammable projectiles below 400 grams	1868	1868 St. Petersburg Declaration	legally binding	banned use	superseded by technology, but adhered to in spirit	unnecessary suffering
	expanding bullets	1899	1899 Hague Declaration	legally binding	banned use	successful in limiting battlefield use, although lawful in civilian applications	unnecessary suffering
	asphyxiating gases (from projectiles)	1899	1899 Hague Declaration	legally binding	banned use	failed - used in WW I	unnecessary suffering
	poison	1899; 1907	1899 and 1907 Hague Declarations	legally binding	banned use	successful	unnecessary suffering
	weapons that cause superfluous injury	1899; 1907	1899 and 1907 Hague Declarations	legally binding	banned use	mixed, but generally successful	unnecessary suffering
	balloon-delivered projectiles or explosives	1899; 1907	1899 and 1907 Hague Declarations	legally binding	banned use	short-lived	civilian casualties
	aerial bombardment against undefended cities	1907	1907 Hague Declaration	legally binding	banned use	failed	civilian casualties

Era	Weapon	Year	Regulation or Treaty	Legally binding?	Type of Regulation	Successful?	Motivation
WORLD WAR I TO WORLD WAR II	sawback bayonets	World War I	tacit cooperation on the battlefield	no explicit agreement	norm against possession	successful	unnecessary suffering
	chemical and bacteriological weapons	1925	1925 Geneva Gas and Bacteriological Protocol	legally binding	banned use	largely successful in restraining battlefield use in WWII	unnecessary suffering
	submarines	1899; 1921–1922	1899 Hague convention; 1921–1922 Washington Naval Conference	never ratified	attempted bans - never ratified	failed - treaty never ratified	civilian casualties
	submarines	1907; 1930; 1936	1907 Hague Declaration; 1930 London Naval Treaty; 1936 London Protocol	legally binding	regulated use	failed - compliance collapsed in war	civilian casualties
	size of navies	1922; 1930; 1936	1922 Washington Naval Treaty; 1930 London Naval Treaty; 1936 Second London Naval Treaty	legally binding	limited quantities and size of ships	short-lived	limit arms races

continued

Era	Weapon	Year	Regulation or Treaty	Legally binding?	Type of Regulation	Successful?	Motivation
COLD WAR	nuclear tests	1963; 1967; 1985; 1995; 1996	Limited Test Ban Treaty; Treaty of Tlatelolco; Treaty of Rarotonga; Treaty of Bangkok; Treaty of Pelindaba; Comprehensive Nuclear Test Ban Treaty	legally binding	restricted testing	generally successful, with some exceptions	effects on civilians; limit arms races
	weapons in Antarctica	1959	Antarctic Treaty	legally binding	banned deployment	successful	limit arms races
	weapons of mass destruction in space	1967	Outer Space Treaty	legally binding	banned deployment	successful	strategic stability
	weapons on the moon	1967	Outer Space Treaty	legally binding	banned deployment	successful	limit arms races
	nuclear-free zones	1967; 1985; 1995; 1996	Treaty of Tlatelolco; Treaty of Rarotonga; Treaty of Bangkok; Treaty of Pelindaba	legally binding	banned developing, manufacturing, possessing, or stationing	successful	limit arms races
	nuclear weapons	1970	Nuclear Non-Proliferation Treaty	legally binding	banned proliferation	generally successful, with some exceptions	strategic stability

Era	Weapon	Year	Regulation or Treaty	Legally binding?	Type of Regulation	Successful?	Motivation
COLD WAR	nuclear weapons on the seabed	1971	Seabed Treaty	legally binding	banned deployment	successful	strategic stability
	ballistic missile defenses	1972	Anti-ballistic Missile Treaty	legally binding	limited deployment	successful during Cold War; collapsed in multipolar world	strategic stability
	biological weapons	1972	Biological Weapons Convention	legally binding	banned development, production, stockpiling, and use	generally successful, with some exceptions	unnecessary suffering; civilian casualties; prevent arms race
	using the environment as a weapon	1976	Environmental Modification Convention	legally binding	banned use	successful	civilian casualties; prevent arms race
	anti-satellite weapons	1970s & 1980s	tacit cooperation between U.S. and U.S.S.R.	no explicit agreement	norm against deployment	successful, but currently threatened in multipolar world	strategic stability
	neutron bombs	1970s	tacit cooperation between U.S. and U.S.S.R.	no explicit agreement	norm against deployment	successful	strategic stability

continued

Era	Weapon	Year	Regulation or Treaty	Legally binding?	Type of Regulation	Successful?	Motivation
	non-x-ray-detectable fragments	1980	Convention on Certain Conventional Weapons (CCW) Protocol I	legally binding	banned use	successful	unnecessary suffering
	land mines	1980	CCW Protocol II	legally binding	regulated use	unsuccessful	civilian casualties
	incendiary weapons	1980	CCW Protocol III	legally binding	regulated use	mixed success	civilian casualties
	chemical and biological weapons	1985	Australia Group	not legally binding	banned proliferation	mixed success	unnecessary suffering; civilian casualties
COLD WAR	ballistic and cruise missiles	1987; 2002	Missile Technology Control Regime; Hague Code of Conduct	not legally binding	limited proliferation	has had some success	strategic stability
	intermediate-range missiles	1987	Intermediate-Range Nuclear Forces (INF) Treaty	legally binding	banned possession	successful, but currently threatened in multipolar world	strategic stability
	nuclear weapons and launcher quantities	1972; 1979; 1991; 2002; 2011	SALT I; SALT II; START; SORT; New START	legally binding	limited quantities	successful	limit arms races

Era	Weapon	Year	Regulation or Treaty	Legally binding?	Type of Regulation	Successful?	Motivation
POST-COLD WAR	conventional air and ground forces	1991	Conventional Forces in Europe	legally binding	limited quantities	collapsed in multipolar world	limit arms races
	chemical weapons	1993	Chemical Weapons Convention	legally binding	banned development, production, stockpiling, and use	generally successful, with some exceptions	unnecessary suffering; civilian casualties
	blinding lasers	1995	CCW Protocol IV	legally binding	banned use	successful	unnecessary suffering
	conventional weapons	1996	Wassenaar Arrangement	not legally binding	limited proliferation	has had some success	political control
	land mines	1997	Mine Ban Treaty (Ottawa Treaty)	legally binding	banned development, production, stockpiling, and use	generally successful, with some exceptions	civilian casualties
	cluster munitions	2008	Convention on Cluster Munitions	legally binding	banned development, production, stockpiling, and use	generally successful, with some exceptions	civilian casualties

a decisive advantage on the battlefield, though, which is why the Nuclear Non-Proliferation Treaty's goals of global nuclear disarmament remain unrealized. Chemical weapons, on the other hand, have some battlefield advantages, but are far from decisive. Had Saddam Hussein used them against the United States, the result might have more U.S. casualties, but it would not have changed the course of the first Gulf War or the 2003 Iraq War.

The result of this dynamic is that many ineffective weapons have been banned. But it is overly simplistic to say that if a weapon has value, then a ban is doomed to fail. If the only factor that mattered was the battlefield utility of a weapon, then militaries would almost certainly use poison gas. It has value in disrupting enemy operations and terrorizing enemy troops. Expanding bullets and blinding lasers—both of which are banned by treaties—also have some military utility. In these cases, though, the perceived value is low enough that states have generally not considered them important enough to break these prohibitions.

The number of countries that need to participate for a ban to succeed is also a critical factor. Arms control was easier during the Cold War when there were only two great powers. It was far more difficult in the early twentieth century, when all powers needed to agree. A single defector could cause an arms control agreement to unravel. Since the end of the Cold War, this dynamic has begun to reemerge.

Interestingly, the legal status of a treaty seems to have little to no bearing on its success. Legally-binding treaties have been routinely violated and restraint has existed in some cases without any formal agreements. International agreements, legally binding or not, primarily serve as a focal point for coordination. What actually deters countries from violating bans is not a treaty, since by default there are no legal consequences if one wins the war, but rather reciprocity. Countries show restraint when they fear that another country might retaliate in kind. When fighting nations who do not have the ability to retaliate, they have shown less restraint. For example, during World War II Japan used chemical weapons in small amounts against China, who did not have them, and Germany killed millions of people in gas chambers during the Holocaust. Neither country used poison gas against adversaries who could retaliate in kind.

For mutual restraint to occur, there must be a clear focal point for coordination. In his books *Strategy of Conflict* and *Arms and Influence,* Thomas Schelling explained that "the most powerful limitations, the most appealing ones, are those that have a conspicuousness and simplicity, that are qualitative and not a matter of degree, that provide recognizable boundaries." Schelling observed:

> "Some gas" raises complicated questions of how much where, under what circumstances: "no gas" is simple and unambiguous. Gas only on military personnel; gas used only by defending forces; gas only when carried by vehicle or projectile; no gas without warning—a variety of limits is conceivable; some may make sense, and many might have been more impartial to the outcome of the war. But there is a simplicity to "no gas" that makes it almost uniquely a focus of agreement when each side can only conjecture at what rules the other side would propose and when failure to coordinate on the first try may spoil the chances for acquiescence in any limits at all.

This simplicity undoubtedly played a role in making it possible for European nations to refrain from using poison gas against each other in World War II, in spite of a total war that devastated the continent.

Germany and the United Kingdom also attempted to mutually avoid bombing attacks on civilian targets. These failed, but not necessarily because aerial bombing was more effective than gas or less horrible. Aerial bombing of cities was largely ineffective and universally reviled. The main purpose of Britain and Germany launching these attacks seemed to be relation for the other having done so.

The chief difference between aerial bombardment and gas, and what made restraint with aerial bombardment so difficult, is that restraint against civilian targets lacked the clarity and simplicity of the "no gas" rule. Bombers were already used in other capacities outside of attacks on cities. First they were used against ships, then land-based military targets (which inevitably had civilian casualties), then eventually cities. Each of these steps was gradual. Escalation from one step to another could even happen by accident. In fact, the final step toward full-scale aerial bombardment seems to have occurred because of an accident. Early in the war, Hitler gave explicit instructions to the Luftwaffe to avoid attacks on cities and stick to military targets, because he was worried about British reprisals. On August 24, 1940, however, several German bombers strayed in the dark and bombed central London by mistake. The British retaliated by hitting Berlin. Hitler was incensed. In a public speech, he declared, "If they declare that they will attack our cities on a large scale—we will eradicate their cities." Germany launched the London Blitz, and all attempts at restraint were gone. Gas was different. Moving from "no gas" to suddenly using it crossed a clear threshold. It was an unambiguous decision to escalate. Had gas been used on the battlefield against military targets, it likely would have expanded to attacks on cities as well.

Treaties that completely ban a weapon tend to be more successful than complicated rules governing a weapon's use. Other attempts to regulate how weapons are used on the battlefield in order to avoid civilian casualties—such as restrictions on submarine warfare, incendiary weapons, and the CCW land mine protocol—have had a similarly poor track record of success. Complete bans on weapons—such as those on exploding bullets, expanding bullets, chemical weapons, biological weapons, environmental-modification weapons, and blinding lasers—have fared better.

Two interesting exceptions that seem to prove this rule are the bans on land mines and cluster munitions. Both treaties articulate a simple and straightforward prohibition in their text. States who sign the treaties pledge "never under any circumstances to use" land mines and cluster munitions. That's about as straightforward as it gets, a clear and simple prohibition. The complicating details are buried in the definitions. In both cases, the definitions are written in such a way to carve out loopholes for certain existing weapons. The definition of "antipersonnel land mine" permits anti-vehicle mines, including those that have anti-handling devices (which are lethal to people). The cluster munitions convention has an even more complicated definition that covers the number and weight of submunitions. The effect is to permit many weapon systems that, to an ordinary person, would appear to be cluster munitions. This is no accident. The definition was crafted in such a way during negotiations to permit some countries to retain their existing inventories of now-not-quite-cluster-munitions. During signing, Australia made clear that the treaty would not cover its SMArt 155 artillery shells, which dispense two antitank submunitions. By burying these complicated rules in the definitions, though, the ban has the appearance of clarity, making it a stronger ban from a normative perspective. It's easier to stigmatize a weapon if it is perceived as illegitimate in all circumstances. "No cluster munitions" is simpler and easier to justify than "these cluster munitions, but not those," even though in practice that's what the ban does.

Carving out exceptions can make it easier to get more countries to sign on to a ban, but can be problematic if the technology is still evolving. One lesson from history is that it is very hard to predict the future path of technology. The 1899 Hague declarations banned gas-filled projectiles, but not poison gas in canisters, a technicality that Germany exploited in World War I in defense of its first large-scale poison gas attack at Ypres. On the other hand, the 1899 declarations also banned expanding bullets, a technology that turned out not to be particularly terrible. Expanding bullets are widely available for purchase by civilians in the United States for personal self-defense, although militaries have generally refrained from their use.

Hague delegates were aware of these challenges and tried to mitigate them, particularly for rapidly-evolving aerial weapons. The 1899 declarations banned projectiles from balloons or "or by other new methods of a similar nature," anticipating the possibility of aircraft, which came only four years later. The 1907 Hague rules attempted to solve the problem of evolving technology by prohibiting "attack or bombardment, by whatever means, of towns, villages, dwellings, or buildings which are undefended." This still fell short, however. The focus on "undefended" targets failed to anticipate the futility of defending against air attack, and the reality that even with defenses, "the bomber will always get through."

Technology will evolve in unforeseen ways. Successful preemptive bans focus on the intent behind a technology, rather than specific restrictions. For example, the ban on blinding lasers prohibits lasers specifically designed to cause permanent blindness, rather than limit a certain power level in lasers. The United States takes a similar intent-based interpretation of the ban on expanding bullets, that they are prohibit only to the extent that they are intended to cause unnecessary suffering.

Preemptive bans pose unique challenges and opportunities. Because they are not yet in states' inventories, the military utility of a new weapon, such as blinding lasers or environmental modification, may be amorphous. This can sometimes make it easier for a ban to succeed. States may not be willing to run the risk of sparking an arms race if the military utility of a new weapon seems uncertain. On the other hand, states often may not fully understand how terrible a weapon is until they see it on the battlefield. States correctly anticipated the harm that air-delivered weapons could cause in unprotected cities, but poison gas and nuclear weapons shocked the conscience in ways that contemporaries were not prepared for.

VERIFICATION

One topic that frequently arises in discussions about autonomous weapons is the role of verification regimes in treaties. Here the track record is mixed. A number of treaties, such as the Nuclear Non-Proliferation Treaty, Chemical Weapons Convention, INF Treaty, START, and New START have formal inspections to verify compliance. Others, such as the Outer Space Treaty's prohibition against military installations on the moon, have de facto inspection regimes. The land mine and cluster munitions bans do not have inspection regimes, but do require transparency from states on their stockpile elimination.

Not all successful bans include verification. The 1899 ban on expanding bullets, 1925 Geneva Gas Protocol, CCW, SORT, and the Outer Space Treaty's ban on putting weapons of mass destruction (WMD) in orbit all do not have verification regimes. The Environmental Modification Convention and Biological Weapons Convention (BWC) only say that states who are concerned that another is cheating should lodge a complaint with the UN Security Council. (The Soviet Union reportedly had a secret biological weapons program, making the BWC a mixed case.)

In general, verification regimes are useful if there is a reason to believe that countries might be developing the prohibited weapon in secret. That could be the case if they already have it (chemical weapons, land mines, or cluster munitions) or if they might be close (nuclear weapons). Inspection regimes are not always essential. What is required is transparency. Countries need to know whether other nations are complying or not for mutual restraint to succeed. In some cases, the need for transparency can be met by the simple fact that some weapons are difficult to keep secret. Anti-ballistic missile facilities and ships cannot be easily hidden. Other weapons can be.

WHY BAN?

Finally, the motivation behind a ban seems to matter in terms of the likelihood of success. Successful bans fall into a few categories. The first is weapons that are perceived to cause unnecessary suffering. By definition, these are weapons that harm combatants excessive to their military value. Restraint with these weapons is self-reinforcing. Combatants have little incentive to use these weapons and strong incentives not to, since the enemy would almost certainly retaliate.

Bans on weapons that were seen as causing excessive civilian harm have also succeeded, but only when those bans prohibit possessing the weapon at all (cluster munitions and the Ottawa land mine ban), not when they permit use in some circumstances (air-delivered weapons, submarine warfare, incendiary weapons, and the CCW land mine protocol). Bans on weapons that are seen as destabilizing (Seabed Treaty, Outer Space Treaty, ABM Treaty, INF Treaty) have generally succeeded, at least when only a few parties are needed for cooperation. Arms limitation has been exceptionally difficult, even when there are only a few parties, but has some record of success. Prohibiting the expansion of war into new geographic areas has only worked when the focal point for cooperation is clear and there is low military utility in doing so, such as banning weapons on the moon or in Antarctica. Attempts to regulate or restrict warfare from undersea or the air failed, most likely because the regulations were too nuanced. "No submarines" or "no aircraft" would have been clearer, for example.

Ultimately, even in the best of cases, bans aren't perfect. Even for highly successful bans, there will be some nations who don't comply. This makes military utility a decisive factor. Nations want to know they aren't giving up a potentially war-winning weapon. This is a profound challenge for those seeking a ban on autonomous weapons.

ARE AUTONOMOUS WEAPONS INEVITABLE?

THE SEARCH FOR LETHAL LAWS OF ROBOTICS

In the nearly ten years I have spent working on the issue of autonomous weapons, almost every person I have spoken with has argued there ought to be some limits on what actions machines can take in war, although they draw this line in very different places. Ron Arkin said he could potentially be convinced to support a ban on unsupervised machine learning to generate new targets in the field. Bob Work drew the line at a weapon with artificial general intelligence. There are clearly applications of autonomy and machine intelligence in war that would be dangerous, unethical, or downright illegal. Whether nations can cooperate to avoid those harmful outcomes is another matter.

Since 2014, countries have met annually at the United Nations Convention on Certain Conventional Weapons (CCW) in Geneva to discuss autonomous weapons. The glacial progress of diplomacy is in marked contrast to the rapid pace of technology development. After three years of informal meetings, the CCW agreed in 2016 to establish a Group of Governmental Experts (GGE) to discuss autonomous weapons. The GGE is a more formal forum, but has no mandate to negotiate a multinational treaty. Its main charge is to establish a working definition for autonomous weapons, a sign of how little progress countries have made.

Definitions matter, though. Some envision autonomous weapons as simple robotic systems that could search over a wide area and attack targets on their own. Such weapons could be built today, but compliance with the law of war in many settings would be difficult. For others, "autonomous weapon" is a general term that applies to any kind of missile or weapon that uses autonomy in any fashion, from an LRASM to a torpedo. From this perspective, concern about autonomous weapons is ill-founded (since they've been around for seventy years!). Some equate "autonomy" with self-learning and adapting systems, which although possible today, have yet to be incorporated into weapons. Others hear the term "autonomous weapons" and envision machines with human-level intelligence, a development that is unlikely to happen any time soon and would raise a host of other problems if it did. Without a common lexicon, countries can have heated disagreements talking about completely different things.

The second problem is common to any discussions about emerging technologies, which is that it is hard to foresee how these weapons might be used, under what conditions, and to what effect in future wars. Some envision autonomous weapons as more reliable and precise than humans, the next logical evolution of precision-guided weapons, leading to more-humane wars with fewer civilian casualties. Others envision calamity, with rogue robot death machines killing multitudes. It's hard to know which vision is more likely. It is entirely possible that both come true, with autonomous weapons making war more precise and humane when they function properly, but causing mass lethality when they fail.

The third problem is politics. Countries view autonomous weapons through the lens of their own security interests. Nations have very different positions depending on whether or not they think autonomous weapons might benefit them. It would be a mistake to assume that discussions are generating momentum toward a ban.

Still, international discussions have made some progress. An early consensus has begun to form around the notion that the use of force requires some human involvement. This concept has been articulated in different ways, with some NGOs and states calling for "meaningful human control." The United States, drawing on language in DoD Directive 3000.09, has used the term "appropriate human judgment." Reflecting these divergent views, the CCW's final report from its 2016 expert meetings uses the neutral phrase "appropriate human involvement." But no country has suggested that it would be acceptable for there to be no human involvement whatsoever in decisions about the use of lethal force. Weak though it may be, this common ground is a starting point for cooperation.

One of the challenges in current discussions on autonomous weapons is that the push for a ban is being led by NGOs, not states. Only a handful of states have said they support a ban, and none of them are major military powers. When viewed in the context of historical attempts to regulate weapons, this is unusual. Most attempts at restricting weapons have come from great powers.

The fact that the issue's framing has been dominated by NGOs campaigning to ban "killer robots" affects the debate. Potential harm to civilians has been front and center in the discussion. Strategic issues, which have been the rationale for many bans in the past, have taken a back seat. The NGOs campaigning for a ban hope to follow in the footsteps of bans on land mines and cluster munitions, but there are no successful examples of preemptive bans on weapons because of concerns about civilian harm. It is easy to see why this is the case. Bans that are motivated by concern about excessive civilian casualties pit an incidental concern for militaries against a fundamental priority: military necessity. Even when countries genuinely care about avoiding civilian harm, they can justifiably say that law-abiding nations will follow existing rules in IHL while those who do not respect IHL will not. What more would a ban accomplish, other than needlessly tie the hands of those who already respect the law? Advocates for the bans on cluster munitions and land mines could point to actual harm caused by those weapons, but for emerging technologies both sides have only hypotheticals.

When weapons have been seen as causing excessive civilian casualties, the solution has often been to regulate their use, such as avoiding attacks in populated areas. In theory, these regulations allow militaries to use weapons for legitimate purposes while protecting civilians. In practice, these prohibitions have almost always failed in war. In analyzing Robert McNamara's call for a "no cities" nuclear doctrine, Thomas Schelling pointed out the inherent problems with these rules: "How near to a city is a military installation 'part' of a city? If weapons go astray, how many mistakes that hit cities can be allowed for before concluding that cities are 'in' the war? . . . there is no such clean line."

Supporters of an autonomous weapons ban have wisely argued against such an approach, sometimes called a "partition," that would permit them in environments without civilians, such as undersea, but not populated areas. Instead, the Campaign to Stop Killer Robots has called for a complete ban on the development, production, and use of fully autonomous weapons. Opponents of a ban sometimes counter that the technology is too diffuse to be stopped, but this wrongly equates a ban with a nonproliferation regime. There are many examples of successful bans (expanding bullets, environmental modification, chemical and biological weapons, blinding lasers, the Mine Ban Treaty, and cluster munitions) that do not attempt to restrict the underlying technologies that would enable these weapons.

What all these bans have in common and what current discussions on autonomous weapons lack, however, is clarity. Even if no one has yet built a laser intended to cause permanent blindness, the concept is clear. As we've seen, there is no widespread agreement on what an autonomous weapon *is*. Some leaders in the NGO community have actually argued against creating a working definition. Steve Goose from Human Rights Watch told me that it's "not a wise campaign strategy at the very beginning" to come up with a working definition. That's because a definition determines "what's in and what's out." He said, "when you start talking about a definition, you almost always have to begin the conversation of potential exceptions." For prior efforts like land mines and cluster munitions, this was certainly true. Countries defined these terms at the end of negotiations. The difference is that countries could get on board with the general principle of a ban and leave the details to the end because there was a common understanding of what a land mine or a cluster munition was. There is no such common understanding with autonomous weapons. It is entirely reasonable that states and individuals who care a great deal about avoiding civilian casualties are skeptical of endorsing a ban when they have no idea what they would actually be banning. Automation has been used in weapons for decades, and states need to identify which uses of autonomy are truly concerning. Politics gets in the way of solving these definitional problems, though. When the starting point for discussions is that some groups are calling for a ban on "autonomous weapons," then the definition of "autonomous weapons" instantly becomes fraught.

The result is a dynamic that is fundamentally different than other attempted weapons bans. This one isn't being led by great powers, and it isn't being led by democratic nations concerned about civilian harm either, as was the case with land mines and cluster munitions. The list of nations that support a ban on autonomous weapons is telling: Pakistan, Ecuador, Egypt, the Holy See, Cuba, Ghana, Bolivia, Palestine, Zimbabwe, Algeria, Costa Rica, Mexico, Chile, Nicaragua, Panama, Peru, Argentina, Venezuela, Guatemala, Brazil, Iraq, Uganda, Austria, Colombia, and Djibouti (in order of when they endorsed a ban). Do Cuba, Zimbabwe, Algeria, and Pakistan really care more about human rights than countries like Canada, Norway, and Switzerland, who have not endorsed a ban? What the countries supporting a ban have in common is that they are not major military powers. With a few exceptions, like the Holy See, for most of these countries their support for a ban isn't about protecting civilians; it's an attempt to tie the hands of more-powerful nations. Most of the countries on this list don't need to know what autonomous weapons are to be against them. Whatever autonomous weapons may be, these countries know they aren't the ones building them.

The prevailing assumption in international discussions seems to be that autonomous weapons would most benefit advanced militaries. In the short term, this is likely true, but as autonomous technology diffuses across the international system, the dynamic is likely to reverse. Fully autonomous weapons would likely benefit the weak. Keeping a human in the loop in contested environments will require protected communications, which is far more challenging than building a weapon that can hunt targets on its own. Nevertheless, these countries likely have the perception that a ban would benefit them.

This sets up a situation where NGOs and smaller states who are advocating for a ban would asymmetrically benefit, at least in the near term, and would not be giving up anything. This only generates resistance from states who are leaders in military robotics, many of whom see their technology development proceeding in an entirely reasonable and prudent fashion. The more that others want to take them away, the more that autonomous weapons look appealing to the countries that might build them.

This is particularly the case when ban supporters have no answer for how law-abiding nations could defend themselves against those who do develop fully autonomous weapons. Steve Goose acknowledged this problem: "You know you're not going to get every country in the world to sign something immediately, but you can get people to be affected by the stigma that would accompany a comprehensive prohibition," he said. "You have to create this stigma that you don't cross the line." This can be a powerful tool in encouraging restraint, but it isn't foolproof. There is a strong stigma against chemical weapons, but they continue to be used by dictators who care nothing for the rule of law or the suffering of civilians. Thus, for many the case against a ban is simple: it would disarm only the law-abiding states who signed it. This would be the worst of all possible outcomes, empowering the world's most odious regimes with potentially dangerous weapons, while leaving nations who care about international law at a disadvantage. Proponents of a ban have yet to articulate a strategic rationale for why it would be in a leading military power's self-interest to support a ban.

Though they haven't always succeeded in the past, great powers have worked together to avoid weapons that could cause excessive harm. This time, however, leading military powers aren't trying, in part because the issue has been framed as a humanitarian one, not a strategic one. In CCW discussions, countries have heard expert views on the Martens Clause, which has never been used to ban a weapon before, but strategic considerations have gotten short shrift. A few experts have presented on offense-defense balance and arms races, but there has been virtually no discussion of how autonomous weapons might complicate crisis stability, escalation control, and war termination. John Borrie from the UN Institute for Disarmament Research is concerned about the risk of "unintended lethal effects" from autonomous weapons, but he acknowledged, "it's not really a significant feature of the policy debate in the CCW."

This is unfortunate, because autonomous weapons raise important issues for stability. There may be military benefits to using fully autonomous weapons, but it would be facile and wrong to suggest that overall they are safer and more humane than semiautonomous weapons that retain a human in the loop. This argument conflates the benefits of adding automation, which are significant, with completely removing the human from the loop. There may be cases where their use would result in more-humane outcomes, provided they functioned properly, such as hostage rescue in communications-denied environments or destroying mobile missiles launchers armed with WMD. On the whole, though, the net effects of introducing fully autonomous weapons on the battlefield are likely to be increased speed, greater consequences when accidents occur, and reduced human control.

States have every incentive to cooperate to avoid a world where they have less control over the use of force. Mutual restraint is definitely in states' interests. This is especially true for great powers, given the destruction that war among them would bring. Restraint doesn't come from a treaty, though. The fear of reciprocity is what generates restraint. A treaty is merely a focal point for coordination. Is restraint possible? History suggests any attempt to restrain autonomous weapons must meet three essential conditions to succeed.

First, a clear focal point for coordination is needed. The simpler and clearer the line, the better. This means that some rules like "no general intelligence" are dead on arrival. The open letter signed by 3,000 AI scientists called for a ban on "offensive autonomous weapons beyond meaningful human control." Every single one of those words is a morass of ambiguity. If states could agree on the difference between "offensive" and "defensive" weapons, they would have banned offensive weapons long ago. "Meaningful human control" is even more vague. Preemptive bans that try to specify the exact shape of the technology don't work either. The best preemptive bans focus on the key prohibited concept, like banning lasers intended to cause permanent blindness.

Second, the horribleness of a weapon must outweigh its military utility for a ban to succeed. Regardless of whether the weapon is seen as destabilizing, a danger to civilians, or causing unnecessary suffering, it must be perceived as bad enough—or sufficiently useless militarily—that states are not tempted to breach the ban.

Third, transparency is essential. States must trust that others are not secretly developing the weapon they themselves have foresworn. Bob Work told me that he thought countries "will move toward some type of broad international discussion on how far we should go on autonomous weapons." The problem he saw was verification: "The verification of this regime is going to be very, very difficult because it's just—it's ubiquitous. It's now exploding around us." This is a fundamental problem for autonomous weapons. The essence of autonomy is software, not hardware, making transparency very difficult.

Are there models for restraint with autonomous weapons that meet these criteria? Is there a military equivalent to Asimov's Laws, a "lethal laws of robotics" that states could agree on? There are many possible places countries could draw a line. States could focus on physical characteristics of autonomous weapons: size, range, payload, etc. States could agree to refrain from certain types of machine intelligence, such as unsupervised machine learning on the battlefield. To illustrate the range of possibilities, here are four very different ways that nations could approach this problem.

OPTION I: BAN FULLY AUTONOMOUS WEAPONS

The Campaign to Stop Killer Robots has called for "a comprehensive, pre-emptive prohibition on the development, production and use of fully autonomous weapons." Assuming that states found it in their interests to do so, could they create a ban that is likely to result in successful restraint?

Any prohibition would need to clearly distinguish between banned weapons and the many existing weapons that already use autonomy. It should be possible to clearly differentiate between the kind of defensive human-supervised autonomous weapons in use today and fully autonomous weapons that would have no human supervision. "Offensive" and "defensive" are distinctions that wouldn't work, but "fixed" and "mobile" autonomous weapons could. The types of systems in use today are all fixed in place. They are either static (immobile) or affixed to a vehicle occupied by people.

Distinguishing between mobile, fully autonomous weapons and advanced missiles would be harder. The chief difference between the semiautonomous HARM and the fully autonomous Harpy is the Harpy's ability to loiter over a wide area and search for targets. Debates over weapons like the LRASM and Brimstone show how difficult it can be to make this distinction without understanding details about not only the weapon's functionality, but also its intended use. Drawing a distinction between recoverable robotic vehicles and nonrecoverable munitions would be easier.

From the perspective of balancing military necessity against the horribleness of the weapon, these distinctions would be sensible. The most troubling applications of autonomy would be fully autonomous weapons on mobile robotic vehicles. Fixed autonomous weapons would primarily be defensive. They also would be lower risk, since humans could supervise engagements and physically disable the system if it malfunctioned. Non-recoverable fully autonomous weapons (e.g., loitering munitions) would be permitted, but their risks would be mitigated by the fact that they can't be sent on patrol. Militaries would want to have some indication that there is an enemy in the vicinity before launching them. There are other ways nations could draw lines on what is and isn't allowed, but this is one set of choices that would seem sensible.

Regardless of where nations draw the line, there are a number of factors that make restraint challenging. How would nations know that others were complying? The United States, the United Kingdom, France, Russia, China, and Israel are already developing experimental stealth drones. Operational versions of these aircraft would be sent into areas in which communications might be jammed. Even if nations agreed that these combat drones should not attack targets unless authorized by a human, there would be no way for them to verify each other's compliance. Delegating full autonomy would likely be valuable in some settings. Even if in peacetime nations genuinely desired mutual restraint, in wartime the temptation might be great enough to overcome any reservations. After all, it's hard to argue that weapons like the Harpy, TASM, or a radar-hunting combat drone shock the conscience. Using them may entail accepting a different level of risk, but it's hard to see them as inherently immoral. Further complicating restraint, it might be difficult to even know whether nations were complying with the rules during wartime. If a robot destroyed a target, how would others know whether a human had authorized the target or the robot itself?

All of these factors: clarity, military utility, horribleness of the weapon, and transparency suggest that a ban on fully autonomous weapons is unlikely to succeed. It is almost certain not to pass in the CCW, where consensus is needed, but even if it did, it is hard to see how such rules would remain viable in wartime. Armed robots that had a person in the loop would need only a flip of the switch, or perhaps a software patch, to become fully autonomous. Once a war begins, history suggests that nations will flip the switch, and quickly.

OPTION 2: BAN ANTIPERSONNEL AUTONOMOUS WEAPONS

A ban on autonomous weapons that targeted people may be another matter. The ban is clearer, the horribleness of the weapon greater, and the military utility lower. These factors may make restraint more feasible for antipersonnel autonomous weapons.

It would be easier for states to distinguish between antipersonnel autonomous weapons and existing systems. There are no antipersonnel equivalents of homing missiles or automated defensive systems in use around the world. This could allow states to sidestep the tricky business of carving out exceptions for existing uses.

The balance between military utility and the weapon's perceived horribleness is also very different for antipersonnel autonomous weapons. Targeting people is much more problematic than targeting objects for a variety of reasons. Antipersonnel autonomous weapons are also significantly more hazardous than anti-matériel autonomous weapons. If the weapon malfunctions, humans cannot simply climb out of a tank to escape being targeted. A person can't stop being human. Antipersonnel autonomous weapons also pose a greater risk of abuse by those deliberating wanting to attack civilians.

Finally, the public may see machines that target and kill people on their own as genuinely horrific. Weapons that autonomously targeted people would tap into an age-old fear of machines rising up against their makers. Public revulsion could be a decisive factor in achieving political support for a ban. There is something clean and satisfying to the rule, to paraphrase Navy engineer John Canning: "let machines target machines; let people target people."

The military utility of antipersonnel autonomous weapons is also far lower that anti-matériel autonomous weapons. The reasons for moving to supervised autonomy (speed) or full autonomy (no communications) don't generally apply when targeting people. Defensive systems like Aegis need a supervised autonomous mode to defend against salvos of high-speed missiles, but overwhelming defensive positions through waves of human attackers has not been an effective tactic since the invention of the machine gun. The additional half second it would take to keep a human in the loop for a weapon like the South Korean sentry gun is marginal. Antipersonnel autonomous weapons in communications-denied environments are also likely to be of marginal value for militaries. At the early stages of a war when communications are contested, militaries will be targeting objects such as radars, missile launchers, bases, airplanes, and ships, not people. Militaries would want the ability to use small, discriminating antipersonnel weapons to target specific individuals, such as terrorist leaders, but those would be semiautonomous weapons; a human would be choosing the target.

Transparency would still be challenging. As is the case for weapons like the South Korean sentry gun, others would have to essentially trust countries when they say they have a human in the loop. Many nations are already fielding armed robotic ground vehicles, and they are likely to become a common feature of future militaries. It would be impossible to verify that these robotic weapons do not have a mode or software patch waiting on the shelf that would enable them to autonomously target people. Given the ubiquity of autonomous technology, it would also be impossible to prevent terrorists from creating homemade autonomous weapons. Large-scale industrial production of the kinds of antipersonnel weapons that Stuart Russell fears, however, would be hard to hide. If the military utility of these weapons were low enough, it isn't clear that the risk of small scale uses would compel other nations to violate a prohibition.

Russell has argued that a treaty could be effective in "stopping an arms race and preventing large-scale manufacturing of such weapons." The combination of low military utility and high potential harm may make restraint possible for antipersonnel autonomous weapons.

OPTION 3: ESTABLISH "RULES OF THE ROAD" FOR AUTONOMOUS WEAPONS

Different problems with autonomous weapons lend themselves to different solutions. A ban on antipersonnel autonomous weapons would reduce the risk of harm to civilians, but would not address the problems autonomous weapons pose for crisis stability, escalation control, and war termination. These are very real concerns, and nations will want to cooperate to ensure their robotic systems do not interact in ways that lead to unintended outcomes.

Rather than a treaty, one solution could be to adopt a non-legally-binding code of conduct to establish a "rules of the road" for autonomous weapons. The main goal of such a set of rules would be to reduce the potential for unintended interactions between autonomous systems in crises. The best rules would be simple and self-enforcing, like "robotic vehicles should not fire unless fired upon" and "return fire must be limited, discriminating, and proportionate." Like maritime law, these rules would be intended to govern how autonomous agents interact when they encounter one another in unstructured environments, respecting the right of self-defense but also a desire to avoid unwanted escalation.

Any ruleset could undoubtedly be manipulated by clever adversaries spoiling for a fight, but the main purpose would be to ensure predictable reactions from robotic systems among nations seeking to control escalation. The rules wouldn't need to be legally binding, since it would be in states' best interests to cooperate. These rules would likely collapse in war, as rules on submarine warfare did, but that wouldn't matter since the intent would be to control escalation in circumstances short of war. Once a full-blown war is under way, the rules wouldn't be needed.

OPTION 4: CREATE A GENERAL PRINCIPLE ABOUT THE ROLE OF HUMAN JUDGMENT IN WAR

The problem with the above approaches is that technology is always changing. Even the most thoughtful regulations or prohibitions will not be able to foresee all the ways that autonomous weapons could evolve over time. An alternative approach would be to focus on the unchanging element in war: the human.

The laws of war do not specify what role(s) humans should play in lethal force decisions, but perhaps they should. Is there a place for human judgment in war, even if we had all the technology we could imagine? Should there be limits on what decisions machines make in war, not because they can't, but because they shouldn't?

One approach would be to articulate a positive requirement for human involvement in the use of force. Phrases like "meaningful human control," "appropriate human judgment," and "appropriate human involvement" all seem to get at this concept. While these terms are not yet defined, they suggest broad agreement that there is some irreducible role for humans in lethal force decisions on the battlefield. Setting aside for the moment the specific label, what would be the underlying idea behind a principle of "_____ human _____"?

IHL may help give us some purchase on the problem, if one adopts the viewpoint that the laws of war apply to people, not machines. This was the view captured in the *U.S. Department of Defense Law of War Manual*:

> The law of war rules on conducting attacks (such as the rules relating to discrimination and proportionality) impose obligations on persons. These rules do not impose obligations on the weapons themselves; . . . Rather, it is persons who must comply with the law of war.

Humans are obligated under IHL to make a determination about the lawfulness of an attack and cannot delegate this obligation to a machine. This means that the human must have some information about the specific attack in order to make a determination about whether it complies with the principles of distinction, proportionality, and precautions in attack. The human must have sufficient information about the target(s), the weapon, the environment, and the context for the attack to determine whether that particular attack is lawful. The attack also must be bounded in time, space, targets, and means of attack for the determination about the lawfulness of that attack to be meaningful. There would presumably be some conditions (time elapsed, geographic boundaries crossed, circumstances changed) under which the human's determination about the lawfulness of the attack might no longer be valid.

How much information the person needs and what those bounds are on autonomy is open for debate. This perspective would seem to suggest, though, that IHL requires some minimum degree of human involvement in lethal force decisions: (1) human judgment about the lawfulness of an attack; (2) some specific information about the target(s), weapon, environment, and context for attack in order to make a determination about lawfulness of that particular attack; and (3) the weapon's autonomy be bounded in space, time, possible targets, and means of attack.

There may be other ways of phrasing this principle and reasonable people might disagree, but there could be merit in countries reaching agreement on a common standard for human involvement in lethal force. While an overarching principle along these lines would not tell states which weapons are permitted and which are not, it could be a common starting point for evaluating technology as it evolves. Many principles in IHL are open to interpretation: unnecessary suffering, proportionality, and precautions in attack, for example. These terms do not tell states which weapons cause unnecessary suffering or how much collateral damage is proportionate, but they still have value. Similarly, a broad principle outlining the role of human judgment in war could be a valuable benchmark against which to evaluate future weapons.

HARD PROBLEMS, IMPERFECT INSTITUTIONS

Humanity is at the threshold of a new technology that could fundamentally change our relationship with war. The institutions that human society has to deal with these challenges are imperfect. Getting agreement in the CCW is challenging, given its structure as a consensus-based organization. It's possible that fully autonomous weapons are a bad idea, whether for legal, moral, or strategic reasons, but that restraint among nations is doomed to fail. It wouldn't be the first time. For now, nations, NGOs, and international organizations like the ICRC continue to meet in the CCW to discuss the challenges of autonomous weapons. Meanwhile, technology races forward.

Conclusion

NO FATE BUT WHAT WE MAKE

In the *Terminator* films, Sarah Connor and her son John, who will eventually lead the resistance against the machines, are hounded by an enemy even worse than the Terminators: fate. No matter how many times Sarah and John defeat Skynet, it still returns to haunt them in yet another film. Part of this is good Hollywood business. Sarah and John are victims of being in a film series where a sequel is a surefire moneymaker. But their trap of fate is also essential to the storytelling of the *Terminator* movies. In film after film, Sarah and John are perpetually hunted by Terminators sent back from the future to kill them and prevent John from eventually leading the human resistance against the machines. The essence of the events that propel the story are a time-travel paradox: if the Terminators succeeded in killing Sarah or John, then John couldn't lead the human resistance, which would negate the reason for killing them in the first place. Meanwhile, Sarah and John attempt to destroy Skynet before it can come into existence. It's another paradox; if they succeeded, Skynet would never send a Terminator back in time to attack them, giving them the motivation to destroy Skynet.

Sarah and John are forever trapped in a battle against Skynet across the past, present, and future. Judgment Day continues to occur, no matter their actions, although the date keeps shifting (conveniently, to just a few years after each film's release date, keeping Judgment Day forever in the audience's future). In spite of this, Sarah and John fight against fate, never wavering in their faith that this time they will be able to defeat Skynet for good and avert Judgment Day. In *Terminator 2: Judgment Day,* John Connor quotes his mother as saying, "The future's not set. There's no fate but what we make for ourselves." The line returns again and again in subsequent movies: "There is no fate but what we make."

Of course, Sarah Conner is right. In the real world, the future isn't written. The visions of possible futures presented in this book—scary visions, good visions—are only wisps of imagination. The real future unfolds one step at a time, one day at a time, one line of code at a time. Will the future be shaped by technology? Of course. But that technology is made by people. It's being crafted by people like Duane Davis, Bradford Tousley, and Brandon Tseng. It's shaped by government officials like Larry Schuette, Frank Kendall, and Bob Work. Their decisions are guided by voices like Stuart Russell, Jody Williams, and Ron Arkin. Each of these individuals has choices and, collectively, humanity has choices.

The technology to enable machines that can take life on their own, without human judgment or decision-making, is upon us. What we do with that technology is up to us. We can use artificial intelligence to build a safer world, one with less human suffering, fewer accidents, fewer atrocities, and one that keeps human judgment where it is needed. We can preserve a space for empathy and compassion, however rare they may be in war, and leave the door open to our better angels. Or we can become seduced by the allure of machines—their speed, their seeming perfection, their cold precision. We can delegate power to the machines, trusting that they will perform their assigned tasks correctly and without hesitation, and hope that we haven't got it wrong, that there are no flaws lurking in the code for unforeseen events to trigger or enemies to exploit.

There are no easy answers. If war could be averted and nations could secure their peace through treaties and not force of arms, they would have done so long ago. Militaries exist as a means to defend people from those who are not deterred by laws or international goodwill. To ask nations to surrender a potential means to defend themselves is to ask them to take a grave and weighty gamble.

And yet . . .

Despite this—despite the reality that there are no police to enforce the laws of war and that only the victors decide who stands trial. . . . Despite the reality that might, not right, decides who wins and dies on the battle-field. . . . Despite all this, codes of conduct have governed human behavior in war for millennia. Even the earliest of these codes contain guidance for which weapons could be used in war and which were beyond the pale. Barbed and poison-tipped arrows were surely useful in war. Yet they were *wrong* nevertheless.

Human societies have cooperated time and again to restrain the worst excesses in war, to place some actions or means of killing out of bounds, even when life and death are at stake. Sometimes this cooperation has failed, but the miracle is that sometimes it hasn't. In the modern era, mili-taries have largely stepped away from chemical weapons, biological weap-ons, blinding lasers, land mines, and cluster munitions as weapons of war. Not all militaries, but most of them. Nuclear powers have further agreed to limit how nuclear weapons are deployed in order to improve strategic stability. These rules are sometimes broken, but the fact that restraint exists at all among states that otherwise fear each other shows that there is hope for a better world.

This restraint—the conscious choice to pull back from weapons that are too dangerous, too inhumane—is what is needed today. No piece of paper can prevent a state from building autonomous weapons if they desire it. At the same time, a pell-mell race forward in autonomy, with no clear sense of where it leads us, benefits no one. States must come together to develop an understanding of which uses of autonomy are appropriate and which go too far and surrender human judgment where it is needed in war. These rules must preserve what we value about human decision-making, while attempting to improve on the many human failings in war. Weighing these human values is a debate that requires all members of society, not just academics, lawyers, and military professionals. Average citizens are needed too, because ultimately autonomous military robots will live—and fight—in our world.

Machines can do many things, but they cannot create meaning. They cannot answer these questions for us. Machines cannot tell us what we value, what choices we should make. The world we are creating is one that will have intelligent machines in it, but it is not for them. It is a world for us.

AFTERWORD

HOW ROBOTIC WEAPONS ARE TRANSFORMING
THE BATTLEFIELD TODAY

Come, check this out," the robotics professor said, urging me toward the back of his office. On a table in the back of the room, tucked behind a filing cabinet, an open laptop sprouted wires that snaked to a control board with a small robot arm on top. The robot arm gripped a plastic airsoft pistol. On the laptop screen, I saw the feed from a camera looking at us.

"It has facial recognition software," he explained. I saw an outlined box appear on the screen near my face. Without even really thinking about what I was doing, I tilted my head to line it up in the box. The box changed color as the computer recognized a face inside the box.

BANG!

With a jolt, the bolt slammed forward on the airsoft pistol as the computer pulled the trigger—at me. I jumped, startled that this computer had just tried to shoot me in the face. There was a cover on the front of the gun, so nothing had actually come out. Still, it felt a bit . . . rude.

"I had to put the cover on because I kept shooting myself in the face," the professor explained.

"Wow, it just . . . did that." There was a certain irony in a robot trying to shoot me. I was visiting the university on a book tour for *Army of None* and here I'd just been "shot" with a homemade autonomous weapon.

I shouldn't have been surprised. I had spent the last several years researching this technology and I knew how widely accessible it was. All the pieces existed to make a crude autonomous weapon; all it took was for someone to put them together. The professor explained that he had

whipped this little contraption together in an afternoon. "Not that difficult," he said. He had downloaded the face tracking software for free from
GitHub. He made one minor modification so it would send a signal to a
controller to fire the gun when the face was centered. And that was it. The
software to make it autonomous was actually easier to get working than
the electronics of making the gun fire, he told me. I'd written an entire
chapter on the ease with which a person with rudimentary programming
skills could build an autonomous weapon. Yet knowing it in the abstract
was different from seeing it right in front of me, with the barrel of an
autonomous weapon literally staring me in the face.

The future is coming, and we aren't ready.

I spend my job following emerging technologies, yet I am continually
surprised by the pace of advancement in artificial intelligence and autonomy. In the past year, a number of events pointed to the increasing rele-
vance of robotic weapons. In January 2018, Syrian rebels launched a mass
drone attack against a Russian air base using thirteen drones. The Russians managed to take out all thirteen drones, shooting down seven with
anti-aircraft missiles and grounding six with electronic warfare countermeasures, but it was an ominous harbinger of the dangers of widely
proliferated robotic technology. Not to be outdone, in May 2018 Russia
deployed their own robotic weapons to Syria, sending the heavily armed
Uran-9 ground robotic vehicle into combat. According to reports, the
vehicle did not perform well. Operators frequently lost communications
with the Uran-9 and the deployment was short-lived. But the Russian military will undoubtedly learn from this experience and use it to improve
future robotic weapons with greater capacity for autonomous operation.
The Russians aren't alone. In August 2018, the ability of non-state actors
to field simple robotic weapons was vividly demonstrated in a drone-
based assassination attempt on Venezuelan President Nicolas Maduro.
The attackers used DJI M600 drones, which retail for $5,000, each car-
rying a kilogram of explosives. With each passing month, we enter deeper
into a world of ubiquitous robotic weapons.

The basic technology underlying artificial intelligence also continues
to advance. In August 2018, the AI research company OpenAI fielded a
team of five bots against five humans in the real-time computer strategy
game, Dota 2. The bots narrowly lost to some of the top professional Dota

2 players in the world but played well, demonstrating superhuman precision and aggression in the match. One observer remarked, "Often, the humans would win a fight and then let their guard down slightly, expecting the enemy team to retreat and regroup. But the bots don't do that. If they can see a kill, they take it." The military advantage of relentless killing machines in real-world combat is obvious. Humans beat the bots at Dota 2 because of advantages in long-term strategy, but for real-world applications militaries might overcome these weaknesses by combining human and machine intelligence through "centaur" approaches. Or it's possible that OpenAI's Dota 2 bots will improve to surpass humans. One major advantage is that computers can rapidly gain experience playing simulated games, racking up the equivalent of "100 human lifetimes of experience every single day."

Meanwhile, international diplomacy on autonomous weapons continues to plod forward. Countries met again at the United Nations Convention on Certain Conventional Weapons in 2018. China threw diplomats for a loop when they offered a proposal for a partial ban on the use of fully autonomous weapons, but not their development and production. As of October 2018, there remain unanswered questions about the Chinese proposal—what specifically it covered and the sincerity of the Chinese offer. France and Germany argued for a politically-binding resolution, a statement expressing views on autonomous weapons but not a legally-binding treaty. The United States and Russia opposed any new regulatory efforts, instead arguing that existing international law was sufficient. One of the most potentially consequential shifts was Austria joining the set of nations—none of them major military powers or robotics developers—calling for a ban, the first Western democratic nation to do so. Prior bans on land mines and cluster munitions required a deep-pocketed Western state to act as a champion for the issue, funding a separate process for like-minded states to draft a treaty outside the UN. Canada played this role for land mines and Norway did the same for cluster munitions. It remains to be seen whether Austria will play a similar role for autonomous weapons and whether their support for a ban could lead other European nations to follow suit.

Despite the wide range of views on outcome, the August 2018 meetings concluded with a consensus document that asserted: "Human responsibility for decisions on the use of weapons systems must be retained since accountability cannot be transferred to machines." These kinds of principles, while not easily translated directly to engineering specifications, could be the type of broad overarching guidance about the role of humans in warfare that could help inform how nations use autonomy in future weapons. This sort of diplomatic achievement, while incremental and not legally binding, is nevertheless an important step toward achieving a common understanding about how to deal with the challenges of autonomous weapons.

In the background of these developments has been a broadening of the discussion of the changes artificial intelligence is bringing to society. AI is sweeping across society and bringing with it a tsunami of change to industry, transportation, medicine, the future of work, cybersecurity, surveillance, and other areas. It is not just the speed of change but its scale that is overwhelming. Autonomous weapons are one slice of a much bigger issue about how artificial intelligence will affect global security. Russia is using bots to spread disinformation and disrupt Western democracies. China is building a techno-dystopian surveillance state to control its citizens through facial recognition technology and a social credit system. The Pentagon is using artificial intelligence to process drone video feeds, much to the dismay of thousands of AI researchers at Google who protested the company's involvement. And private companies are building the platforms that sift and sort the information citizens access, often using opaque algorithms designed to maximize profit, not the public good.

These and other actors fight not just to control the technology, but to shape the very course of our future. China has launched a national-level AI development plan, increasing their research investment and aggressively courting top experts from Silicon Valley. The U.S. government has belatedly followed suit, creating internal committees to better understand what could be done to ensure U.S. competitiveness and cracking down on foreign investment in U.S. tech firms. In the absence of federal government leadership, some states are taking action on their own. In September 2018, California approved a "Blade Runner" law that would require bots to disclose that they're not human, effectively making it illegal for machines to impersonate a human. Europe is also shaping the regulatory space surrounding artificial intelligence, despite lagging behind the United States and China in AI research. Data is central to machine learning systems and Europe's new data privacy laws are shaping how companies access and control user data, which can be used to train machine learning algorithms. Tech companies themselves are responding to public pressure following the disclosure of hacks or disinformation campaigns on their platforms by changing their practices, such as tweaking their algorithms to "de-rank" certain content. Few private actors have embraced regulation, though, with Microsoft's call for regulation of facial recognition technology a rare exception.

As Russian President Vladimir Putin has stated, "Whoever becomes the leader in [artificial intelligence] will become the ruler of the world." The race is underway to shape that future. Those who are building AI technology will have tremendous power to set the terms of how it is used—who has access to it, how data is managed, and what if any regulations govern its use. Decisions made in the coming years will have a long-lasting impact on human society as we see the emergence of a new digital order.

Notes

Introduction: The Power Over Life and Death

1 **shot down a commercial airliner:** Thom Patterson, "The downing of Flight 007: 30 years later, a Cold War tragedy still seems surreal," CNN.com, August 31, 2013, http://www.cnn.com/2013 /08/31/us/kal-fight-007-anniversary/index.html.

1 **Stanislav Petrov:** David Hoffman, "'I Had a Funny Feeling in My Gut,'" *Washington Post,* February 10, 1999, http://www.washingtonpost.com/wp-srv/inatl/longterm/coldwar/shatter021099b.htm.

1 **red backlit screen:** Pavel Aksenov, "Stanislav Petrov: The Man Who May Have Saved the World," *BBC.com,* September 26, 2013, http://www.bbc.com/news/world-europe-24280831.

2 **five altogether:** Ibid.

2 **Petrov had a funny feeling:** Hoffman, "I Had a Funny Feeling in My Gut.'"

2 **Petrov put the odds:** Aksenov, "Stanislav Petrov: The Man Who May Have Saved the World."

5 **Sixteen nations already have armed drones:** The United States, United Kingdom, Israel, China, Nigeria, Iran, Iraq, Jordan, Egypt, United Arab Emirates, Saudi Arabia, Kazakhstan, Turkmenistan, Pakistan, Myanmar, Turkey. Matt Fuhrmann and Michael C. Horowitz, "Droning On: Explaining the Proliferation of Unmanned Aerial Vehicles," *International Organization,* 71 no. 2 (Spring 2017), 397–418.

5 **"next industrial revolution":** "Robot Revolution—Global Robot & AI Primer," *Bank of America Merrill Lynch,* December 16, 2015, http://www.bofaml.com/content/dam/boamlimages/documents/PDFs/robotics_and_ai_condensed_primer.pdf.

5 **Kevin Kelly:** Kevin Kelly, "The Three Breakthroughs That Have Finally Unleashed AI on the World," *Wired,* October 27, 2014, http://www.wired.com/2014/10/future-of-artificial-intelligence/.

5 *cognitization* **of machines:** Antonio Manzalini, "Cognitization is Upon Us!," *5G Network Softwarization,* May 21, 2015, http://ieee-sdn.blogspot.com/2015/05/cognitization-is-upon-us.html.

6 **"fully roboticized ... military operations":** Robert Coalson, "Top Russian General Lays Bare Putin's Plan for Ukraine," *Huffington Post,* September 2, 2014, http://www

.huffingtonpost.com/robert-coalson/valery-gerasimov-putin-ukraine_b_5748480
.html.

6   **Department of Defense officials state:** Deputy Assistant Secretary of Defense for
      Research Melissa Flagg, as quoted in Stew Magnuson, "Autonomous, Lethal Robot
      Concepts Must Be 'On the Table,' DoD Official Says," March 3, 2016, http://www
      .nationaldefensemagazine.org/blog/Lists/Posts/Post.aspx?ID=2110.

6   **AI programs today:** For an overview of machine capabilities and limitations today
      for image recognition and understanding, see JASON, "Perspectives on Research in
      Artificial Intelligence and Artificial General Intelligence Relevant to DoD, JSR-16-
      Task-003," The Mitre Corporation, January 2017, 10–11, https://fas.org/irp/agency/
      dod/jason/ai-dod.pdf.

7   **Over 3,000 robotics and artificial intelligence experts:** "Autonomous Weapons:
      An Open Letter From AI & Robotics Researchers," *Future of Life Institute,* http://
      futureoflife.org/open-letter-autonomous-weapons/. Additionally, over one hun-
      dred robotics and AI company founders and CEOs signed an open letter in 2017
      warning of the dangers of autonomous weapons. This second letter had a more
      muted call to action, however. Rather than calling for a ban as the 2015 letter did,
      the 2017 letter simply implored countries engaged in discussions at the United
      Nations to "find a way to protect us from all these dangers." "An Open Letter to the
      United Nations Convention on Certain Conventional Weapons," accessed August
      24, 2017, https://www.dropbox.com/s/g4ijcaqq6ivq19d/2017%20Open%20Letter
      %20to%20the%20United%20Nations%20Convention%20on%20Certain%20
      Conventional%20Weapons.pdf?dl=0.

7   **Campaign to Stop Killer Robots:** "Who We Are," *Campaign to Stop Killer Robots,*
      http://www.stopkillerrobots.org/coalition/.

7   **"global AI arms race":** Autonomous Weapons: An Open Letter From AI & Robotics
      Researchers," *Future of Life Institute.*

8   **"If our competitors go to Terminators":** Bob Work, remarks at the Atlantic Council
      Global Strategy Forum, Washington, DC, May 2, 2016, http://www.atlanticcouncil
      .org/events/webcasts/2016-global-strategy-forum.

8   **"The Terminator Conundrum":** Andrew Clevenger, " 'The Terminator Conun-
      drum': Pentagon Weighs Ethics of Pairing Deadly Force, AI," *Defense News,*
      January 23, 2016, http://www.defensenews.com/story/defense/policy-budget/
      budget/2016/01/23/terminator-conundrum-pentagon-weighs-ethics-pairing
      -deadly-force-ai/79205722/.

1   The Coming Swarm: The Military Robotics Revolution

13   **Global spending on military robotics:** "Robot Revolution–Global Robot & AI
      Primer," *Bank of America Merrill Lynch.*

14   **increasing sixfold to over $2 billion per year:** Office of the Secretary of Defense,
      "Unmanned Aircraft Systems Roadmap, 2005–2030," August 4, 2005, 37, Figure
      2.6-1, http://fas.org:8080/irp/program/collect/uav_roadmap2005.pdf.

14   **"over-the-hill reconnaissance":** Dyke Weatherington, "Unmanned Aircraft Sys-
      tems Roadmap, 2005–2030," presentation, 2005, http://www.uadrones.net/military/
      research/acrobat/050713.pdf.

14   **Hundreds of drones:** Ibid.

14 **Drones weren't new:** "Thomas P. Ehrhard, "Air Force UAVs: The Secret History," July 2010 Mitchell Institute Study, 28.

14 **over $6 billion per year:** Office of the Secretary of Defense, "Unmanned Systems Integrated Roadmap, FY2011–2036," 13, http://www.acq.osd.mil/sts/docs/Unmanned%20Systems%20Integrated%20Roadmap%20FY2011-2036.pdf.

14 **DoD had over 7,000 drones:** Ibid, 21.

14 **over 6,000 ground robots:** Office of the Secretary of Defense, "Unmanned Systems Integrated Roadmap, FY2009–2034," 3, http://www.acq.osd.mil/sts/docs/UMSIntegratedRoadmap2009.pdf.

15 **"For unmanned systems to fully realize":** Office of the Secretary of Defense, "Unmanned Systems Integrated Roadmap, FY2011–2036," 45.

15 **ways of communicating that are more resistant to jamming:** Kelley Sayler, "Talk Stealthy to Me," *War on the Rocks*, December 4, 2014, https://warontherocks.com/2014/12/talk-stealthy-to-me/.

16 **human pilot could remain effective sitting in the cockpit:** Graham Warwick, "Aurora Claims Endurance Record For Orion UAS," *Aviation Week & Space Technology*, January 22, 2015, http://aviationweek.com/defense/aurora-claims-endurance-record-orion-uas. Paul Scharre, "The Value of Endurance," Center for a New American Security, Washington, DC, November 12, 2015, https://www.cnas.org/publications/blog/infographic-the-value-of-endurance.

16 **"Autonomy reduces the human workload":** Office of the Secretary of Defense, "Unmanned Systems Integrated Roadmap, FY2011–2036," 45.

17 **"fully autonomous swarms":** Office of the Secretary of Defense, "Unmanned Aircraft Systems Roadmap, 2005–2030," 48.

17 **2011 roadmap articulated . . . four levels of autonomy:** Office of the Secretary of Defense, "Unmanned Systems Integrated Roadmap, FY2011–2036," 46.

17 **"single greatest theme":** Office of the U.S. Air Force Chief Scientist, "Technology Horizons: A Vision for Air Force Science and Technology During 2010-30," May 15, 2010, *xx* http://www.defenseinnovationmarketplace.mil/resources/AF_TechnologyHorizons2010-2030.pdf.

18 **"If I have fifty planes":** Duane Davis, interview, May 10, 2016.

19 **"fifty humans and fifty balls":** Ibid.

21 **the colony converges on the fastest route:** Goss S., Beckers R., Deneubourg J. L., Aron S., Pasteels J. M., "How Trail Laying and Trail Following Can Solve Foraging Problems for Ant Colonies," in Hughes R. N., ed. *Behavioural Mechanisms of Food Selection, NATO ASI Series* (Series G: Ecological Sciences), vol 20 (Berlin, Heidelberg: Springer, 1990) http://www.ulb.ac.be/sciences/use/publications/JLD/77.pdf.

21 *stigmergy:* For an excellent overview of animal swarming, see Eric Bonabeau, Guy Theraulaz, Marco Dorigo, *Swarm Intelligence: From Natural to Artificial Systems* (New York: Oxford University Press, 1999).

21 **"a collective organism . . . like swarms in nature":** Chris Baraniuk, "US Military Tests Swarm of Mini-Drones Launched from Jets," *BBC News*, January 10, 2017, http://www.bbc.com/news/technology-38569027. Shawn Snow, "Pentagon Successfully Tests World's Largest Micro-Drone Swarm," *Military Times*, January 9, 2017, http://www.militarytimes.com/articles/pentagon-successfully-tests-worlds-largest-micro-drone-swarm.

21   **China demonstrated a 119-drone swarm:** Emily Feng and Charles Clover, "Drone Swarms vs. Conventional Arms: China's Military Debate," *Financial Times*, August 24, 2017.

22   **a swarm of small boats:** Office of Naval Research, "Autonomous Swarm," video, October 4, 2014, https://www.youtube.com/watch?v=ITTvgkO2Xw4.

22   **five boats working together:** Ibid.

22   **"game changer":** Bob Brizzolara as quoted in Office of Naval Research, "Autonomous Swarm."

22   **"always a human in the loop":** Sydney J. Freedberg, Jr., "Naval Drones 'Swarm,' But Who Pulls The Trigger?" *Breaking Defense,* October 5, 2015, http://breakingdefense .com/2014/10/who-pulls-trigger-for-new-navy-drone-swarm-boats/.

23   **"Goal: Collapse adversary's system":** John Boyd, "Patterns of Conflict," eds. Chet Richards and Chuck Spinney, presentation, slide 87, January 2007.

24   **"without necessarily requiring human input":** United States Air Force, "Unmanned Aircraft Systems Flight Plan, 2009–2047," May 18, 2009, 41, https:// fas.org/irp/program/collect/uas_2009.pdf.

24   **"Authorizing a machine":** Ibid, 41.

25   **"Policy guidelines":** Office of the Secretary of Defense, "Unmanned Systems Integrated Roadmap, FY2011–2036," 50.

2   The Terminator and the Roomba: What Is Autonomy?

26   **Three Laws of Robotics:** Asimov's Three Laws of Robotics first appear in their original form in the short story "Runaround" in his 1950 collection, *I, Robot.* Isaac Asimov, *I, Robot,* (New York: Grome Press, 1950).

28   **rescuing a U.S. F-16 in Syria:** Guy Norris, "Ground Collision Avoidance System 'Saves' First F-16 In Syria," *Aerospace Daily & Defense Report,* February 5, 2015, http:// aviationweek.com/defense/ground-collision-avoidance-system-saves-first-f-16 -syria.

3   Machines That Kill: What Is an Autonomous Weapon?

35   **Well-trained troops could fire:** George Knapp, "Rifled Musket, Springfield, Model 1861," in Jerold E. Brown, *Historical Dictionary of the U.S. Army* (Santa Barbara, CA: Greenwood Publishing Group, 2001), 401.

35   **"[T]he Gatling gun . . . I liked it very much.":** As quoted in Julia Keller, *Mr. Gatling's Terrible Marvel* (New York: Penguin Books, 2009).

36   **as more than a hundred men:** The Gatling gun could fire 350–400 rounds per minute. A well-trained Civil War–era soldier could fire three rounds per minute with a Springfield rifle, allowing a hundred men to fire 300 rounds per minute. George Knapp, "Rifled Musket, Springfield, Model 1861."

36   **Richard Gatling's motivation:** Keller, *Mr. Gatling's Terrible Marvel*, 7.

36   **"It occurred to me":** Ibid, 27.

36   **Gatling was an accomplished inventor:** Ibid, 71.

36   **"bears the same relation":** Ibid, 43.

38   **they were mowed down:** Ibid, 9.

39   **B. F. Skinner:** C. V. Glines, "Top Secret WWII Bat and Bird Bomber Program," His-

toryNet.com, June 12, 2006, http://www.historynet.com/top-secret-wwii-bat-and-bird-bomber-program.htm.

39 G7e/T4 *Falke*: "The Torpedoes," uboat.net, http://uboat.net/technical/torpedoes.htm, accessed June 19, 2017; "Torpedoes of Germany," NavWeaps, http://www.navweaps.com/Weapons/WTGER_WWII.php, accessed June 19, 2017; "TV (G7es) Acoustic Homing Torpedo," German U-Boat, http://www.uboataces.com/torpedo-tv.shtml, accessed June 19, 2017.

39 G7es/T5 *Zaunkönig*: Ibid.

39 tactic of diving immediately after launch: Ibid.

40 an improved acoustic seeker: "The Torpedoes." "Torpedoes of Germany." "TV (G7es) Acoustic Homing Torpedo."

41 Bat anti-ship glide bomb: Jim Sweeney, "Restoration: The Bat," *Air & Space Magazine* (January 2002), http://www.airspacemag.com/military-aviation/restoration-the-bat-2925632/.

41 going "maddog": Air Land Sea Application Center, "Brevity: Multi-Service Brevity Codes," February, 2002, http://www.dtic.mil/dtic/tr/fulltext/u2/a404426.pdf, I–19.

42 allows the missile to fly past other ships: "Harpoon Missiles," United States Navy Fact File, http://www.navy.mil/navydata/fact_display.asp?cid=2200&tid=200&ct=2, accessed June 19, 2017. "Harpoon," WeaponSystems.net, http://weaponsystems.net/weaponsystem/HH10+-+Harpoon.html, accessed June 19, 2017.

43 A weapon system consists of: The official DoD definition of "weapon system" takes a somewhat broader view, including the personnel and support equipment needed for a weapon to function: "A combination of one or more weapons with all related equipment, materials, services, personnel, and means of delivery and deployment (if applicable) required for self-sufficiency." U.S. Department of Defense, "weapon system," in "DoD Dictionary of Military and Associated Terms," March 2017, http://www.dtic.mil/doctrine/dod_dictionary/data/w/7965.html.

43 "Because 'precision munitions' require detailed data": Barry D. Watts, "Six Decades of Guided Munitions and Battle Networks: Progress and Prospects," Center for Strategic and Budgetary Assessments, Washington, DC, March, 2007, http://csbaonline.org/uploads/documents/2007.03.01-Six-Decades-Of-Guided-Weapons.pdf, ix.

45 At least thirty nations currently employ: Australia, Bahrain, Belgium, Canada, Chile, China, Egypt, France, Germany, Greece, India, Israel, Japan, Kuwait, the Netherlands, New Zealand, Norway, Pakistan, Poland, Portugal, Qatar, Russia, Saudi Arabia, South Africa, South Korea, Spain, Taiwan, the United Arab Emirates, the United Kingdom, and the United States. Paul Scharre and Michael Horowitz, "An Introduction to Autonomy in Weapon Systems," Center for a New American Security, February 2015, Appendix B, https://s3.amazonaws.com/files.cnas.org/documents/Ethical-Autonomy-Working-Paper_021015_v02.pdf.

46 Some loitering munitions keep humans in the loop: Dan Gettinger and Arthur Holland Michel, "Loitering Munitions," Center for the Study of the Drone at Bard College, February 10, 2017, http://dronecenter.bard.edu/files/2017/02/CSD-Loitering-Munitions.pdf.

47 Harpy has been sold to several countries: Israel Aerospace Industries, "Harpy NG," http://www.iai.co.il/Sip_Storage//FILES/5/41655.pdf. Israel Aerospace Industries, "Harpy NG," http://www.iai.co.il/2013/36694-16153-en/IAI.aspx. "Harpy," Israeli

-Weapons.com, http://www.israeli-weapons.com/weapons/aircraft/uav/harpy/HARPY
.html. "Harpy Air Defense Suppression System," Defense Update, http://defense
-update.com/directory/harpy.htm. Tamir Eshel, "IAI Introduces New Loitering
Weapons for Anti-Radiation, Precision strike," Defense-Update.com, February 15,
2016, http://defense-update.com/20160215_loitering-weapons.html.

47   **airborne for approximately four and a half minutes:** United States Navy,
"AGM-88 HARM Missile," February 20, 2009, http://www.navy.mil/navydata/
fact_display. asp?cid=2200&tid=300&ct=2.

48   **stay aloft for over two and a half hours:** Robert O'Gorman and Chriss Abbott,
"Remote Control War: Unmanned Combat Air Vehicles in China, India, Israel,
Iran, Russia, and Turkey" (Open Briefing, September 2013), 75, http://www
.oxfordresearchgroup.org.uk/sites/default/files/Remote%20Control%20War.pdf.

49   **Tomahawk Anti-Ship Missile (TASM):** Carlo Kopp, "Tomahawk Cruise Missile
Variants," Air Power Australia, http://www.ausairpower.net/Tomahawk-Subtypes
.html.

49   **Tomahawk Land Attack Missile [TLAM]:** U.S. Navy, "Tactical Tomahawk,"
http://www.navair.navy.mil/index.cfm?fuseaction=home.display&key=F4E98B0F
-33F5-413B-9FAE-8B8F7C5F0766.

49   **TASM was taken out of Navy service:** The Tomahawk Anti-Ship Missile (TASM)
refers to the BGM/RGM-109B, a Tomahawk variant that was employed by the U.S.
Navy from 1982 to 1994. In 2016, the Navy decided to reconfigure existing Tomahawk
Land Attack Missiles (TLAMs) to a new anti-ship version of the Tomahawk, which
would enter the force in 2021. Sam LaGrone, "WEST: U.S. Navy Anti-Ship Tomahawk Set for Surface Ships, Subs Starting in 2021," *USNI News*, February 18, 2016,
https://news.usni.org/2016/02/18/west-u-s-navy-anti-ship-tomahawk-set-for
-surface-ships-subs-starting-in-2021.

49   **Tacit Rainbow:** Carlo Kopp, "Rockwell AGM-130A/B and Northrop AGM-136A
Tacit Rainbow," Air Power Australia, last updated January 27, 2014, http://www
.ausairpower.net/TE-AGM-130-136.html. Andreas Parsch, "AGM/BGM-136,"
Designation-Systems.net, 2002, http://www.designation-systems.net/dusrm/m-136
.html.

49   **LOCAAS:** Andreas Parsch, "LOCAAS," Designation-Systems.net, 2006, http://
www.designation-systems.net/dusrm/app4/locaas.html

51   **110 million land mines:** Graca Machel, "Impact of Conflict on Armed Children,"
UNICEF, 1996, http://www.unicef.org/graca/.

51   **Ottawa Treaty:** International Committee of the Red Cross, "Convention on
the Prohibition of the Use, Stockpiling, Production and Transfer of Anti-Per-
sonnel Mines and on their Destruction, 18 September 1997," ICRC.org, https://
ihl-databases.icrc.org/applic/ihl/ihl.nsf/States.xsp?xp_viewStates=XPages
_NORMStatesParties&xp_treatySelected=580.

51   **Antitank land mines and naval mines are still permitted:** From 2004 to 2014,
the U.S. land mine policy was to use only self-destructing/self-deactivating mines.
U.S. Department of State, "New United States Policy on Landmines: Reducing
Humanitarian Risk and Saving Lives of United States Soldiers," U.S. Department
of State Archive, February 27, 2004, http://2001-2009.state.gov/t/pm/rls/fs/30044
.htm. In 2014, the United States established a new policy of aligning itself with the
requirements of the Ottawa Treaty, with the exception of the Korean Peninsula.
U.S. Department of State, "U.S. Landmine Policy," State.gov, http://www.state
.gov/t/pm/wra/c11735.htm.

51   Encapsulated torpedo mines: "Mine Warfare Trends," Mine Warfare Association, presentation, May 10, 2011, www.minwara.org/Meetings/2011_05/Presentations/tuespdf/WMason_0900/MWTrends.pdf.

51   Mk 60 CAPTOR: Federation of American Scientists, "MK60 Encapsulated Torpedo (CAPTOR)," FAS Military Analysis Network, December 13, 1998, https://web.archive.org/web/20160902152533/http://fas.org:80/man/dod-101/sys/dumb/mk60.htm.

51   Russian PMK-2: Scott C. Truver, "Taking Mines Seriously: Mine Warfare in China's Near Seas," Naval War College Review, 65 no. 2 (Spring 2012), 40–41, https://www.usnwc.edu/getattachment/19669a3b-6795-406c-8924-106d7a5adb93/Taking-Mines-Seriously--Mine-Warfare-in-China-s-Ne. Andrew S. Erickson, Lyle J. Goldstein, and William S. Murray, "Chinese Mine Warfare," China Maritime Studies Institute, (Newport, RI: Naval War College, 2009), 16, 20, 28–30, 44, 90.

51   Sensor Fuzed Weapon: Textron Systems, "SFW: Sensor Fuzed Weapon," video, published on November 21, 2015, https://www.youtube.com/watch?v=AEXMHf2Usso.

53   TASM was in service in the U.S. Navy from 1982 to 1994: "Harpoon," NavSource Online, http://www.navsource.org/archives/01/57s1.htm. James C. O'Halloran, ed., "RGM/UGM-109 Tomahawk," IHS Jane's Weapons: Strategic, 2015-2016 (United Kingdom, 2015), 219-223. Carlo Kopp, "Tomahawk Cruise Missile Variants." "AGM/BGM/RGM/UGM-109," Designation-Systems.net, http://www.designation-systems.net/dusrm/m-109.html.

53   "lack of confidence in how the targeting picture": Bryan McGrath, interview, May 19, 2016.

54   "a weapon we just didn't want to fire": Ibid.

54   "Because the weapons cost money": Ibid.

55   Harpy 2, or Harop: Publicly available documents are unclear on whether the Harop retains the Harpy's ability to conduct fully autonomous anti-radar engagements as one mode of operation. Israel Aerospace Industries, developer of the Harpy and Harop, declined to comment on details of the Harpy and Harop functionality. Israel Aerospace Industries, "Harop," http://www.iai.co.il/2013/36694-46079-EN/Business_Areas_Land.aspx.

55   "You've got to talk to the missile:": Bryan McGrath, interview, May 19, 2016.

56   At least sixteen countries already possess: The United States, United Kingdom, Israel, China, Nigeria, Iran, Iraq, Jordan, Egypt, United Arab Emirates, Saudi Arabia, Kazakhstan, Turkmenistan, Pakistan, Myanmar, Turkey. Fuhrmann and Horowitz, "Droning On: Explaining the Proliferation of Unmanned Aerial Vehicles."

4   The Future Being Built Today:
Autonomous Missiles, Drones, and Robot Swarms

60   the MQ-25 is envisioned primarily as a tanker: Sydney J. Freedburg, Jr., "CBARS Drone Under OSD Review; Can A Tanker Become A Bomber?" Breaking Defense, February 19, 2016, http://breakingdefense.com/2016/02/cbars-drone-under-osd-review-can-a-tanker-become-a-bomber/. Richard Whittle, "Navy Refueling Drone May Tie Into F-35s," Breaking Defense, March 22, 2016, http://breakingdefense.com/2016/03/navy-refueling-drone-may-tie-into-f-35s-f-22s/.

61   2013 Remotely Piloted Aircraft Vector: United States Air Force, "RPA Vector: Vision and Enabling Concepts, 2013–2038," February 17, 2014, http://www

.defenseinnovationmarketplace.mil/resources/USAF-RPA_VectorVision
EnablingConcepts2013-2038_ForPublicRelease.pdf.

62   **cultural resistance to combat drones:** Jeremiah Gertler, "History of the Navy
     UCLASS Program Requirements: In Brief," Congressional Research Service,
     August 3, 2015, https://www.fas.org/sgp/crs/weapons/R44131.pdf.

62   **uninhabited combat aerial vehicle:** "UCAV" specifically refers to the air vehicle
     whereas "UCAS" refers to the entire system: air vehicle, communications links, and
     ground control station. In practice, the terms are often used interchangeably.

62   **China has developed anti-ship ballistic and cruise missiles:** Kelley Sayler,
     "Red Alert: The Growing Threat to U.S. Aircraft Carriers," Center for a New
     American Security, Washington, DC, February 22, 2016, https://www.cnas.org/
     publications/reports/red-alert-the-growing-threat-to-u-s-aircraft-carriers. Jerry
     Hendrix, "Retreat from Range: The Rise and Fall of Carrier Aviation," Center for a
     New American Security, Washington, DC, October 19, 2015, https://www.cnas.org/
     publications/reports/retreat-from-range-the-rise-and-fall-of-carrier-aviation.

62   **Sea power advocates outside the Navy:** Sam LaGrone, "Compromise Defense
     Bill Restricts Navy UCLASS Funds," *USNI News,* December 3, 2014, https://
     news.usni.org/2014/12/03/compromise-defense-bill-restricts-navy-uclass
     -funds. Sam LaGrone, "McCain Weighs in on UCLASS Debate, Current Navy
     Requirements 'Strategically Misguided,'" *USNI News,* March 24, 2015, https://
     news.usni.org/2015/03/24/mccain-weighs-in-on-uclass-debate-current-navy
     -requirements-strategically-misguided.

62   **the Navy is deferring any plans for a future UCAV:** Sydney J. Freedburg, Jr.,
     "Navy Hits Gas On Flying Gas Truck, CBARS: Will It Be Armed?" *Breaking
     Defense,* March 11, 2016, http://breakingdefense.com/2016/03/navy-hits-gas-on
     -flying-gas-truck-cbars-will-it-be-armed/.

62   **a range of only 67 nautical miles:** 67 nautical miles equals 124 kilometers.

62   **can fly up to 500 nautical miles:** 500 nautical miles equals 930 kilometers.

63   **three *New York Times* articles:** John Markoff, "Fearing Bombs That Can Pick
     Whom to Kill," *New York Times,* November 11, 2014, http://www.nytimes.com/
     2014/11/12/science/weapons-directed-by-robots-not-humans-raise-ethical
     -questions.html?_r=0. John Markoff, "Report Cites Dangers of Autonomous
     Weapons," *New York Times,* February 26, 2016, http://www.nytimes.com/2016/
     02/29/technology/report-cites-dangers-of-autonomous-weapons.html.
     John Markoff, "Arms Control Groups Urge Human Control of Robot Weap-
     onry," *New York Times,* April 11, 2016, http://www.nytimes.com/2016/04/12/
     technology/arms-control-groups-urge-human-control-of-robot-weaponry
     .html.

63   **"artificial intelligence outside human control":** Markoff, "Fearing Bombs That
     Can Pick Whom to Kill."

63   **"an autonomous weapons arms race":** Ibid.

63   **"LRASM employed precision routing and guidance":** Lockheed Martin, "Long
     Range Anti-Ship Missile," http://www.lockheedmartin.com/us/products/LRASM/
     overview.html (accessed on May 15, 2017).

63   **Lockheed's description of LRASM:** Lockheed Martin, "Long Range Anti-Ship Mis-
     sile," as of October 20, 2014, https://web.archive.org/web/20141020231650/http://
     www.lockheedmartin.com/us/products/LRASM.html.

64   **"The semi-autonomous guidance capability gets LRASM":** Lockheed Martin,
     "Long Range Anti-Ship Missile," as of December 16, 2014, https://web.archive

.org/web/20141216100706/http://www.lockheedmartin.com/us/products/LRASM.html.

64   **video online that explains LRASM's functionality:** The video is no longer available on the Lockheed Martin website. Lockheed Martin, "LRASM: Long Range Anti-Ship Missile," published on May 3, 2016, archived on December 15, 2016, https://web.archive.org/web/20160504083941/https://www.youtube.com/watch?v=6eFGPIg05q0&gl=US&hl=en.

68   **literally wrote the textbook:** Stuart Russell and Peter Norvig, *Artificial Intelligence: A Modern Approach,* 3rd ed. (Boston: Pearson, 2009).

68   **"offensive autonomous weapons beyond meaningful human control":** "Autonomous Weapons: An Open Letter From AI & Robotics Researchers," Future of Life Institute, https://futureoflife.org/open-letter-autonomous-weapons/.

69   **"The challenge for the teams now":** DARPA, "FLA Program Takes Flight," DARPA.mil, February 12, 2016, http://www.darpa.mil/news-events/2016-02-12.

69   **"FLA technologies could be especially useful":** Ibid.

70   **Lee explained:** Daniel Lee, email to author, June 3, 2016.

70   **"localization, mapping, obstacle detection":** Ibid.

70   **"applications to search and rescue":** Vijay Kumar, email to author, June 3, 2016.

71   **"foreshadow planned uses":** Stuart Russell, "Take a Stand on AI Weapons," Nature.com, May 27, 2015, http://www.nature.com/news/robotics-ethics-of-artificial-intelligence-1.17611.

71   **wasn't "cleanly directed only at":** Stuart Russell, interview, June 23, 2016.

71   **"You can make small, lethal quadcopters":** Ibid.

71   **"if you were wanting to develop autonomous weapons":** Ibid.

71   **"certainly think twice" about working on:** Ibid.

72   **"collaborative autonomy—the capability of groups":** DARPA, "Collaborate Operations in Denied Environments," DARPA.com, http://www.darpa.mil/program/collaborative-operations-in-denied-environment.

72   **"just as wolves hunt in coordinated packs":** DARPA, "Establishing the CODE for Unmanned Aircraft to Fly as Collaborative Teams," DARPA.com, http://www.darpa.mil/news-events/2015-01-21.

72   **"multiple CODE-enabled unmanned aircraft":** Ibid.

72   **Graphics on DARPA's website:** DARPA, "Collaborate Operations in Denied Environments."

72   **"contested electromagnetic environments":** Ibid.

73   **methods of communicating stealthily:** Sayler, "Talk Stealthy to Me." Amy Butler, "5th-To-4th Gen Fighter Comms Competition Eyed In Fiscal 2015," *AWIN First,* June 18, 2014, http://aviationweek.com/defense/5th-4th-gen-fighter-comms-competition-eyed-fiscal-2015.

73   **56K dial-up modem:** DARPA, "Broad Agency Announcement: Collaborative Operations in Denied Environment (CODE) Program," DARPA-BAA-14-33, April 25, 2014, 13, available at https://www.fbo.gov/index?s=opportunity&mode=form&id=2f2733be59230cf2ddaa46498fe5765a&tab=core&_cview=1.

73   **"under a single person's supervisory control":** DARPA, "Collaborative Operations in Denied Environment."

73   **A May 2016 video released online:** DARPA, "Collaborative Operations in Denied Environment (CODE): Test of Phase 1 Human-System Interface," https://www.youtube.com/watch?v=o8AFuiO6ZSs&feature=youtu.be.

74   **under the supervision of the human commander:** DARPA, "Collaborative

Operations in Denied Environment (CODE): Phase 2 Concept Video," https://www.youtube.com/watch?v=BPBuE6fMBnE.

75   **The CODE website says:** DARPA, "Collaborative Operations in Denied Environment."

75   **"Provide a concise but comprehensive targeting chipset":** DARPA, "Broad Agency Announcement: Collaborative Operations in Denied Environment (CODE) Program," 10.

75   **"Autonomous and semi-autonomous weapon systems shall be":** Department of Defense, "Department of Defense Directive Number 3000.09: Autonomy in Weapon Systems," November 21, 2012, http://www.dtic.mil/whs/directives/corres/pdf/300009p.pdf, 2.

75   **approval to build and deploy autonomous weapons:** Ibid, 7–8.

76   **"providing multi-modal sensors":** DARPA, "Broad Agency Announcement: Collaborative Operations in Denied Environment (CODE) Program," 6.

77   **Stuart Russell said that he found these projects concerning:** Stuart Russell, interview, June 23, 2016.

5   Inside the Puzzle Palace:
Is the Pentagon Building Autonomous Weapons?

78   **Klingon Bird of Prey:** Bob Work, "Remarks at the ACTUV 'Seahunter' Christening Ceremony," April 7, 2016, https://www.defense.gov/News/Speeches/Speech-View/Article/779197/remarks-at-the-actuv-seahunter-christening-ceremony/.

79   **"fighting ship":** Ibid.

79   **"You can imagine anti-submarine warfare pickets":** Ibid.

80   **"Our fundamental job":** Bradford Tousley, interview, April 27, 2016.

80   **"That final decision is with humans, period":** Ibid.

81   **"Until the machine processors equal or surpass":** Ibid.

81   **"Groups of platforms that are unmanned":** Ibid.

81   **"We're using physical machines and electronics":** Ibid.

82   **"As humans ascend to the higher-level":** Ibid.

82   **"We expect that there will be jamming":** Ibid.

82   **"I think that will be a rule of engagement-dependent decision":** Ibid.

83   **"If [CODE] enables software":** Ibid.

84   **"In a target-dense environment":** DARPA, "Target Recognition and Adaptation in Contested Environments," http://www.darpa.mil/program/trace.

84   **"develop algorithms and techniques":** DARPA, "Broad Agency Announcement: Target Recognition and Adaptation in Contested Environments (TRACE)," DARPA-BAA-15-09, December 1, 2014, 6, https://www.fbo.gov/index?s=opportunity&mode=form&id=087d9fba51700a89d154e8c9d9fdd93d&tab=core&_cview=1.

87   *Deep* **neural networks:** Alex Krizhevsky, Ilya Sutskever, and Geoffrey E. Hinton, "ImageNet Classification with Deep Convolutional Neural Networks," https://papers.nips.cc/paper/4824-imagenet-classification-with-deep-convolutional-neural-networks.pdf.

87   **over a hundred layers:** Christian Szegedy et al., "Going Deeper With Convolutions," https://www.cs.unc.edu/~wliu/papers/GoogLeNet.pdf.

87   **error rate of only 4.94 percent:** Richard Eckel, "Microsoft Researchers' Algorithm Sets ImageNet Challenge Milestone," Microsoft Research Blog, February 10, 2015,

https://www.microsoft.com/en-us/research/microsoft-researchers-algorithm
-sets-imagenet-challenge-milestone/. Kaiming He et al., "Delving Deep into Rectifiers: Surpassing Human-Level Performance on ImageNet Classification," https://arxiv
.org/pdf/1502.01852.pdf.

87   **estimated 5.1 percent error rate:** Olga Russakovsky et al., "ImageNet Large Scale
Visual Recognition Challenge," January 20, 2015, https://arxiv.org/pdf/1409.0575
.pdf.

87   **3.57 percent rate:** Kaiming He et al., "Deep Residual Learning for Image Recognition," December 10, 2015, https://arxiv.org/pdf/1512.03385v1.pdf.

6   Crossing the Threshold: Approving Autonomous Weapons

89   **delineation of three classes of systems:** Department of Defense, "Department of
Defense Directive Number 3000.09."

90   **"minimize the probability and consequences":** Ibid, 1.

91   **"We haven't had anything that was even remotely close":** Frank Kendall, interview, November 7, 2016.

91   **"We had an automatic mode":** Ibid.

91   **"relatively soon":** Ibid.

91   **"sort through all that":** Ibid.

91   **"Are you just driving down":** Ibid.

92   **"other side of the equation":** Ibid.

92   **"a reasonable question to ask":** Ibid.

92   **"where technology supports it":** Ibid.

92   **"principles and obey them":** Ibid.

93   **"Automation and artificial intelligence are":** Ibid.

93   **Work explained in a 2014 monograph:** Robert O. Work and Shawn Brimley, "20YY:
Preparing for War in the Robotic Age," *Center for a New American Security*, January
2014, 7–8, http://www.cnas.org/sites/default/files/publications-pdf/CNAS_20YY_
WorkBrimley.pdf.

95   **"We will not delegate lethal authority":** Bob Work, interviewed by David Ignatius,
"Securing Tomorrow," March 30, 2016, https://static.dvidshub.net/media/video/
1603/DOD_103167280/DOD_103167280-512x288-442k.mp4. Comments on autonomy start around 29:00.

95   **"We might be going up against":** Ibid.

95   **"our potential competitors may not":** Bob Work, interviewed by August Cole,
"Global Strategy Forum," Atlantic Council, http://www.atlanticcouncil.org/events/
webcasts/2016-global-strategy-forum. Comments starting around 38:00.

96   **"We, the United States, have had":** Bob Work, interview, June 22, 2016.

97   **"We are moving to a world":** Ibid.

97   **"The thing that people worry about":** Ibid.

98   **"same determination that we have right now":** Ibid.

98   **"What is your comfort level on target":** Ibid.

98   **"I hear people say":** Ibid.

98   **Work contrasted these narrow AI systems:** Ibid.

99   **"People are going to use AI":** Ibid.

100 **Schuette made it clear to me:** Larry Schuette, interview, May 5, 2016.

100 **"The man pushes a button":** Ibid.

100 **"History is full of innovations":** Ibid.

101 **"We've had these debates before":** Ibid.

101 **"EXECUTE AGAINST JAPAN":** Joel Ira Holwitt, "'EXECUTE AGAINST JAPAN':
Freedom-of-the-Seas, The U.S. Navy, Fleet Submarines, and the U.S. Decision to
Conduct Unrestricted Submarine Warfare, 1919–1941," Dissertation, Ohio State
University, 2005, https://etd.ohiolink.edu/rws_etd/document/get/osu1127506553/
inline.

101 **"Is it December eighth or December sixth?":** Larry Schuette, interview, May 5,
2016.

7 World War R: Robotic Weapons around the World

102 **"A growing number of countries":** The United States, United Kingdom, Israel,
China, Nigeria, Iran, Iraq, Jordan, Egypt, United Arab Emirates, Saudi Arabia,
Kazakhstan, Turkmenistan, Pakistan, Myanmar, Turkey. Fuhrmann and Horo-
witz, "Droning On: Explaining the Proliferation of Unmanned Aerial Vehicles."

102 **Even Shiite militias in Iraq:** David Axe, "An Iraqi Shi'ite Militia Now Has Ground
Combat Robots," *War Is Boring*, March 23, 2015, https://warisboring.com/an-iraqi
-shi-ite-militia-now-has-ground-combat-robots-68ed69121d21#.hj0vxomjl.

103 **armed uninhabited boat, the Protector:** Berenice Baker, "No Hands on Deck—
Arming Unmanned Surface Vessels," naval-technology.com, November 23, 2012,
http://www.naval-technology.com/features/featurehands-on-deck-armed
-unmanned-surface-vessels/.

103 **Singapore has purchased the Protector:** "Protector Unmanned Surface Vessel,
Israel," naval-technology.com, http://www.naval-technology.com/projects/
protector-unmanned-surface-vehicle/.

103 **ESGRUM:** "BAE ESGRUM USV," NavalDrones.com, http://www.navaldrones
.com/BAE-ESGRUM.html.

103 **Only twenty-two nations have said they support:** Campaign to Stop Killer Robots,
"Country Views on Killer Robots," November 16, 2017, http://www.stopkillerrobots
.org/wp-content/uploads/2013/03/KRC_CountryViews_16Nov2017.pdf.

104 **These include the United Kingdom's Taranis:** Nicholas de Larringa, "France Begins
Naval Testing of Neuron UCAV," IHS Jane's Defence Weekly, May 19, 2016; archived at
https://web.archive.org/web/20161104112421/http://www.janes.com/article/60482/
france-begins-naval-testing-of-neuron-ucav. "New Imagery Details Indian Aura
UCAV," *Aviation Week & Space Technology,* July 16, 2012, http://aviationweek.com/
awin/new-imagery-details-indian-aura-ucav. "Israel Working on Low-Observable
UAV," FlightGlobal, November 28, 2012, https://www.flightglobal.com/news/articles/
israel-working-on-low-observable-uav-379564/.

104 **a handful of countries already possess the fully autonomous Harpy:** Tamir Eshel,
"IAI Introduces New Loitering Weapons for Anti-Radiation, Precision Strike."

105 **"the ultimate decision about shooting":** Jean Kumagai, "A Robotic Sentry For
Korea's Demilitarized Zone," *IEEE Spectrum: Technology, Engineering, and Science News,* March 1, 2007, http://spectrum.ieee.org/robotics/military-robots/
a-robotic-sentry-for-koreas-demilitarized-zone.

105 **SGR-A1 cited as an example:** Christopher Moyer, "How Google's AlphaGo Beat a
Go World Champion," *The Atlantic,* March 28, 2016, https://www.theatlantic.com/
technology/archive/2016/03/the-invisible-opponent/475611/. Adrianne Jeffries,

"Should a Robot Decide When to Kill?," *The Verge*, January 28, 2014, https://www.theverge.com/2014/1/28/5339246/war-machines-ethics-of-robots-on-the-battlefield. "Future Tech? Autonomous Killer Robots Are Already Here," *NBC News*, May 16, 2014, http://www.nbcnews.com/tech/security/future-tech-autonomous-killer-robots-are-already-here-n105656. Sharon Weinberger, "Next Generation Military Robots Have Minds of Their Own," accessed June 18, 2017, http://www.bbc.com/future/story/20120928-battle-bots-think-for-themselves. "The Scariest Ideas in Science," *Popular Science*, accessed June 18, 2017, http://www.popsci.com/scitech/article/2007-02/scariest-ideas-science.

105 **"WHY, GOD? WHY?":** "The Scariest Ideas in Science."

105 **cited the SGR-A1 as fully autonomous:** Ronald C. Arkin, *Governing Lethal Behavior in Autonomous Robots* (Boca Raton, FL: Taylor & Francis Group, 2009), 10. Patrick Lin, George Bekey, and Keith Abney, "Autonomous Military Robotics: Risk, Ethics, and Design," California Polytechnic State University, San Luis Obispo, CA, December 20, 2008, http://digitalcommons.calpoly.edu/cgi/viewcontent.cgi?article=1001&context=phil_fac; Hin-Yan Liu, "Categorization and legality of autonomous and remote weapons systems," *International Review of the Red Cross* 94, no. 886 (Summer 2012), https://www.icrc.org/eng/assets/files/review/2012/irrc-886-liu.pdf.

105 **In 2010, a spokesperson for Samsung:** "Machine Gun-Toting Robots Deployed on DMZ," *Stars and Stripes*, July 12, 2010, https://www.stripes.com/machine-gun-toting-robots-deployed-on-dmz-1.110809.

105 **"the SGR-1 can and will prevent wars":** Ibid.

105 **critics who have questioned whether it has too much:** Markoff, "Fearing Bombs That Can Pick Whom to Kill."

106 **Brimstone has two primary modes of operation:** MBDA, "Brimstone 2 Data Sheet November 2015," accessed June 7, 2017, https://mbdainc.com/wp-content/uploads/2015/11/Brimstone2-Data-Sheet_Nov-2015.pdf.

106 **"This mode provides through-weather targeting":** Ibid.

106 **"It can identify, track, and lock on":** David Hambling Dec 4 and 2015, "The U.K. Fights ISIS With a Missile the U.S. Lacks," *Popular Mechanics*, December 4, 2015, http://www.popularmechanics.com/military/weapons/a18410/brimstone-missile-uk-david-cameron-isis/.

107 **"In May 2013, multiple Brimstone":** MBDA, "Brimstone 2 Data Sheet."

107 **reported range in excess of 20 kilometers:** "Dual-Mode Brimstone Missile Proves Itself in Combat," *Defense Media Network*, April 26, 2012, http://www.defensemedianetwork.com/stories/dual-mode-brimstone-missile-proves-itself-in-combat/.

107 **Brimstone can engage these targets:** I am grateful to MBDA Missile Systems for agreeing to an interview on background in 2016 to discuss the Brimstone's functionality.

109 **"1. Taranis would reach the search area":** BAE Systems, "Taranis: Looking to the Future," http://www.baesystems.com/en/download-en/20151124120336/1434555376407.pdf.

109 **BAE Chairman Sir Roger Carr:** Sir Roger Carr, "What If: Robots Go to War?—World Economic Forum Annual Meeting 2016 | World Economic Forum," video, accessed June 7, 2017, https://www.weforum.org/events/world-economic-forum-annual-meeting-2016/sessions/what-if-robots-go-to-war.

109 **"decisions to release a lethal mechanism"**: John Ingham, "WATCH: Unmanned Test Plane Can Seek and Destroy Heavily Defended Targets," Express.co .uk, June 9, 2016, http://www.express.co.uk/news/uk/678514/WATCH-Video -Unmanned-test-plane-Taranis.

109 **"The UK does not possess"**: UK Ministry of Defence, "Defence in the Media: 10 June 2016," June 10, 2016, https://modmedia.blog.gov.uk/2016/06/10/defence-in-the -media-10-june-2016/.

110 **"must be capable of achieving"**: UK Ministry of Defence, "Joint Doctrine Note 2/11: The UK Approach to Unmanned Aircraft Systems," March 30, 2011, https:// www.gov.uk/government/uploads/system/uploads/attachment_data/file/33711/ 20110505JDN_211_UAS_v2U.pdf, 2-3.

110 **"As computing and sensor capability increases"**: Ibid, 2–4.

111 **one short-lived effort during the Iraq war**: "The Inside Story of the SWORDS Armed Robot 'Pullout' in Iraq: Update," *Popular Mechanics*, October 1, 2009, http:// www.popularmechanics.com/technology/gadgets/4258963.

112 **"the military robots were assigned"**: Alexander Korolkov and special to RBTH, "New Combat Robot Is Russian Army's Very Own Deadly WALL-E," *Russia Beyond The Headlines*, July 2, 2014, https://www.rbth.com/defence/2014/07/02/new _combat_robot_is_russian_armys_very_own_deadly_wall-e_37871.html.

112 **"Platform-M . . . is used"**: Ibid.

112 **videos of Russian robots show soldiers**: This video (https://www.youtube.com/ watch?v=RBi977p0plA) is no longer available online.

114 **According to David Hambling of *Popular Mechanics***: David Hambling, "Check Out Russia's Fighting Robots," *Popular Mechanics*, May 12, 2014, http:// www.popularmechanics.com/technology/military/robots/russia-wants-autonomous -fighting-robots-and-lots-of-them-16787165.

114 **amphibious Argo**: "Battle Robotic complex 'Argo'—Military Observer," accessed June 7, 2017, http://warsonline.info/bronetechnika/boevoy-robotizirovanniy-kompleks -argo.html.

114 **Pictures online show Russian soldiers**: Tamir Eshel, "Russian Military to Test Combat Robots in 2016," Defense Update, December 31, 2015, http://defense-update .com/20151231_russian-combat-robots.html.

115 **slo-mo shots of the Uran-9 firing**: Rosoboronexport, *Combat Robotic System Uran-9*, n.d., https://www.youtube.com/watch?v=VBC9BM4-3Ek.

115 **Uran-9 could make the modern battlefield a deadly place**: Rich Smith, "Russia's New Robot Tank Could Disrupt the International Arms Market," *The Motley Fool*, February 7, 2016, https://www.fool.com/investing/general/2016/02/07/russias -new-robot-tank-could-disrupt-international.aspx.

115 **Vikhr . . . "lock onto a target"**: Simon Holmes, "Russian Army Puts New Remote-Controlled Robot Tank to Test," *Mail Online*, April 29, 2017, http://www.dailymail .co.uk/~/article-4457892/index.html.

115 **DJI's base model Spark**: "Spark," DJI.com, http://www.dji.com/spark.

116 **T-14 Armata**: Alex Lockie, "Russia Claims Its Deadly T-14 Armata Tank Is in Full Production," *Business Insider*, March 17, 2016, http://www.businessinsider.com/ russia-claims-t14-armata-tank-is-in-production-2016-3.

116 **"Quite possibly, future wars will be waged"**: "Engineers Envisioned T-14 Tank 'robotization' as They Created Armata Platform," *RT International*, June 1, 2015, https://www.rt.com/news/263757-armata-platform-remote-control/.

116 **"fully automated combat module":** "Kalashnikov Gunmaker Develops Combat Module based on Artificial Intelligence," TASS Russian News Agency, July 5, 2017, http://tass.com/defense/954894.

117 **"Another factor influencing the essence of modern":** Coalson, "Top Russian General Lays Bare Putin's Plan for Ukraine."

117 **Deputy Secretary of Defense Bob Work mentioned:** "Remarks by Defense Deputy Secretary Robert Work at the CNAS Inaugural National Security Forum," December 14, 2015, https://www.cnas.org/publications/transcript/remarks-by-defense-deputy-secretary-robert-work-at-the-cnas-inaugural-national-security-forum.

117 **some have suggested, that a dangerous arms race:** Markoff, "Fearing Bombs That Can Pick Whom to Kill."

118 **Policy discussions may be happening:** Thanks to Sam Bendett for pointing out this important distinction.

118 **NGO Article 36:** Article 36, "The United Kingdom and Lethal Autonomous Weapon Systems," April 2016, http://www.article36.org/wp-content/uploads/2016/04/UK-and-LAWS.pdf

8   Garage Bots: DIY Killer Robots

120 **fifteen-second video clip . . . Connecticut teenager:** Rick Stella, "Update: FAA Launches Investigation into Teenager's Gun-Wielding Drone Video," *Digital Trends*, July 22, 2015, https://www.digitaltrends.com/cool-tech/man-illegally-straps-handgun-to-a-drone/.

120 **For under $500:** "Spark," DJI.com.

122 **Shield AI:** "Shield AI," http://shieldai.com/

122 **grant from the U.S. military:** Mark Prigg, "Special Forces developing 'AI in the sky' drones that can create 3D maps of enemy lairs: Pentagon reveals $1m secretive 'autonomous tactical airborne drone' project," DailyMail.com, http://www.dailymail.co.uk/sciencetech/article-3776601/Special-Forces-developing-AI-sky-drones-create-3D-maps-enemy-lairs-Pentagon-reveals-1m-secretive-autonomous-tactical-airborne-drone-project.html.

123 **"Robotics and artificial intelligence are":** Brandon Tseng, email to author, June 17, 2016.

124 **"fully automated combat module":** "Kalashnikov Gunmaker Develops Combat Module based on Artificial Intelligence."

125 **more possible positions in *go*:** "AlphaGo," *DeepMind*, accessed June 7, 2017, https://deepmind.com/research/alphago/.

125 **"Our goal is to beat the best human players":** "AlphaGo: Using Machine Learning to Master the Ancient Game of Go," *Google*, January 27, 2016, http://blog.google:443/topics/machine-learning/alphago-machine-learning-game-go/.

126 **game 2, on move 37:** Daniel Estrada, "Move 37!! Lee Sedol vs AlphaGo Match 2" video, https://www.youtube.com/watch?v=JNrXgpSEEIE.

126 **"I thought it was a mistake":** Ibid.

126 **"It's not a human move":** Cade Metz, "The Sadness and Beauty of Watching Google's AI Play Go," *WIRED*, March 11, 2016, https://www.wired.com/2016/03/sadness-beauty-watching-googles-ai-play-go/.

126 **1 in 10,000:** Cade Metz, "In Two Moves, AlphaGo and Lee Sedol Redefined

the Future," *WIRED*, accessed June 7, 2017, https://www.wired.com/2016/03/two-moves-alphago-lee-sedol-redefined-future/.

126 **"I kind of felt powerless":** Moyer, "How Google's AlphaGo Beat a Go World Champion."

126 **"AlphaGo isn't just an 'expert' system":** "AlphaGo," January 27, 2016.

127 **AlphaGo Zero:** "AlphaGo Zero: Learning from Scratch," DeepMind, accessed October 22, 2017, https://deepmind.com/blog/alphago=zero=learning=scratch/.

127 **neural network to play Atari games:** Volodymyr Mnih et al., "Human-Level Control through Deep Reinforcement Learning," *Nature* 518, no. 7540 (February 26, 2015): 529–33.

127 **deep neural network:** JASON, "Perspectives on Research in Artificial Intelligence and Artificial General Intelligence Relevant to DoD."

129 **Inception-v3:** Inception-v3 is trained for the Large Scale Visual Recognition Challenge (LSVRC) using the 2012 data. "Image Recognition," *TensorFlow*, accessed June 7, 2017, https://www.tensorflow.org/tutorials/image_recognition.

129 **one of 1,000 categories:** The categories available for Inception-v3 are those from the Large Scale Visual Recognition Challenge (LSVRC) 2012, which is the same as the LSVRC 2014. "ImageNet Large Scale Visual Recognition Competition 2014 (ILSVRC2014)," accessed June 7, 2017, http://image-net.org/challenges/LSVRC/2014/browse-synsets.

129 **Pascal Visual Object Classes database:** "The PASCAL Visual Object Classes Challenge 2012 (VOC2012)," accessed June 7, 2017, http://host.robots.ox.ac.uk/pascal/VOC/voc2012/index.html.

130 **FIRST Robotics Competition:** "FIRST Robotics Competition 2016 Season Facts," FirstInspires.org, http://www.firstinspires.org/sites/default/files/uploads/resource_library/frc-2016-season-facts.pdf.

131 **"They can pretty much program in anything":** Charles Dela Cuesta, interview, May 20, 2016.

132 **"The stuff that was impressive to me":** Ibid.

133 **"I don't think we're ever going to give":** Brandon Tseng, email to author, June 17, 2016.

9   Robots Run Amok: Failure in Autonomous Systems

137 **March 22, 2003:** This account is based on official public records of the fratricide, including an investigation by the UK government in regard to the Tornado fratricide and a Defense Science Board report on both incidents. It is also based, in part, on individuals close to both incidents who asked to remain anonymous as they were not authorized to speak on the record about the incidents. Ministry of Defence, "Aircraft Accident to Royal Air Force Tornado GR MK4A ZG710," https://www.gov.uk/government/uploads/system/uploads/attachment_data/file/82817/maas03_02_tornado_zg710_22mar03.pdf. Defense Science Board, "Report of the Defense Science Board Task Force on Patriot Performance," January 2005, http://www.acq.osd.mil/dsb/reports/2000s/ADA435837.pdf.

138 **All they had was a radio:** Ministry of Defence, "Aircraft Accident to Royal Air Force Tornado GR MK4A ZG710."

138 **attack on a nearby base:** "Army: U.S. Soldier Acted Out of Resentment in Grenade Attack," *Associated Press*, (March 24, 2003), http://www.foxnews.com/

story/2003/03/24/army-us-soldier-acted-out-resentment-in-grenade-attack
.html.

138 **Tornado GR4A fighter jet:** Staff and Agencies, "'Glaring Failures' Caused US to
Kill RAF Crew," *The Guardian*, October 31, 2006, https://www.theguardian.com/
uk/2006/oct/31/military.iraq.

138 **It could be because the system simply broke:** A UK board of inquiry discounted
the possibility that Main and Williams had intentionally turned off the IFF,
although without explaining why. In the absence of other possible explanations, it
concluded the IFF system "had a fault." Ministry of Defence, "Aircraft Accident to
Royal Air Force Tornado GR MK4A ZG710," 4–5.

139 **Their screen showed a radar-hunting enemy missile:** Ministry of Defence, "Aircraft Accident to Royal Air Force Tornado GR MK4A ZG710."

139 **parabolic trajectory:** This is an approximation. Lior M. Burko and Richard H.
Price, "Ballistic Trajectory: Parabola, Ellipse, or What?," May 17, 2004, https://
arxiv.org/pdf/physics/0310049.pdf.

139 **The Tornado's IFF signal:** There are actually two relevant IFF modes of operation
that might have prevented the Tornado fratricide, Mode 1 and Mode 4. More 4 is the
standard encrypted military IFF used by coalition aircraft in theater. It was tested
on the Tornado prior to starting engines on the ground and found functional, but
there is no evidence it was broadcasting at any point during the flight. The reasons
why are unclear. Mode 1 is an unencrypted mode that all coalition aircraft were
supposed to be using as a backup. The UK accident investigation report does not
specify whether this mode was broadcasting or not. In any case, the Patriot battery
did not have the codes for Mode 1, so they couldn't have received a Mode 1 signal in
any case. This is likely because their equipment was not interoperable and so they
were not on the network. The Mode 1 codes would have had to have been delivered
by hand to be loaded, and apparently they were not. Ministry of Defence, "Aircraft
Accident to Royal Air Force Tornado GR MK4A ZG710."

139 **They had seconds to decide:** John K. Hawley, "Looking Back at 20 Years of MAN-
PRINT on Patriot: Observations and Lessons," Army Research Laboratory, September, 2007, 7, http://www.dtic.mil/docs/citations/ADA472740.

139 **Main and Williams' wingman landed in Kuwait:** Stewart Payne, "US Colonel Says
Sorry for Tornado Missile Blunder," March 25, 2003, http://www.telegraph.co.uk/
news/worldnews/middleeast/iraq/1425545/US-colonel-says-sorry-for-Tornado
-missile-blunder.html.

141 **Patriot crew was unharmed:** "F-16 Fires on Patriot Missile Battery," *Associated
Press*, March 25, 2003, http://www.foxnews.com/story/2003/03/25/f-16-fires-on-
patriot-missile-battery.html.

143 **Two PAC-3 missiles launched automatically:** Pamela Hess, "Feature: The Patriot's
Fratricide Record," *UPI*, accessed June 7, 2017, http://www.upi.com/Feature-
The-Patriots-fratricide-record/63991051224638/. "The Patriot Flawed?," April 24,
2003, http://www.cbsnews.com/news/the-patriot-flawed-19-02-2004/.

143 **U.S. Navy F/A-18C Hornet:** Ibid.

143 **both missiles struck his aircraft:** Ibid.

143 **"substantial success":** The Defense Science Board Task Force assessed the Patriot's
performance as a "substantial success." This seems perhaps overstated. It's worth
asking at what point a system's fratricide rate negates its operational advantages.
Defense Science Board, "Report of the Defense Science Board Task Force on Patriot
Performance," 1.

144 **"unacceptable" fratricide rate:** Hawley, "Looking Back at 20 Years of MANPRINT on Patriot."

144 **IFF was well understood:** Defense Science Board, "Report of the Defense Science Board Task Force on Patriot Performance."

144 **"trusting the system without question":** Hawley, "Looking Back at 20 Years of MANPRINT on Patriot."

144 **"unwarranted and uncritical trust":** John K. Hawley, "Not by Widgets Alone: The Human Challenge of Technology-intensive Military Systems," *Armed Forces Journal*, February 1, 2011, http://www.armedforcesjournal.com/not-by-widgets-alone/. Patriot operators now train on this and other similar scenarios to avoid this problem of unwarranted trust in the automation.

145 **more than 30,000 people a year:** "Accidents or Unintentional Injuries," Centers for Disease Control and Prevention, http://www.cdc.gov/nchs/fastats/accidental -injury.htm.

145 **advanced vehicle autopilots:** For example "Intelligent Drive," Mercedes-Benz, https://www.mbusa.com/mercedes/technology/videos/detail/title-safety/videoId -fc0835ab8d127410VgnVCM100000ccec1e35RCRD.

146 **"No, Ken said that":** Bin Kenney, "Jeopardy!—The IBM Challenge (Day 1—February 14)," video, https://www.youtube.com/watch?v=i-vMW_Ce51w.

146 **Watson hadn't been programmed:** Casey Johnston, "Jeopardy: IBM's Watson Almost Sneaks Wrong Answer by Trebek," *Ars Technica*, February 15, 2011, https:// arstechnica.com/media/news/2011/02/ibms-watson-tied-for-1st-in-jeopardy-almost -sneaks-wrong-answer-by-trebek.ars.

146 **"We just didn't think it would ever happen":** Ibid.

147 **2016 fatality involving a Tesla Model S:** Neither the autopilot nor driver applied the brake when a tractor-trailer turned in front of the vehicle. Anjali Singhvi and Karl Russell, "Inside the Self-Driving Tesla Fatal Accident," *New York Times*, July 1, 2016, https://www.nytimes.com/interactive/2016/07/01/business/inside -tesla-accident.html. "A Tragic Loss," June 30, 2016, https://www.tesla.com/blog/ tragic-loss.

148 *The Sorcerer's Apprentice*: *Sorcerer's Apprentice—Fantasia*, accessed June 7, 2017, http://video.disney.com/watch/sorcerer-s-apprentice-fantasia -4ea9ebc01a74ea59a5867853.

148 **German poem written in 1797:** Johann Wolfgang von Goethe, "The Sorcerer's Apprentice," accessed June 7, 2017, http://germanstories.vcu.edu/goethe/zauber _e4.html.

149 **"When you delegate authority to a machine":** Bob Work, interview, June 22, 2016.

150 **"Traditional methods . . . fail to address":** U.S. Air Force Office of the Chief Scientist, *Autonomous Horizons: System Autonomy in the Air Force—A Path to the Future* (June 2015), 23, http://www.af.mil/Portals/1/documents/SECAF/AutonomousHorizons .pdf?timestamp=1435068339702.

150 **"We had seen it once before":** Interestingly, this random move may have played a key role in shaking Kasparov's confidence. Unlike AlphaGo's 1 in 10,000 surprise move that later turned out to be a stroke of brilliance, Kasparov could see right away that Deep Blue's 44th move was tactically nonsensical. Deep Blue resigned the game one move later. Later that evening while pouring over a recreation of the final moves, Kasparov discovered that in 20 moves he would have checkmated Deep Blue. The implication was that Deep Blue made a nonsense move and resigned because it could see 20 moves ahead, a staggering advantage in chess. Nate Silver reports that

this bug may have irreparably shaken Kasparov's confidence. Nate Silver, *The Signal and the Noise: Why So Many Predictions Fail* (New York: Penguin, 2015), 276–289.

150 **recent UNIDIR report on autonomous weapons and risk:** UN Institute for Disarmament Research, "Safety, Unintentional Risk and Accidents in the Weaponization of Increasingly Autonomous Technologies," 2016, http://www.unidir.org/files/publications/pdfs/safety-unintentional-risk-and-accidents-en-668.pdf. (I was a participant in a UNIDIR-hosted workshop that helped inform this project and I spoke at a UNIDIR-hosted panel in 2016.)

151 **"With very complex technological systems":** John Borrie, interview, April 12, 2016.

151 **"Why would autonomous systems be any different?":** Ibid.

151 **Three Mile Island incident:** This description is taken from Charles Perrow, *Normal Accidents: Living with High-Risk Technologies* (Princeton, NJ: Princeton University Press, 1999), 15–31; and United States Nuclear Regulatory Commission, "Backgrounder on the Three Mile Island Accident," https://www.nrc.gov/reading-rm/doc-collections/fact-sheets/3mile-isle.html.

153 *Apollo 13:* For a very brief summary of the incident, see National Aeronautics and Space Administration, "Apollo 13," https://www.nasa.gov/mission_pages/apollo/missions/apollo13.html. NASA's full report on the Apollo 13 disaster can be found at National Aeronautics and Space Administration, "Report of the Apollo 13 Review Board," June 15, 1970, http://nssdc.gsfc.nasa.gov/planetary/lunar/apollo_13_review_board.txt. See also Perrow, *Normal Accidents,* 271–281.

154 **"failures . . . we hadn't anticipated":** John Borrie, interview, April 12, 2016.

154 *Challenger* **(1986) and** *Columbia* **(2003):** On *Challenger,* see National Aeronautics and Space Administration, "Report of the Presidential Commission on the Space Shuttle Challenger Accident," June 6, 1986, http://history.nasa.gov/rogersrep/51lcover.htm. On the *Columbia* accident, see National Aeronautics and Space Administration, "Columbia Accident Investigation Board, Volume 1," August 2003, http://spaceflight.nasa.gov/shuttle/archives/sts-107/investigation/CAIB_medres_full.pdf.

154 **"never been encountered before":** Matt Burgess, "Elon Musk Confirms SpaceX's Falcon 9 Explosion Was Caused by 'Frozen Oxygen,'" *WIRED,* November 8, 2016, http://www.wired.co.uk/article/elon-musk-universal-basic-income-falcon-9-explosion. "Musk: SpaceX Explosion Toughest Puzzle We've Ever Had to Solve," *CNBC,* video accessed June 7, 2017, http://video.cnbc.com/gallery/?video=3000565513.

154 **Fukushima Daiichi:** Phillip Y. Lipscy, Kenji E. Kushida, and Trevor Incerti, "The Fukushima Disaster and Japan's Nuclear Plant Vulnerability in Comparative Perspective," *Environmental Science and Technology* 47 (2013), http://web.stanford.edu/~plipscy/LipscyKushidaIncertiEST2013.pdf.

156 **"A significant message for the":** William Kennedy, interview, December 8, 2015.

156 **"almost never occur individually":** Ibid.

156 **"The automated systems":** Ibid.

156 **"Both sides have strengths and weaknesses":** Ibid.

156 **F-16 fighter aircraft:** Guy Norris, "Ground Collision Avoidance System 'Saves' First F-16 In Syria," February 5, 2015, http://aviationweek.com/defense/ground-collision-avoidance-system-saves-first-f-16-syria.

156 **software-based limits on its flight controls:** Dan Canin, "Semper Lightning: F-35 Flight Control System," Code One, December 9, 2015, http://www.codeonemagazine.com/f35_article.html?item_id=187.

157   **software with millions of lines of code:** Robert N. Charette, "This Car Runs on Code,"
      IEEE Spectrum: Technology, Engineering, and Science News, February 1, 2009, http://
      spectrum.ieee.org/transportation/systems/this-car-runs-on-code. Joey Cheng,
      "Army Lab to Provide Software Analysis for Joint Strike Fighter," *Defense Systems*,
      August 12, 2014, https://defensesystems.com/articles/2014/08/14/army-f-35-joint
      -strike-fighter-software-tests.aspx. Robert N. Charette, "F-35 Program Continues
      to Struggle with Software," *IEEE Spectrum: Technology, Engineering, and Science
      News*, September 19, 2012, http://spectrum.ieee.org/riskfactor/aerospace/military/
      f35-program-continues-to-struggle-with-software.

157   **0.1 to 0.5 errors per 1,000 lines of code:** Steve McConnell, *Code Complete: A Prac-
      tical Handbook of Software Construction* (Redmond, WA: Microsoft Press, 2004),
      http://www.amazon.com/Code-Complete-Practical-Handbook-Construction/
      dp/0735619670.

157   **some errors are inevitable:** The space shuttle is an interesting exception that
      proves the rule. NASA has been able to drive the number of errors on space shuttle
      code down to zero through a labor-intensive process employing teams of engineers.
      However, the space shuttle has only approximately 500,000 lines of code, and this
      process would be entirely unfeasible for more complex systems using millions of
      lines of code. The F-35 Joint Strike Fighter, for example, has over 20 million lines of
      code. Charles Fishman, "They Write the Right Stuff," FastCompany.com, December
      31, 1996, http://www.fastcompany.com/28121/they-write-right-stuff.

157   **F-22 fighter jets:** Remarks by Air Force retired Major General Don Sheppard
      on "This Week at War," *CNN*, February 24, 2007, http://transcripts.cnn.com/
      TRANSCRIPTS/0702/24/tww.01.html.

157   **hack certain automobiles:** Andy Greenberg, "Hackers Remotely Kill a Jeep on the
      Highway – With Me in It," *Wired*, July 21, 2015, http://www.wired.com/2015/07/
      hackers-remotely-kill-jeep-highway/.

158   **A study of Nest users:** Rayoung Yang and Mark W. Newman, "Learning from a
      Learning Thermostat: Lessons for Intelligent Systems for the Home," UbiComp'13,
      September 8–12, 2013.

158   **"As systems get increasingly complex":** John Borrie, interview, April 12, 2016.

159   **Air France Flight 447:** "Final Report: On the accidents of 1st June 2009 to the
      Airbus A330-203 registered F-GZCP operated by Air France flight 447 Rio de
      Janeiro—Paris," Bureau d'Enquêtes et d'Analyses pour la sécurité de l'aviation civile,
      [English translation], 2012, http://www.bea.aero/docspa/2009/f-cp090601.en/
      pdf/f-cp090601.en.pdf. William Langewiesche, "The Human Factor," *Vanity Fair,*
      October 2014, http://www.vanityfair.com/news/business/2014/10/air-france-
      flight-447-crash. Nick Ross and Neil Tweedie, "Air France Flight 447: 'Damn it,
      We're Going to Crash,'" *The Telegraph*, April 28, 2012, http://www.telegraph.co.uk/
      technology/9231855/Air-France-Flight-447-Damn-it-were-going-to-crash.html.

159   **Normal accident theory sheds light:** In fact, Army researchers specifically cited
      the Three Mile Island incident as having much in common with the Patriot fratri-
      cides. Hawley, "Looking Back at 20 Years of MANPRINT on Patriot."

160   **"even very-low-probability failures":** Defense Science Board, "Report of the
      Defense Science Board Task Force on Patriot Performance."

10   Command and Decision: Can Autonomous
Weapons Be Used Safely?

161   **aircraft carrier flight decks:** Gene I. Rochlin, Todd R. La Porte, and Karlene H. Roberts, "The Self-DesigningHigh-Reliability Organization: Aircraft Carrier Flight Operations at Sea," *Naval War College Review,* Autumn 1987, https://fas.org/man/dod-101/sys/ship/docs/art7su98.htm. Gene I. Rochlin, Todd R. La Porte, and Karlene H. Roberts, "Aircraft Carrier Operations At Sea: The Challenges of High Reliability Performance," University of California, Berkeley, July 15, 1988, http://www.dtic.mil/dtic/tr/fulltext/u2/a198692.pdf.

161   **High-reliability organizations:** Karl E. Weick and Kathleen M. Sutcliffe, *Managing the Unexpected: Sustained Performance in a Complex World,* 3rd ed. (San Francisco: Jossey-Bass, 2015).

161   **militaries as a whole would not be considered:** Scott A Snook, *Friendly Fire: The Accidental Shootdown of U.S. Black Hawks over Northern Iraq* (Princeton, NJ: Princeton University Press, 2002).

162   **"The SUBSAFE Program":** Paul E. Sullivan, "Statement before the House Science Committee on the SUBSAFE Program," October 29, 2003, http://www.navy.mil/navydata/testimony/safety/sullivan031029.txt.

162   **seventy submarines in its force:** Department of Defense, "Quadrennial Defense Review 2014," http://archive.defense.gov/pubs/2014_Quadrennial_Defense_Review.pdf.

164   **"You can mix and match":** Peter Galluch, interview, July 15, 2016.

164   **"kill or be killed":** Ibid.

165   **"there is no voltage that can be applied":** Ibid.

165   **"Absolutely, it's automated":** Ibid.

166   **"You're never driving around":** Ibid.

167   **"there is a conscious decision to fire":** Ibid.

167   **"ROLLGREEN":** The command, as reported on the Navy's website, is "roll FIS green." U.S. Navy, "Naval Terminology," http://www.public.navy.mil/surfor/Pages/Navy-Terminology.aspx.

169   **"terrible, painful lesson":** Peter Galluch, interview, July 15, 2016.

169   **"tanker war":** Ronald O'Rourke, "The Tanker War," Proceedings, May 1988, https://www.usni.org/magazines/proceedings/1988-05/tanker-war.

170   **Iran Air 655:** This account comes from an ABC special on the *Vincennes* incident which relies on first-hand interviews and video and audio recordings from the *Vincennes* during the incident. ABC Four Corners, "Shooting Down of Iran Air 655," 2000, https://www.youtube.com/watch?v=Onk_wI3ZVME.

171   **"unwarranted and uncritical trust":** Hawley, "Not by Widgets Alone."

171   **"spent a lot of money looking into":** John Hawley, interview, December 5, 2016.

171   **"If you make the [training]":** Ibid.

171   **"sham environment . . . the Army deceives":** Ibid.

172   **"consistent objective feedback":** Ibid.

172   **"Even when the Army guys":** Ibid.

172   **"Navy brass in the Aegis":** Ibid.

172   **"too sloppy an organization":** Ibid.

172   **"Judging from history":** Ibid.

173   **training tape left in a computer:** Lewis et al., "Too Close for Comfort: Cases of Near Nuclear Use and Options for Policy," The Royal Institute of International Affairs,

London, April 2014, https://www.chathamhouse.org/sites/files/chathamhouse/field/
field_document/20140428TooCloseforComfortNuclearUseLewisWilliamsPelopidas
Aghlani.pdf, 12–13.

173 **faulty computer chip:** Ibid, 13. William Burr, "The 3 A.M. Phone Call," The National
Security Archive, March 1, 2012, http://nsarchive.gwu.edu/nukevault/ebb371/.

173 **brought President Boris Yeltsin the nuclear briefcase:** Lewis, "Too Close for
Comfort," 16–17.

174 **"erosion" of adherence:** Defense Science Board Permanent Task Force on
Nuclear Weapons Surety, "Report on the Unauthorized Movement of Nuclear
Weapons," February 2008, http://web.archive.org/web/20110509185852/http://
www.nti.org/e_research/source_docs/us/department_defense/reports/11.pdf.
Richard Newton, "Press Briefing with Maj. Gen. Newton from the Pentagon, Arling-
ton, Va.," October 19, 2007, http://web.archive.org/web/20071023092652/http://
www.defenselink.mil/transcripts/transcript.aspx?transcriptid=4067.

174 **thirteen near-use nuclear incidents:** Lewis, "Too Close for Comfort."

174 **"When I began this book":** Scott D. Sagan, *The Limits of Safety: Organizations,
Accidents, and Nuclear Weapons* (Princeton, NJ: Princeton University Press, 1993),
251.

174 **"the historical evidence . . . nuclear weapon systems":** Ibid, 252.

175 **"the inherent limits of organizational safety":** Ibid, 279.

175 **"always/never dilemma":** Ibid, 278.

175 **this is effectively "impossible":** Ibid, 278.

175 **Autonomous weapons have an analogous problem to the always/never dilemma:**
Special thanks to Heather Roff at Arizona State University for pointing out this
parallel.

177 **"You can go through all of the kinds of training":** John Hawley, interview, December 5, 2016.

178 **"planned actions":** William Kennedy, interview, December 8, 2015.

11   Black Box: The Weird, Alien World of Deep Neural Networks

180 **object recognition, performing as well or better than humans:** Kaiming He et al.,
"Delving Deep into Rectifiers: Surpassing Human-Level Performance on ImageNet
Classification."

180 **Adversarial images:** Christian Szegedy et al., "Intriguing properties of neural
networks," February 2014, https://arxiv.org/pdf/1312.6199v4.pdf.

180 **usually created by researchers intentionally:** In at least one case, the researchers
were intentionally evolving the images, but they were not attempting to fool the
deep neural network by making nonsensical images. The researchers explained,
"we were trying to produce recognizable images, but these unrecognizable images
emerged." "Deep neural networks are easily fooled: High confidence predictions for
unrecognizable images," Evolving Artificial Intelligence Laboratory, University of
Wyoming, http://www.evolvingai.org/fooling. For more, see Nguyen A, Yosinski J,
Clune J (2015) "Deep neural networks are easily fooled: High confidence predictions for unrecognizable images," *Computer Vision and Pattern Recognition* (CVPR
'15), IEEE, 2015.

182 specific internal structure of the network: Szegedy et al., "Intriguing properties of neural networks."

182 "huge, weird, alien world of imagery": Jeff Clune, interview, September 28, 2016.

182 surreptitiously embedded into normal images: Ibid.

183 "hidden exploit": Ibid.

183 linear methods to interpret data: Ian J. Goodfellow, Jonathan Shlens, and Christian Szegedy, "Explaining and Harnessing Adversarial Examples," March 20, 2015, https://arxiv.org/abs/1412.6572; Ian Goodfellow, Presentation at Re-Work Deep Learning Summit, 2015, https://www.youtube.com/watch?v=Pq4A2mPCB0Y.

184 "infinitely far to the left": Jeff Clune, interview, September 28, 2016.

184 "real-world images are a very, very small": Ibid.

184 present in essentially every deep neural network: "Deep neural networks are easily fooled." Goodfellow et al., "Explaining and Harnessing Adversarial Examples."

184 specially evolved noise: Corey Kereliuk, Bob L. Sturm, and Jan Larsen, "Deep Learning and Music Adversaries," http://www2.imm.dtu.dk/pubdb/views/edoc_download.php/6904/pdf/imm6904.pdf.

184 News-reading trading bots: John Carney, "The Trading Robots Really Are Reading Twitter," April 23, 2013, http://www.cnbc.com/id/100666302. Patti Domm, "False Rumor of Explosion at White House Causes Stocks to Briefly Plunge; AP Confirms Its Twitter Feed Was Hacked," April 23, 2013, http://www.cnbc.com/id/100646197.

185 deep neural networks to understand text: Xiang Zhang and Yann LeCun, "Text Understanding from Scratch," April 4, 2016, https://arxiv.org/pdf/1502.01710v5.pdf.

185 Associated Press Twitter account was hacked: Domm, "False Rumor of Explosion at White House Causes Stocks to Briefly Plunge; AP Confirms Its Twitter Feed Was Hacked."

186 design deep neural networks that aren't vulnerable: "Deep neural networks are easily fooled."

186 "counterintuitive, weird" vulnerability: Jeff Clune, interview, September 28, 2016.

186 "[T]he sheer magnitude, millions or billions": JASON, "Perspectives on Research in Artificial Intelligence and Artificial General Intelligence Relevant to DoD," 28–29.

186 "the very nature of [deep neural networks]": Ibid, 28.

186 "As deep learning gets even more powerful": Jeff Clune, interview, September 28, 2016.

186 "super complicated and big and weird": Ibid.

187 "sobering message . . . tragic extremely quickly": Ibid.

187 "[I]t is not clear that the existing AI paradigm": JASON, "Perspectives on Research in Artificial Intelligence and Artificial General Intelligence Relevant to DoD," Ibid, 27.

188 "nonintuitive characteristics": Szegedy et al., "Intriguing Properties of Neural Networks."

188 we don't really understand how it happens: For a readable explanation of this broader problem, see David Berreby, "Artificial Intelligence Is Already Weirdly Inhuman," Nautilus, August 6, 2015, http://nautil.us/issue/27/dark-matter/artificial-intelligence-is-already-weirdly-inhuman.

12   Failing Deadly: The Risk of Autonomous Weapons

189   **"I think that we're being overly optimistic"**: John Borrie, interview, April 12, 2016.

189   **"If you're going to turn these things loose"**: John Hawley, interview, December 5, 2016.

189   **"[E]ven with our improved knowledge"**: Perrow, *Normal Accidents*, 354.

191   **"robo-cannon rampage"**: Noah Shachtman, "Inside the Robo-Cannon Rampage (Updated)," *WIRED*, October 19, 2007, https://www.wired.com/2007/10/inside-the-robo/.

191   **bad luck, not deliberate targeting**: "'Robotic Rampage' Unlikely Reason for Deaths," *New Scientist*, accessed June 12, 2017, https://www.newscientist.com/article/dn12812-robotic-rampage-unlikely-reason-for-deaths/.

191   **35 mm rounds into a neighboring gun position**: "Robot Cannon Kills 9, Wounds 14," *WIRED*, accessed June 12, 2017, https://www.wired.com/2007/10/robot-cannon-ki/.

191   **"The machine doesn't know it's making a mistake"**: John Hawley, interview, December 5, 2016.

193   **"incidents of mass lethality"**: John Borrie, interview, April 12, 2016.

193   **"If you put someone else"**: John Hawley, interview, December 5, 2016.

194   **"I don't have a lot of good answers for that"**: Peter Galluch, interview, July 15, 2016.

13   Bot vs. Bot: An Arms Race in Speed

199   **May 6, 2010**: U.S. Commodity Futures Trading Commission and U.S. Securities and Exchange Commission, "Findings Regarding the Market Events of May 6, 2010," September 30, 2010), 2, http://www.sec.gov/news/studies/2010/marketevents-report .pdf.

199   **"horrifying" and "absolute chaos"**: Tom Lauricella and Peter A. McKay, "Dow Takes a Harrowing 1,010.14-Point Trip," *Wall Street Journal*, May 7, 2010, http://www.wsj.com/articles/SB10001424052748704370704575227754131412596. Alexandra Twin, "Glitches Send Dow on Wild Ride," *CNN Money*, May 6, 2010, http://money.cnn.com/2010/05/06/markets/markets_newyork/.

200   **Gone are the days of floor trading**: D. M. Levine, "A Day in the Quiet Life of a NYSE Floor Trader," *Fortune*, May 20, 2013, http://fortune.com/2013/05/29/a-day-in-the-quiet-life-of-a-nyse-floor-trader/.

200   **three-quarters of all trades**: "Rocky Markets Test the Rise of Amateur 'Algo' Traders," *Reuters*, January 28, 2016, http://www.reuters.com/article/us-europe-stocks-algos -insight-idUSKCN0V61T6. The percentage of trades that are automated has varied over time and is subject to some uncertainty. Estimates for the past several years have ranged from 50 percent to 80 percent of all U.S. equity trades. WashingtonsBlog, "What Percentage of U.S. Equity Trades Are High Frequency Trades?," October 28, 2010, http://www.washingtonsblog.com/2010/10/what-percentage-of -u-s-equity-trades-are-high-frequency-trades.html.

200   **sometimes called algorithmic trading**: Tom C. W. Lin, "The New Investor," SSRN Scholarly Paper (Rochester, NY: Social Science Research Network, March 3, 2013), https://papers.ssrn.com/abstract=2227498.

200 **automated trading decisions to buy or sell:** Some writers on automated stock trading differentiate between automated trading and algorithmic trading, using the term algorithmic trading only to refer to the practice of breaking up large orders to execute via algorithm by price, time, or volume, and referring to other practices such as seeking arbitrage opportunities as automated trading. Others treat algorithmic trading and automated trading as effectively synonymous.

200 **Automated trading offers the advantage:** Shobhit Seth, "Basics of Algorithmic Trading: Concepts and Examples," *Investopedia*, October 10, 2014, http://www.investopedia.com/articles/active-trading/101014/basics-algorithmic-trading-concepts-and-examples.asp.

201 **blink of an eye:** "Average Duration of a Single Eye Blink—Human Homo Sapiens—BNID 100706," accessed June 12, 2017, http://bionumbers.hms.harvard.edu//bionumber.aspx?id=100706&ver=0.

201 **speeds measured in microseconds:** Michael Lewis, *Flash Boys: A Wall Street Revolt* (New York: W. W. Norton, 2015), 63, 69, 74, 81.

201 **shortest route for their cables:** Ibid, 62–63.

201 **optimizing every part of their hardware for speed:** Ibid, 63–64.

201 **test them against actual stock market data:** D7, "Knightmare: A DevOps Cautionary Tale," *Doug Seven*, April 17, 2014, https://dougseven.com/2014/04/17/knightmare-a-devops-cautionary-tale/.

202 **"The Science of Trading, the Standard of Trust":** Jeff Cox, " 'Knight-Mare': Trading Glitches May Just Get Worse," August 2, 2012, http://www.cnbc.com/id/48464725.

202 **Knight's trading system began flooding the market:** "Knight Shows How to Lose $440 Million in 30 Minutes," *Bloomberg.com*, August 2, 2012, https://www.bloomberg.com/news/articles/2012-08-02/knight-shows-how-to-lose-440-million-in-30-minutes.

202 **neglected to install a "kill switch":** D7, "Knightmare."

202 **executed 4 million trades:** "How the Robots Lost: High-Frequency Trading's Rise and Fall," *Bloomberg.com*, June 7, 2013, https://www.bloomberg.com/news/articles/2013-06-06/how-the-robots-lost-high-frequency-tradings-rise-and-fall.

202 **Knight was bankrupt:** D7, "Knightmare."

202 **"Knightmare on Wall Street":** For a theory on what happened, see Nanex Research, "03-Aug-2012—The Knightmare Explained," http://www.nanex.net/aqck2/3525.html.

203 **Waddell & Reed:** Waddell & Reed was not named in the official SEC and CFTC report, which referred only to a "large fundamental trader (a mutual fund complex)." U.S. Commodity Futures Trading Commission and U.S. Securities and Exchange Commission, "Findings Regarding the Market Events of May 6, 2010," 2. However, multiple news sources later identified the firm as Waddell & Reed Financial Inc. E. S. Browning and Jenny Strasburg, "The Mutual Fund in the 'Flash Crash,'" *Wall Street Journal*, October 6, 2010, http://www.wsj.com/articles/SB10001424052748704689804575536513798579500. "Flash-Crash Spotlight on Kansas Manager Avery," *Reuters*, May 18, 2010, http://www.reuters.com/article/us-selloff-ivyasset-analysis-idUSTRE64H6G620100518.

203 **E-minis:** "What Are Emini Futures? Why Trade Emini Futures?," *Emini-Watch.com*, accessed June 13, 2017, http://emini-watch.com/emini-trading/emini-futures/.

203 **9 percent of the trading volume over the previous minute:** U.S. Commodity Futures Trading Commission and U.S. Securities and Exchange Commission, "Findings Regarding the Market Events of May 6, 2010," 2.

203 **no instructions with regard to time or price:** Both Waddell & Reed and the Chicago Mercantile Exchange contest the argument, made principally by the CFTF & SEC, that the large E-mini sale was a factor in precipitating the Flash Crash. The independent market research firm Nanex has similarly criticized the suggestion that the Waddell & Reed sell algorithm led to the crash. "Waddell & Reed Responds to 'Flash Crash' Reports—May. 14, 2010," accessed June 13, 2017, http://money.cnn.com/2010/05/14/markets/flash_crash_waddell_reed/. "CME Group Statement on the Joint CFTC/SEC Report Regarding the Events of May 6 - CME Investor Relations," October 1, 2010, http://investor.cmegroup.com/investor-relations/releasedetail.cfm?ReleaseID=513388. "Nanex - Flash Crash Analysis - Continuing Developments," accessed June 13, 2017, http://www.nanex.net/FlashCrashFinal/FlashCrashAnalysis_WR_Update.html.

203 **"unusually turbulent":** U.S. Commodity Futures Trading Commission and U.S. Securities and Exchange Commission, "Findings Regarding the Market Events of May 6, 2010," 1.

203 **"unusually high volatility":** Ibid, 2.

204 **"hot potato" effect:** Ibid, 3.

204 **27,000 E-mini contracts:** Ibid, 3.

204 **automated "stop logic" safety:** Ibid, 4.

204 **"irrational prices":** Ibid, 5.

204 **"clearly erroneous" trades:** Ibid, 6.

205 **Michael Eisen:** Michael Eisen, "Amazon's $23,698,655.93 book about flies," *it is NOT junk*, April 22, 2011, http://www.michaeleisen.org/blog/?p=358.

205 **Eisen hypothesized:** "Amazon.com At a Glance: bordeebook," accessed on October 1, 2016.

206 **Navinder Singh Sarao:** Department of Justice, "Futures Trader Pleads Guilty to Illegally Manipulating the Futures Market in Connection With 2010 'Flash Crash,'" November 9, 2016, https://www.justice.gov/opa/pr/futures-trader-pleads-guilty-illegally-manipulating-futures-market-connection-2010-flash.

206 **By deliberately manipulating the price:** Department of Justice, "Futures Trader Charged with Illegally Manipulating Stock Market, Contributing to the May 2010 Market 'Flash Crash,'" April 21, 2015, https://www.justice.gov/opa/pr/futures-trader-charged-illegally-manipulating-stock-market-contributing-may-2010-market-flash. United States District Court Northern District of Illinois Eastern Division, "Criminal Complaint: United States of America v. Navinder Singh Sarao," February 11, 2015, https://www.justice.gov/sites/default/files/opa/press-releases/attachments/2015/04/21/sarao_criminal_complaint.pdf.

206 **pin the blame for the Flash Crash:** "Post Flash Crash, Regulators Still Use Bicycles To Catch Ferraris," *Traders Magazine Online News*, accessed June 13, 2017, http://www.tradersmagazine.com/news/technology/post-flash-crash-regulators-still-use-bicycles-to-catch-ferraris-113762-1.html.

206 **spoofing algorithm was reportedly turned *off*:** "Guy Trading at Home Caused the Flash Crash," *Bloomberg.com*, April 21, 2015, https://www.bloomberg.com/view/articles/2015-04-21/guy-trading-at-home-caused-the-flash-crash.

206 **exacerbated instability in the E-mini market:** Department of Justice, "Futures

Trader Charged with Illegally Manipulating Stock Market, Contributing to the May 2010 Market 'Flash Crash,'" April 21, 2015, https://www.justice.gov/opa/pr/futures -trader-charged-illegally-manipulating-stock-market-contributing-may-2010 -market-flash.

206 **"circuit breakers":** The first tranche of individual stock circuit breakers, implemented in the immediate aftermath of the Flash Crash, initiated a five-minute pause if a stock's price moved up or down more than 10 percent in the preceding five minutes. U.S. Commodity Futures Trading Commission and U.S. Securities and Exchange Commission, "Findings Regarding the Market Events of May 6, 2010," 7.

206 **Market-wide circuit breakers:** A 7 percent or 13 percent drop halts trading for 15 minutes. A 20 percent drop stops trading for the rest of the day.

206 **"limit up–limit down":** Securities and Exchange Commission, "Investor Bulletin: Measures to Address Market Volatility," July 1, 2012, https://www.sec.gov/oiea/ investor-alerts-bulletins/investor-alerts-circuitbreakersbulletinhtm.html.Specific price bands are listed here: Investopedia Staff, "Circuit Breaker," *Investopedia*, November 18, 2003, http://www.investopedia.com/terms/c/circuitbreaker.asp.

207 **An average day sees a handful of circuit breakers tripped:** Maureen Farrell, "Mini Flash Crashes: A Dozen a Day," *CNNMoney*, March 20, 2013, http:// money.cnn.com/2013/03/20/investing/mini-flash-crash/index.html.

207 **over 1,200 circuit breakers:** Matt Egan, "Trading Was Halted 1,200 Times Monday," *CNNMoney*, August 24, 2015, http://money.cnn.com/2015/08/24/investing/stocks -markets-selloff-circuit-breakers-1200-times/index.html. Todd C. Frankel, "Mini Flash Crash? Trading Anomalies on Manic Monday Hit Small Investors," *Washington Post*, August 26, 2015, https://www.washingtonpost.com/business/economy/mini-flash -crash-trading-anomalies-on-manic-monday-hit-small-investors/2015/08/26/ 6bdc57b0-4c22-11e5-bfb9-9736d04fc8e4_story.html?utm_term=.749eb0bbbf5b.

207 **simple human error:** Joshua Jamerson and Aruna Viswanatha, "Merrill Lynch to Pay $12.5 Million Fine for Mini-Flash Crashes," *Wall Street Journal*, September 26, 2016, http://www.wsj.com/articles/merrill-lynch-to-pay-12-5-million -fine-for-mini-flash-crashes-1474906677.

207 **underlying conditions for flash crashes remain:** Bob Pisani, "A Year after the 1,000-Point Plunge, Changes Cut Trading Halts on Volatile Days," August 23, 2016, http://www.cnbc.com/2016/08/23/a-year-after-the-1000-point-plunge-changes -cut-trading-halts-on-volatile-days.html.

207 **"Circuit breakers don't prevent":** Farrell, "Mini Flash Crashes."

208 **Gulf of Tonkin incident:** Historians have since argued that there was ample evidence at the time that the purported naval battle on August 4, 1964, never happened, that Secretary of Defense Robert McNamara deliberately misled Congress in order to push for war in Vietnam, and that President Lyndon Johnson was aware that the battle never took place. Pat Paterson, "The Truth About Tonkin," *Naval History* magazine, February 2008, https://www.usni.org/magazines/navalhistory/ 2008-02/truth-about-tonkin. John Prados, "The Gulf of Tonkin Incident, 40 Years Later," The National Security Archive, August 4, 2004, http://nsarchive.gwu.edu/ NSAEBB/NSAEBB132/. John Prados, "Tonkin Gulf Intelligence 'Skewed' According to Official History and Intercepts," The National Security Archive, December 1, 2005. Robert J. Hanyok, "Skunks, Bogies, Silent Hounds, and the Flying Fish: The Gulf of Tonkin Mystery, 2–4 August 1964," *Cryptologic Quarterly*, http:// nsarchive.gwu.edu/NSAEBB/NSAEBB132/relea00012.pdf.

208 **an "act of war"**: Dan Gettinger, "'An Act of War': Drones Are Testing China-Japan Relations," Center for the Study of the Drone, November 8, 2013, http://dronecenter.bard.edu/act-war-drones-testing-china-japan-relations/. "Japan to Shoot down Foreign Drones That Invade Its Airspace," *The Japan Times Online*, October 20, 2013, http://www.japantimes.co.jp/news/2013/10/20/national/japan-to-shoot-down-foreign-drones-that-invade-its-airspace/. "China Warns Japan against Shooting down Drones over Islands," *The Times of India*, October 27, 2013, http://timesofindia.indiatimes.com/world/china/China-warns-Japan-against-shooting-down-drones-over-islands/articleshow/24779422.cms.

208 **broach other nations' sovereignty**: "Kashmir Firing: Five Civilians Killed after Drone Downed," *BBC News*, July 16, 2015, sec. India, http://www.bbc.com/news/world-asia-india-33546468.

208 **Pakistan shot down**: "Pakistan Shoots Down Indian Drone 'Trespassing' into Kashmir," *AP News*, accessed June 13, 2017, https://apnews.com/b67d689fb1f7410e9d0be01daaded3a7/pakistani-and-indian-troops-trade-fire-kashmir.

208 **Israel has shot down drones**: Gili Cohen, "Israeli Fighter Jet Shoots Down Hamas Drone Over Gaza," *Haaretz*, September 20, 2016, http://www.haaretz.com/israel-news/1.743169.

208 **Syria shot down**: Missy Ryan, "U.S. Drone Believed Shot down in Syria Ventured into New Area, Official Says," *Washington Post*, March 19, 2015, https://www.washingtonpost.com/world/national-security/us-drone-believed-shot-down-in-syria-ventured-into-new-area-official-says/2015/03/19/891a3d08-ce5d-11e4-a2a7-9517a3a70506_story.html.

208 **Turkey shot down**: "Turkey Shoots down Drone near Syria, U.S. Suspects Russian Origin," *Reuters*, October 16, 2015, http://www.reuters.com/article/us-mideast-crisis-turkey-warplane-idUSKCN0SA15K20151016.

209 **China seized a small underwater robot**: "China to Return Seized U.S. Drone, Says Washington 'Hyping Up' Incident," *Reuters*, December 18, 2016, http://www.reuters.com/article/us-usa-china-drone-idUSKBN14526J.

209 **Navy Fire Scout drone**: Elisabeth Bumiller, "Navy Drone Wanders Into Restricted Airspace Around Washington," *New York Times*, August 25, 2010, https://www.nytimes.com/2010/08/26/us/26drone.html.

209 **Army Shadow drone**: Alex Horton, "Questions Hover over Army Drone's 630-Mile Odyssey across Western US," *Stars and Stripes*, March 1, 2017, https://www.stripes.com/news/questions-hover-over-army-drone-s-630-mile-odyssey-across-western-us-1.456505#.WLby0hLyv5Z.

209 **RQ-170**: Greg Jaffe and Thomas Erdbrink, "Iran Says It Downed U.S. Stealth Drone; Pentagon Acknowledges Aircraft Downing," *Washington Post*, December 4, 2011, https://www.washingtonpost.com/world/national-security/iran-says-it-downed-us-stealth-drone-pentagon-acknowledges-aircraft-downing/2011/12/04/gIQAyxa8TO_story.html.

209 **Reports swirled online**: Scott Peterson and Payam Faramarzi, "Exclusive: Iran Hijacked US Drone, Says Iranian Engineer," *Christian Science Monitor*, December 15, 2011, http://www.csmonitor.com/World/Middle-East/2011/1215/Exclusive-Iran-hijacked-US-drone-says-Iranian-engineer.

209 **"complete bullshit"**: David Axe, "Did Iran Hack a Captured U.S. Stealth Drone?" *WIRED*, April 24, 2012, https://www.wired.com/2012/04/iran-drone-hack/.

209 **United States did awkwardly confirm:** Bob Orr, "U.S. Official: Iran Does Have Our Drone," *CBS News*, December 8, 2011, http://www.cbsnews.com/news/us-official-iran-does-have-our-drone/.

210 **"networks of systems":** Heather Roff, interview, October 26, 2016.

210 **"If my autonomous agent":** Ibid.

210 **"What are the unexpected side effects":** Bradford Tousley, interview, April 27, 2016.

210 **"I don't know that large-scale military impacts":** Ibid.

210 **"machine speed . . . milliseconds":** Ibid.

14 The Invisible War: Autonomy in Cyberspace

212 **Internet Worm of 1988:** Ted Eisenberg et al., "The Cornell Commission: On Morris and the Worm," *Communications of the ACM* 32, 6 (June 1989), 706–709, http://www.cs.cornell.edu/courses/cs1110/2009sp/assignments/a1/p706-eisenberg.pdf;

212 **over 70,000 reported cybersecurity incidents:** Government Accountability Office, "Information Security: Agencies Need to Improve Controls over Selected High-Impact Systems," GAO-16-501, Washington, DC, May 2016, http://www.gao.gov/assets/680/677293.pdf.

212 **most frequent and most serious attacks:** Ibid, 11.

212 **exposed security clearance investigation data:** James Eng, "OPM Hack: Government Finally Starts Notifying 21.5 Million Victims," *NBC News*, October 1, 2015, http://www.nbcnews.com/tech/security/opm-hack-government-finally-starts-notifying-21-5-million-victims-n437126. "Why the OPM Hack Is Far Worse Than You Imagine," *Lawfare*, March 11, 2016, https://www.lawfareblog.com/why-opm-hack-far-worse-you-imagine.

212 **Chinese government claimed:** David E. Sanger, "U.S. Decides to Retaliate Against China's Hacking," *New York Times*, July 31, 2015, https://www.nytimes.com/2015/08/01/world/asia/us-decides-to-retaliate-against-chinas-hacking.html. Sean Gallagher, "At First Cyber Meeting, China Claims OPM Hack Is 'criminal Case' [Updated]," *Ars Technica*, December 3, 2015, https://arstechnica.com/tech-policy/2015/12/at-first-cyber-meeting-china-claims-opm-hack-is-criminal-case/. David E. Sanger and Julie Hirschfeld Davis, "Hacking Linked to China Exposes Millions of U.S. Workers," *New York Times*, June 4, 2015, https://www.nytimes.com/2015/06/05/us/breach-in-a-federal-computer-system-exposes-personnel-data.html.

212 **affected Estonia's entire electronic infrastructure:** Dan Holden, "Estonia, Six Years Later," Arbor Networks, May 16, 2013, https://www.arbornetworks.com/blog/asert/estonia-six-years-later/.

213 **over a million botnet-infected computers:** "Hackers Take Down the Most Wired Country in Europe," *WIRED*, accessed June 14, 2017, https://www.wired.com/2007/08/ff-estonia/. "Denial-of-Service: The Estonian Cyberwar and Its Implications for U.S. National Security," *International Affairs Review*, accessed June 14, 2017, http://www.iar-gwu.org/node/65.

213 **"disastrous" consequences if Estonia removed the monument:** "Hackers Take Down the Most Wired Country in Europe."

213 **Russian Duma official confirmed:** "Russia Confirms Involvement with Estonia

DDoS Attacks," *SC Media US*, March 12, 2009, https://www.scmagazine.com/news/russia-confirms-involvement-with-estonia-ddos-attacks/article/555577/.

213 **many alleged or confirmed cyberattacks:** "Estonia, Six Years Later."

213 **cyberattacks against Saudi Arabia and the United States:** Keith B. Alexander, "Prepared Statement of GEN (Ret) Keith B. Alexander* on the Future of Warfare before the Senate Armed Services Committee," November 3, 2015, http://www.armed-services.senate.gov/imo/media/doc/Alexander_11-03-15.pdf.

213 **team of professional hackers months if not years:** David Kushner, "The Real Story of Stuxnet," *IEEE Spectrum: Technology, Engineering, and Science News*, February 26, 2013, http://spectrum.ieee.org/telecom/security/the-real-story-of-stuxnet.

213 **"zero days":** Kim Zetter, "Hacker Lexicon: What Is a Zero Day?," *WIRED*, November 11, 2014, https://www.wired.com/2014/11/what-is-a-zero-day/.

213 **Stuxnet had four:** Michael Joseph Gross, "A Declaration of Cyber War." *Vanity Fair*, March 2011, https://www.vanityfair.com/news/2011/03/stuxnet-201104.

214 **programmable logic controllers:** Gross, "A Declaration of Cyber War." Nicolas Falliere, Liam O Murchu, and Eric Chien, "W32.Stuxnet Dossier," Symantec Security Response, February 2011, https://www.symantec.com/content/en/us/enterprise/media/security_response/whitepapers/w32_stuxnet_dossier.pdf.

214 **two encrypted "warheads":** Gross, "A Declaration of Cyber War."

214 **Computer security specialists widely agree:** Falliere et al., "W32.Stuxnet Dossier," 2, 7.

214 **Natanz nuclear enrichment facility:** Gross, "A Declaration of Cyber War." Ralph Langner, "Stuxnet Deep Dive," S4x12, https://vimeopro.com/s42012/s4-2012/video/35806770. Kushner, imeopro.com/s42012/Stuxnet.t

214 **Nearly 60 percent of Stuxnet infections:** Falliere et al., "W32.Stuxnet Dossier," 5–7. Kim Zetter, "An Unprecedented Look at Stuxnet, the World's First Digital Weapon," *WIRED*, November 3, 2014, https://www.wired.com/2014/11/countdown-to-zero-day-stuxnet/.

214 **sharp decline in the number of centrifuges:** John Markoff and David E. Sanger, "In a Computer Worm, a Possible Biblical Clue," *New York Times*, September 24, 2010, http://www.nytimes.com/2010/09/30/world/middleeast/30worm.html.

214 **Security specialists have further speculated:** Ibid. Gross, "A Declaration of Cyber War."

215 **"While attackers could control Stuxnet":** Falliere et al., "W32.Stuxnet Dossier," 3.

215 **"collateral damage":** Ibid, 7.

215 **spread via USB to only three other machines:** Ibid, 10.

215 **self-terminate date:** Ibid, 18.

215 **Some experts saw these features as further evidence:** Gross, "A Declaration of Cyber War."

215 **"open-source weapon":** Patrick Clair, "Stuxnet: Anatomy of a Computer Virus," video, 2011, https://vimeo.com/25118844.

215 **blueprint for cyber-weapons to come:** Josh Homan, Sean McBride, and Rob Caldwell, "IRONGATE ICS Malware: Nothing to See Here … Masking Malicious Activity on SCADA Systems," FireEye, June 2, 2016, https://www.fireeye.com/blog/threat-research/2016/06/irongate_ics_malware.html.

216 **"It should be automated":** Keith B. Alexander, Testimony on the Future of War-

fare, Senate Armed Services Committee, November 3, 2015, http://www.armed
-services.senate.gov/hearings/15-11-03-future-of-warfare. Alexander's comments
on automation come up in the question-and-answer period, starting at 1:14:00.

217 **DARPA held a Robotics Challenge:** DARPA, "DARPA Robotics Challenge (DRC),"
accessed June 14, 2017, http://www.darpa.mil/program/darpa-robotics-challenge.
DARPA, "Home | DRC Finals," accessed June 14, 2017, http://archive.darpa.mil/
roboticschallenge/.

217 **"automatically check the world's software":** David Brumley, "Why CGC Matters
to Me," ForAllSecure, July 26, 2016, https://forallsecure.com/blog/2016/07/26/
why-cgc-matters-to-me/.

217 **"fully autonomous system for finding and fixing":** David Brumley, "Mayhem
Wins DARPA CGC," ForAllSecure, August 6, 2016, https://forallsecure.com/
blog/2016/08/06/mayhem-wins-darpa-cgc/.

217 **vulnerability is analogous to a weak lock:** David Brumley, interview, November
24, 2016.

218 **"There's grades of security":** Ibid.

218 **"an autonomous system that's taking all of those things":** Ibid.

218 **"Our goal was to come up with a skeleton key":** Ibid.

219 **"true autonomy in the cyber domain":** Michael Walker, interview, December 5, 2016.

219 **comparable to a "competent" computer security professional:** David Brumley,
interview, November 24, 2016.

219 **DEF CON hacking conference:** Daniel Tkacik, "CMU Team Wins Fourth 'World
Series of Hacking' Competition," CMU.edu, July 31, 2017.

219 **Brumley's team from Carnegie Mellon:** Ibid.

219 **Mirai:** Brian Krebs, "Who Makes the IoT Things Under Attack?" Krebs on Security, October 3, 2016, https://krebsonsecurity.com/2016/10/who-makes-the-iot
-things-under-attack/.

219 **massive DDoS attack:** Brian Krebs, "KrebsOnSecurity Hit With Record DDoS,"
Krebs on Security, September 21, 2016, https://krebsonsecurity.com/2016/09/
krebsonsecurity-hit-with-record-ddos/.

219 **most IoT devices are "ridiculous vulnerable":** David Brumley, interview, November 24, 2016.

219 **6.4 billion IoT devices:** "Gartner Says 6.4 Billion Connected," Gartner, November
10, 2015, http://www.gartner.com/newsroom/id/3165317.

220 **"check all these locks":** David Brumley, interview, November 24, 2016.

220 **"no difference" between the technology:** Ibid.

220 **"All computer security technologies are dual-use":** Michael Walker, interview,
December 5, 2016.

220 **"you have to trust the researchers":** David Brumley, interview, November 24,
2016.

220 **"It's going to take the same kind":** Michael Walker, interview, December 5, 2016.

221 **"I'm not saying that we can change to a place":** Ibid.

221 **"It's scary to think of Russia":** David Brumley, interview, November 24, 2016.

221 **"counter-autonomy":** David Brumley, "Winning Cyber Battles: The Next 20 Years,"
unpublished working paper, November 2016.

221 **"trying to find vulnerabilities":** David Brumley, interview, November 24, 2016.

221 **"you play the opponent":** Ibid.

221 **"It's a little bit like a Trojan horse":** Ibid.

222 **"computer equivalent to 'the long con'":** Brumley, "Winning Cyber Battles: The Next 20 Years."

222 **"Make no mistake, cyber is a war":** Ibid.

222 **F-35 . . . tens of millions of lines of code:** Jacquelyn Schneider, "Digitally-Enabled Warfare: The Capability-Vulnerability Paradox," Center for a New American Security, Washington DC, August 29, 2016, https://www.cnas.org/publications/reports/digitally-enabled-warfare-the-capability-vulnerability-paradox.

223 **Hacking back is when:** Dorothy E. Denning, "Framework and Principles for Active Cyber Defense," December 2013, 3.

223 **Hacking back can inevitably draw in third parties:** Dan Goodin, "Millions of Dynamic DNS Users Suffer after Microsoft Seizes No-IP Domains," *Ars Technica*, June 30, 2014, https://arstechnica.com/security/2014/06/millions-of-dymanic-dns-users-suffer-after-microsoft-seizes-no-ip-domains/.

223 **Hacking back is controversial:** Hannah Kuchler, "Cyber Insecurity: Hacking Back," *Financial Times*, July 27, 2015, https://www.ft.com/content/c75a0196-2ed6-11e5-8873-775ba7c2ea3d.

223 **"Every action accelerates":** Steve Rosenbush, "Cyber Experts Draw Line Between Active Defense, Illegal Hacking Back," *Wall Street Journal*, July 28, 2016, https://blogs.wsj.com/cio/2016/07/28/cyber-experts-draw-line-between-active-defense-illegal-hacking-back/.

223 **Coreflood botnet:** Denning, 6.

223 **Automated hacking back would delegate:** "Hacking Back: Exploring a New Option of Cyber Defense," *InfoSec Resources*, November 8, 2016, http://resources.infosecinstitute.com/hacking-back-exploring-a-new-option-of-cyber-defense/.

223 **"autonomous cyber weapons could automatically escalate":** Patrick Lin, remarks at conference, "Cyber Weapons and Autonomous Weapons: Potential Overlap, Interaction and Vulnerabilities," UN Institute for Disarmament Research, New York, October 9, 2015, http://www.unidir.org/programmes/emerging-security-threats/the-weaponization-of-increasingly-autonomous-technologies-addressing-competing-narratives-phase-ii/cyber-weapons-and-autonomous-weapons-potential-overlap-interaction-and-vulnerabilities. Comment at 5:10.

224 **Automated hacking back is a theoretical concept:** Alexander Velez-Green, "When 'Killer Robots' Declare War," *Defense One*, April 12, 2015, http://www.defenseone.com/ideas/2015/04/when-killer-robots-declare-war/109882/.

224 **automate "spear phishing" attacks:** Karen Epper Hoffman, "Machine Learning Can Be Used Offensively to Automate Spear Phishing," *Infosecurity Magazine*, August 5, 2016, https://www.infosecurity-magazine.com/news/bhusa-researchers-present-phishing/.

224 **automatically develop "humanlike" tweets:** John Seymour and Philip Tully, "Weaponizing data science for social engineering: Automated E2E spear phishing on Twitter," https://www.blackhat.com/docs/us-16/materials/us-16-Seymour-Tully-Weaponizing-Data-Science-For-Social-Engineering-Automated-E2E-Spear-Phishing-On-Twitter-wp.pdf.

224 **"in offensive cyberwarfare":** Eric Messinger, "Is It Possible to Ban Autonomous Weapons in Cyberwar?," *Just Security*, January 15, 2015, https://www.justsecurity.org/19119/ban-autonomous-weapons-cyberwar/.

225 **estimated 8 to 15 million computers worldwide:** "Virus Strikes 15 Million PCs," *UPI*, January 26, 2009, http://www.upi.com/Top_News/2009/01/26/Virus-strikes-15-million-PCs/19421232924206/.

225 **method to counter Conficker:** "Clock ticking on worm attack code," *BBC News*, January 20, 2009, http://news.bbc.co.uk/2/hi/technology/7832652.stm.

225 **brought Conficker to heel:** *Microsoft Security Intelligence Report: Volume 11* (11), Microsoft, 2011.

226 **"prevent and react to countermeasures":** Alessandro Guarino, "Autonomous Intelligent Agents in Cyber Offence," in K. Podins, J. Stinissen, M. Maybaum, eds., *2013 5th International Conference on Cyber Conflict* (Tallinn, Estonia: NATO CCD COE Publications, 2013), https://ccdcoe.org/cycon/2013/proceedings/d1r1s9_guarino.pdf.

226 **"the synthesis of new logic":** Michael Walker, interview, December 5, 2016.

226 **"those are a possibility and are worrisome":** David Brumley, interview, November 24, 2016.

227 **"Defense is powered by openness":** Michael Walker, interview, December 5, 2016.

227 **"I tend to view everything as a system":** David Brumley, interview, November 24, 2016.

227 **what constitutes a "cyberweapon":** Thomas Rid and Peter McBurney, "Cyber-Weapons," *The RUSI Journal*, 157 (2012):1, 6–13.

227 **specifically exempts cyberweapons:** Department of Defense, "Department of Defense Directive Number 3000.09, 2.

228 **"goal is not offense":** Bradford Tousley, interview, April 27, 2016.

228 **"the narrow cases where we will allow":** Bob Work, interview, June 22, 2016.

228 **"We'll work it through":** Ibid.

230 **"they would just shut it down":** Ibid.

15 "Summoning the Demon": The Rise of Intelligent Machines

231 **cannot piece these objects together:** Machines have been able to caption images with reasonable accuracy, describing in a general sense what the scene depicts. For an overview of current abilities and limitations in scene interpretation, see JASON, "Perspectives on Research in Artificial Intelligence and Artificial General Intelligence Relevant to DoD," 10.

232 **Brain imaging:** "Human Connectome Project | Mapping the Human Brain Connectivity," accessed June 15, 2017, http://www.humanconnectomeproject.org/. "Meet the World's Most Advanced Brain Scanner," *Discover Magazine*, accessed June 15, 2017, http://discovermagazine.com/2013/june/08-meet-the-worlds-most-advanced-brain-scanner.

232 **whole brain emulations:** Anders Sandburg and Nick Bostrom, "Whole Brain Emulation: A Roadmap," Technical Report #2008-3, Oxford, UK, 2008, http://www.fhi.ox.ac.uk/Reports/2008-3.pdf

232 **"When people say a technology":** Andrew Herr, email to the author, October 22, 2016.

232 **"last invention":** Irving J. Good, "Speculations Concerning the First Ultraintelligent Machine", May 1964, https://web.archive.org/web/20010527181244/http://www.aeiveos.com/~bradbury/Authors/Computing/Good-IJ/SCtFUM.html. See also James Barrat, *Our Final Invention* (New York: Thomas Dunne Books, 2013).

232 **"development of full artificial intelligence"**: Rory Cellan-Jones, "Stephen Hawking Warns Artificial Intelligence Could End Mankind," *BBC News*, December 2, 2014, http://www.bbc.com/news/technology-30290540.

232 **"First the machines will"**: Peter Holley, "Bill Gates on Dangers of Artificial Intelligence: 'I Don't Understand Why Some People Are Not Concerned,'" *Washington Post*, January 29, 2015, https://www.washingtonpost.com/news/the-switch/wp/2015/01/28/bill-gates-on-dangers-of-artificial-intelligence-dont-understand-why-some-people-are-not-concerned/.

232 **"summoning the demon"**: Matt McFarland, "Elon Musk: 'With Artificial Intelligence We Are Summoning the Demon,'" *Washington Post*, October 24, 2014, https://www.washingtonpost.com/news/innovations/wp/2014/10/24/elon-musk-with-artificial-intelligence-we-are-summoning-the-demon/.

233 **"I am in the camp that is concerned"**: Holley, "Bill Gates on Dangers of Artificial Intelligence: 'I Don't Understand Why Some People Are Not Concerned.'"

233 **"Let an ultraintelligent machine be defined"**: Good, "Speculations Concerning the First Ultraintelligent Machine."

233 **lift itself up by its own boostraps**: "Intelligence Explosion FAQ," *Machine Intelligence Research Institute*, accessed June 15, 2017, https://intelligence.org/ie-faq/.

233 **"AI FOOM"**: Robin Hanson and Eliezer Yudkowsky, "The Hanson-Yudkowsky AI Foom Debate," http://intelligence.org/files/AIFoomDebate.pdf.

233 **"soft takeoff" scenario**: Müller, Vincent C. and Bostrom, Nick, 'Future progress in artificial intelligence: A Survey of Expert Opinion, in Vincent C. Müller (ed.), *Fundamental Issues of Artificial Intelligence* (Berlin: Springer Synthese Library, 2016), http://www.nickbostrom.com/papers/survey.pdf.

234 **"the dissecting room and the slaughter-house"**: Mary Shelley, *Frankenstein, Or, The Modern Prometheus* (London: Lackington, Hughes, Harding, Mavor & Jones, 1818), 43.

234 **Golem stories**: Executive Committee of the Editorial Board., Ludwig Blau, Joseph Jacobs, Judah David Eisenstein, "Golem," JewishEncylclopedia.com, http://www.jewishencyclopedia.com/articles/6777-golem#1137.

235 **"the dream of AI"**: Micah Clark, interview, May 4, 2016.

235 **"building human-like persons"**: Ibid.

236 **"Why would we expect a silica-based intelligence"**: Ibid.

236 **Turing test**: The Loebner Prize runs the Turing test every year. While no computer has passed the test by fooling all of the judges, some programs have fooled at least one judge in the past. Tracy Staedter, "Chat-Bot Fools Judges Into Thinking It's Human," *Seeker*, June 9, 2014, https://www.seeker.com/chat-bot-fools-judges-into-thinking-its-human-1768649439.html. Every year the Loebner Prize awards a prize to the "most human" AI. You can chat with the 2016 winner, "Rose," here: http://ec2-54-215-197-164.us-west-1.compute.amazonaws.com/speech.php.

236 **AI virtual assistant called "Amy"**: "Amy the Virtual Assistant Is So Human-Like, People Keep Asking It Out on Dates," accessed June 15, 2017, https://mic.com/articles/139512/xai-amy-virtual-assistant-is-so-human-like-people-keep-asking-it-out-on-dates.

236 **"If we presume an intelligent alien life"**: Micah Clark, interview, May 4, 2016.

237 **"any level of intelligence could in principle"**: Nick Bostrom, "The Superintelligent Will: Motivation and Instrumental Rationality in Advanced Artificial Agents," http://www.nickbostrom.com/superintelligentwill.pdf.

237 **"The AI does not hate you"**: Eliezer S. Yudkowsky, "Artificial Intelligence as a Positive and Negative Factor in Global Risk," http://www.yudkowsky.net/singularity/ai-risk.

238 **"[Y]ou build a chess playing robot"**: Stephen M. Omohundro, "The Basic AI Drives," https://selfawaresystems.files.wordpress.com/2008/01/ai_drives_final.pdf.

238 **"Without special precautions"**: Ibid.

238 **lead-lined coffins connected to heroin drips**: Patrick Sawer, "Threat from Artificial Intelligence Not Just Hollywood Fantasy," June 27, 2015, http://www.telegraph.co.uk/news/science/science-news/11703662/Threat-from-Artificial-Intelligence-not-just-Hollywood-fantasy.html.

239 **"its final goal is to make us happy"**: Nick Bostrom, *Superintelligence: Paths, Dangers, Strategies* (Oxford: Oxford University Press, 2014), Chapter 8.

239 **"a system that is optimizing a function"**: Stuart Russell, "Of Myths and Moonshine," *Edge,* November 14, 2014, https://www.edge.org/conversation/the-myth-of-ai#26015.

239 **"perverse instantiation"**: Bostrom, *Superintelligence,* Chapter 8.

239 **learned to pause Tetris**: Tom Murphy VII, "The First Level of Super Mario Bros. is Easy with Lexicographic Orderings and Time Travel ... after that it gets a little tricky," https://www.cs.cmu.edu/~tom7/mario/mario.pdf. The same AI also uncovered and exploited a number of bugs, such as one in Super Mario Brothers that allowed it to stomp goombas from underneath.

239 **EURISKO**: Douglas B. Lenat, "EURISKO: A Program That Learns New Heuristics and Domain Concepts," *Artificial Intelligence* 21 (1983), http://www.cs.northwestern.edu/~mek802/papers/not-mine/Lenat_EURISKO.pdf, 90.

240 **"not to put a specific purpose into the machine"**: Dylan Hadfield-Menell, Anca Dragan, Pieter Abbeel, Stuart Russell, "Cooperative Inverse Reinforcement Learning," 30th Conference on Neural Information Processing Systems (NIPS 2016), Barcelona, Spain, November 12, 2016, https://arxiv.org/pdf/1606.03137.pdf.

240 **correctable by their human programmers**: Nate Soares et al., "Corrigibility," in AAAI Workshops: Workshops at the Twenty-Ninth AAAI Conference on Artificial Intelligence, Austin, TX, January 25–26, 2015, https://intelligence.org/files/Corrigibility.pdf.

240 **indifferent to whether they are turned off**: Laurent Orseau and Stuart Armstrong, "Safely Interruptible Agents," https://intelligence.org/files/Interruptibility.pdf.

240 **designing AIs to be tools**: Holden Karnofsky, "Thoughts on the Singularity Institute," May 11, 2012, http://lesswrong.com/lw/cbs/thoughts_on_the_singularity_institute_si/.

240 **"they might not work"**: Stuart Armstrong, interview, November 18, 2016.

240 **Tool AIs could still slip out of control**: Bostrom, *Superintelligence,* 184–193.

240 **"We also have to consider . . . whether tool AIs"**: Stuart Armstrong, interview, November 18, 2016.

241 **"Just by saying, 'we should only build . . .'"**: Ibid.

241 **AI risk "doesn't concern me"**: Sharon Gaudin, "Ballmer Says Machine Learning Will Be the next Era of Computer Science," *Computerworld,* November 13, 2014, http://www.computerworld.com/article/2847453/ballmer-says-machine-learning-will-be-the-next-era-of-computer-science.html.

241 **"There won't be an intelligence explosion"**: Jeff Hawkins, "The Terminator Is Not Coming. The Future Will Thank Us," *Recode,* March 2, 2015, https://

www.recode.net/2015/3/2/11559576/the-terminator-is-not-coming-the-future-will
-thank-us.

241 **Mark Zuckerberg:** Alanna Petroff, "Elon Musk Says Mark Zuckerberg's Understanding of AI Is 'Limited,'" CNN.com, July 25, 2017.

241 **"not concerned about self-awareness":** David Brumley, interview, November 24,
2016.

242 **"has been completely contradictory":** Stuart Armstrong, interview, November 18,
2016.

242 **poker became the latest game to fall:** Olivia Solon, "Oh the Humanity! Poker
Computer Trounces Humans in Big Step for AI," *The Guardian*, January 30,
2017, sec. Technology, https://www.theguardian.com/technology/2017/jan/30/
libratus-poker-artificial-intelligence-professional-human-players-competition.

242 **"imperfect information" game:** Will Knight, "Why Poker Is a Big Deal for
Artificial Intelligence," *MIT Technology Review*, January 23, 2017, https://www
.technologyreview.com/s/603385/why-poker-is-a-big-deal-for-artificial
-intelligence/.

242 **world's top poker players had handily beaten:** Cameron Tung, "Humans Out-Play
an AI at Texas Hold 'Em—For Now," *WIRED*, May 21, 2015, https://www.wired.com/
2015/05/humans-play-ai-texas-hold-em-now/.

242 **upgraded AI "crushed":** Cade Metz, "A Mystery AI Just Crushed the Best Human
Players at Poker," *WIRED*, January 31, 2017, https://www.wired.com/2017/01/
mystery-ai-just-crushed-best-human-players-poker/.

242 **"as soon as something works":** Micah Clark, interview, May 4, 2016.

242 **"as soon as a computer can do it":** Stuart Armstrong, interview, November 18,
2016. This point was also made by authors of a Stanford study of AI. Peter Stone,
Rodney Brooks, Erik Brynjolfsson, Ryan Calo, Oren Etzioni, Greg Hager, Julia
Hirschberg, Shivaram Kalyanakrishnan, Ece Kamar, Sarit Kraus, Kevin Leyton-
Brown, David Parkes, William Press, AnnaLee Saxenian, Julie Shah, Milind Tambe,
and Astro Teller. "Artificial Intelligence and Life in 2030." One Hundred Year Study
on Artificial Intelligence: Report of the 2015–2016 Study Panel, Stanford University,
Stanford, CA, September 2016, 13. http://ai100.stanford.edu/2016-report.

243 **"responsible use":** AAAI.org, http://www.aaai.org/home.html.

243 **"most of the discussion about superintelligence":** Tom Dietterich, interview,
April 27, 2016.

243 **"runs counter to our current understandings":** Thomas G. Dietterich and Eric J.
Horvitz, "Viewpoint Rise of Concerns about AI: Reflections and Directions," *Communications of the ACM* 58, no. 10 (October 2015): 38–40, http://web.engr.oregonstate
.edu/~tgd/publications/dietterich-horvitz-rise-of-concerns-about-ai-reflections-
and-directions-CACM_Oct_2015-VP.pdf.

243 **"The increasing abilities of AI":** Tom Dietterich, interview, April 27, 2016.

244 **"robust to adversarial attack":** Ibid.

244 **"The human should be taking the actions":** Ibid.

244 **"The whole goal in military doctrine":** Ibid.

245 **AGI as "dangerous":** Bob Work, interview, June 22, 2016.

245 **more Iron Man than Terminator:** Sydney J. Freedburg Jr., "Iron Man, Not Terminator: The Pentagon's Sci-Fi Inspirations," *Breaking Defense*, May 3, 2016, http://
breakingdefense.com/2016/05/iron-man-not-terminator-the-pentagons-sci-fi
-inspirations/. Matthew Rosenberg and John Markoff, "The Pentagon's 'Terminator
Conundrum': Robots That Could Kill on Their Own," *New York Times*, October 25,

2016, https://www.nytimes.com/2016/10/26/us/pentagon-artificial-intelligence
-terminator.html.

245 **"impose obligations on persons"**: Office of General Counsel, Department of Defense, "Department of Defense Law of War Manual," June 2015, https://www.defense.gov/Portals/1/Documents/law_war_manual15.pdf, 330.

245 **"the ultimate goal of AI"**: "The ultimate goal of AI (which we are very far from achieving) is to build a person, or, more humbly, an animal." Eugene Charniak and Drew McDermott, *Introduction to Artificial Intelligence* (Boston: Addison-Wesley Publishing Company, 1985), 7.

245 **"what they're aiming at are human-level"**: Selmer Bringsjord, interview, November 8, 2016.

245 **"we can plan all we want"**: Ibid.

246 **"adversarial AI" and "AI security"**: Stuart Russell, Daniel Dewey, and Max Tegmark, "Research Priorities for Robust and Beneficial Artificial Intelligence," Association for the Advancement of Artificial Intelligence (Winter 2015), http://futureoflife.org/data/documents/research_priorities.pdf.

246 **malicious applications of AI**: One of the few articles to tackle this problem is Federico Pistono and Roman V. Yampolskiy, "Unethical Research: How to Create a Malevolent Artificial Intelligence," September 2016, https://arxiv.org/pdf/1605.02817.pdf.

246 **Elon Musk's reaction**: Elon Musk, Twitter post, July 14, 2016, 2:42am, https://twitter.com/elonmusk/status/753525069553381376.

246 **"adaptive and unpredictable"**: David Brumley, interview, November 24, 2016.

247 **"Faustian bargain"**: Richard Danzig, "Surviving on a Diet of Poisoned Fruit: Reducing the National Security Risks of America's Cyber Dependencies," Center for a New American Security, Washington, DC, July 21, 2014, https://www.cnas.org/publications/reports/surviving-on-a-diet-of-poisoned-fruit-reducing-the-national-security-risks-of-americas-cyber-dependencies, 9.

247 **"placing humans in decision loops"**: Ibid, 21.

247 **"abnegation"**: Ibid, 20.

247 **"ecosystem"**: David Brumley, interview, November 24, 2016.

247 **Armstrong estimated**: Stuart Armstrong, interview, November 18, 2016.

16 Robots on Trial: Autonomous Weapons and the Laws of War

251 **biblical book of Deuteronomy**: Deuteronomy 20:10–19. Laws of Manu 7:90–93.

251 *principle of distinction*: "Article 51: Protection of the Civilian Population" and "Article 52: General Protection of Civilian Objects," Protocol Additional to the Geneva Conventions of 12 August 1949, and relating to the Protection of Victims of International Armed Conflicts (Protocol I), 8 June 1977, https://ihl-databases.icrc.org/ihl/webart/470-750065 and https://ihl-databases.icrc.org/ihl/WebART/470-750067.

251 *principle of proportionality*: Article 51(5)(b), Protocol Additional to the Geneva Conventions of 12 August 1949 (Protocol I); and "Rule 14: Proportionality in Attack," Customary IHL, https://ihl-databases.icrc.org/customary-ihl/eng/docs/v1_cha_chapter4_rule14.

251 *principle of avoiding unnecessary suffering*: "Practice Relating to Rule 70. Weapons of a Nature to Cause Superfluous Injury or Unnecessary Suffering," Customary IHL, https://ihl-databases.icrc.org/customary-ihl/eng/docs/v2_rul_rule70.

252 *precautions in the attack*: "Article 57: Precautions in Attack," Protocol Additional to the Geneva Conventions of 12 August 1949 (Protocol I), https://ihl -databases.icrc.org/applic/ihl/ihl.nsf/9ac284404d38ed2bc1256311002afd89/50f b5579fb098faac12563cd0051dd7c; "Rule 15: Precautions in Attack," Customary IHL, https://ihl-databases.icrc.org/customary-ihl/eng/docs/v1_rul_rule15.

252 *'hors de combat'*: "Article 41: Safeguard of an Enemy Hors de Combat," Protocol Additional to the Geneva Conventions of 12 August 1949 (Protocol I), https:// ihl-databases.icrc.org/applic/ihl/ihl.nsf/WebART/470-750050?OpenDocument; "Rule 41: Attacks Against Persons Hors de Combat," Customary IHL, https://ihl -databases.icrc.org/customary-ihl/eng/docs/v1_rul_rule47.

252 **by their nature, indiscriminate or uncontrollable:** "Article 51: Protection of the Civilian Population," and "Rule 71: Weapons That Are By Nature Indiscriminate," Customary IHL, https://ihl-databases.icrc.org/customary-ihl/eng/docs/ v1_rul_rule71.

252 **"lots of civilian dying":** Steve Goose, interview, October 26, 2016.

255 **"there is no accepted formula":** Kenneth Anderson, Daniel Reisner, and Matthew Waxman, "Adapting the Law of Armed Conflict to Autonomous Weapon Systems," *International Law Studies* 90 (2014): 386–411, https://www.usnwc.edu/getattachment/ a2ce46e7-1c81-4956-a2f3-c8190837afa4/dapting-the-Law-of-Armed-Conflict -to-Autonomous-We.aspx, 403.

257 **Ancient Sanskrit texts:** Dharmaśāstras 1.10.18.8, as quoted in A. Walter Dorn, *The Justifications for War and Peace in World Religions Part III: Comparison of Scriptures from Seven World Religions* (Toronto: Defence R&D Canada, March 2010), 20, http://www.dtic.mil/dtic/tr/fulltext/u2/a535552.pdf. Mahabharata, Book 11, Chapter 841, "Law, Force, and War," verse 96.10, from James L. Fitzgerald, ed., *Mahabharata,* Volume 7, Book 11 and Book 12, Part One, 1st ed. (Chicago: University of Chicago Press, 2003), 411.

257 **"blazing with fire":** Chapter VII: 90, Laws of Manu, translated by G. Buhler, http:// sourcebooks.fordham.edu/halsall/india/manu-full.asp.

257 **"sawback" bayonets:** Sawback bayonets are not illegal, however, provided the purpose is to use the saw as a tool and not for unnecessarily injuring the enemy. Bill Rhodes, *An Introduction to Military Ethics: A Reference Handbook* (Santa Barbara, CA: Praeger, 2009), 13–14.

258 **because of the wounds they cause:** "Protocol on Non-Detectable Fragments (Protocol I), United Nations Conference on Prohibitions or Restrictions on the Use of Certain Conventional Weapons Which May be Deemed to be Excessively Injurious or to Have Indiscriminate Effects," Geneva, 1980, https://ihl-databases .icrc.org/applic/ihl/ihl.nsf/Article.xsp?action=openDocument&documentId =1AF77FFE8082AE07C12563CD0051EDF5; "Rule 79: Weapons Primarily Injuring by Non-Detectable Fragments," Customary IHL, https://ihl-databases.icrc.org/ customary-ihl/eng/docs/v1_rul_rule79.

258 **Is being blinded by a laser really worse:** Charles J. Dunlap, "Is it Really Better to be Dead than Blind?," *Just Security*, January 13, 2015, https://www.justsecurity.org/ 19078/dead-blind/.

258 **"take all feasible precautions":** Article 57(2)(a)(ii), Protocol Additional to the Geneva Conventions of 12 August 1949 (Protocol I); and "Rule 15: Precautions in Attack," Customary IHL.

258 **"feasible" precautions:** Anderson et al., "Adapting the Law of Armed Conflict to Autonomous Weapon Systems," 403–405.

259 **Lieber Code:** "Article 71, General Orders No. 100: The Lieber Code," The Avalon Project, http://avalon.law.yale.edu/19th_century/lieber.asp#art71.

259 **"has been rendered unconscious":** "Article 41: Safeguard of an Enemy Hors de Combat," Protocol Additional to the Geneva Conventions of 12 August 1949 (Protocol I).

259 **"perfidy":** "Practice Relating to Rule 65: Perfidy," Customary IHL, https://ihl-databases.icrc.org/customary-ihl/eng/docs/v2_rul_rule65.

261 **"You've just been disarmed":** John S. Canning, "'You've just been disarmed. Have a nice day!'" *IEEE Technology and Society Magazine* (Spring 2009), 12–15.

261 **"targeting either the bow or the arrow":** John Canning, interview, December 6, 2016.

261 **ultra-precise weapons that would disarm:** John S. Canning, "Weaponized Unmanned Systems: A Transformational Warfighting Opportunity, Government Roles in Making It Happen," http://www.sevenhorizons.org/docs/CanningWeaponizedunmannedsystems.pdf

261 **"let the machines target machines":** John S. Canning, "A Concept of Operations for Armed Autonomous Systems," presentation, http://www.dtic.mil/ndia/2006disruptive_tech/canning.pdf.

261 **"accountability gap":** Bonnie Docherty, "Mind the Gap: The Lack of Accountability for Killer Robots," Human Rights Watch, April 9, 2015, https://www.hrw.org/report/2015/04/09/mind-gap/lack-accountability-killer-robots.

261 **"fair nor legally viable":** Bonnie Docherty, interview, November 18, 2016.

261 **"'punishing' the robot":** Docherty, "Mind the Gap."

262 **generally shielded from civil liability:** Docherty notes: "Immunity for the US military and its defense contractors presents an almost insurmountable hurdle to civil accountability for users or producers of fully autonomous weapons. The military is immune from lawsuits related to: (1) its policy determinations, which would likely include a choice of weapons, (2) the wartime combat activities of military forces, and (3) acts committed in a foreign country. Manufacturers contracted by the military are similarly immune from suit when they design a weapon in accordance with government specifications and without deliberately misleading the military. These same manufacturers are also immune from civil claims relating to acts committed during wartime." Ibid.

262 **"dangerous combination":** Bonnie Docherty, interview, November 18, 2016.

262 **"retributive justice":** Ibid.

262 **"eliminate this accountability gap":** Docherty, "Mind the Gap."

262 **shootdown was a mistake:** Rebecca Crootof, "War Torts: Accountability for Autonomous Weapons," *University of Pennsylvania Law Review* 164, no. 6 (May 2016): 1347-1402, http://scholarship.law.upenn.edu/cgi/viewcontent.cgi?article=9528&context=penn_law_review.

262 **U.S. government paid $61.8 million:** Crootof, "War Torts."

262 **"resonates with everyone":** Bonnie Docherty, interview, November 18, 2016.

262 **must be an individual to hold accountable:** Charles Dunlap Jr., "Accountability and Autonomous Weapons: Much Ado About Nothing?" *Temple International and Comparative Law Journal* 30 (1), Spring 2016, 65–66.

263 **"issue is not with autonomous weapons":** Ibid.

263 **"In cases not covered by the law":** Protocol Additional to the Geneva Conventions of 12 August 1949, and relating to the Protection of Victims of Non-International Armed Conflicts (Protocol II), June 8, 1977, https://ihl-databases.icrc.org/applic/ihl/ihl.nsf/7c4d08d9b287a42141256739003e636b/d67c3971bcff1c10c125641e00 52b545.

264 **"There is no accepted interpretation":** Rupert Ticehurst, "The Martens Clause and the Laws of Armed Conflict," *International Review of the Red Cross* 317 (April 30, 1997), https://www.icrc.org/eng/resources/documents/article/other/57jnhy.htm.

264 **"priming":** Cengiz Erisen, Milton Lodge and Charles S. Taber, "Affective Contagion in Effortful Political Thinking," *Political Psychology* 35, no. 2 (April 2014): 187–206.

264 **Carpenter found:** Charli Carpenter, "How Do Americans Feel About Fully Autonomous Weapons?" Duck of Minerva, June 19, 2013, http://duckofminerva .com/2013/06/how-do-americans-feel-about-fully-autonomous-weapons.html.

265 **sharp arrow in the quiver of ban advocates:** "Q&A on Fully Autonomous Weapons," Human Rights Watch, October 21, 2013, https://www.hrw.org/news/2013/10/21/ qa-fully-autonomous-weapons.

265 **"it is too early to argue that":** Michael C. Horowitz, "Public Opinion and the Politics of the Killer Robots Debate," February 16, 2016, http://rap.sagepub.com/ content/3/1/2053168015627183.

265 **"'Conscience' has an explicitly moral inflection":** Peter Asaro, "Jus nascendi: Robotic Weapons and the Martens Clause," http://www.peterasaro.org/writing/ Asaro%20Jus%20Nascendi%20PROOF.pdf.

265 **"disservice to reduce the 'dictates of public conscience'":** Ibid.

265 **"through public discussion, as well as academic scholarship":** Ibid.

265 **"The bar for claiming to speak for humanity":** Horowitz, "Public Opinion and the Politics of the Killer Robots Debate."

266 **"the clearest manifestation of":** Steve Goose, email to author, November 22, 2016.

266 **"emphasize *effects* rather than weapons":** Charles J. Dunlap et al., "Guest Post: To Ban New Weapons or Regulate Their Use?," *Just Security*, April 3, 2015, https:// www.justsecurity.org/21766/guest-post-ban-weapons-regulate-use/.

266 **modern-day CS gas:** Ibid.

267 **legal for military use *against civilians*:** "Riot Control Agents," Organization for the Prohibition of Chemical Weapons, accessed June 16, 2017, https://www.opcw.org/ about-chemical-weapons/types-of-chemical-agent/riot-control-agents/.

267 **"smart mines":** Dunlap et al., "Guest Post."

267 **"the paradox that requires":** Ibid.

267 **"Given the pace of accelerated scientific development":** Dunlap et al., "Is It Really Better to Be Dead than Blind?"

267 **"*strict* compliance with the core principles":** Dunlap et al., "Guest Post."

268 **"even though there are no victims":** Bonnie Docherty, interview, November 18, 2016.

268 **"grave concern":** Steve Goose, interview, October 26, 2016.

268 **Protocol II:** Protocol on Prohibitions or Restrictions on the Use of Mines, Booby-Traps and Other Devices as amended on 3 May 1996 (Protocol II to the 1980 CCW Convention as amended on 3 May 1996), https://ihl-databases.icrc.org/ihl/ INTRO/575.

268 **"The dangers just far outweigh":** Steve Goose, interview, October 26, 2016.

268 **"where you stand depends on where you sit":** Rufus E. Miles, Jr., "The Origin

and Meaning of Miles' Law" *Public Administration Review* 38, no. 5 (September/October, 1978): 399-403. https://www.jstor.org/stable/975497?seq=1#page_scan_tab_contents.

268 **"Denying such capabilities to nations":** Dunlap et al., "Guest Post."

269 **"The law of war rules on conducting attacks":** Department of Defense, "Department of Defense Law of War Manual," 330.

269 **"acts of violence against the adversary":** "Article 49: Definition of Attacks and Scope of Application," Protocol Additional to the Geneva Conventions of 12 August 1949, and relating to the Protection of Victims of International Armed Conflicts (Protocol I), June 8, 1977, https://ihl-databases.icrc.org/ihl/WebART/470-750062?OpenDocument.

269 **"the size of something that constitutes an attack":** Kenneth Anderson, interview, January 6, 2017.

270 **"is a technical term relating to":** International Committee of the Red Cross, "Commentary of 1987 Protection of the Civilian Population," https://ihl-databases.icrc.org/applic/ihl/ihl.nsf/Comment.xsp?action=openDocument&documentId=2C8494C2FCAF8B27C12563CD0043AA67.

270 **"The notion of the launching of an attack":** Kenneth Anderson, interview, January 6, 2017.

17   Soulless Killers: The Morality of Autonomous Weapons

271 **"morally reprehensible":** Jody Williams, interview, October 27, 2016.

272 **combatants would have an ethical responsibility:** Kenneth Anderson and Matthew Waxman, "Law and Ethics for Autonomous Weapon Systems Why a Ban Won't Work and How the Laws of War Can," Hoover Institution, http://media.hoover.org/sites/default/files/documents/Anderson-Waxman_LawAndEthics_r2_FINAL.pdf, 21-22.

272 **detailed policy guidance:** Department of Defense, "Department of Defense Directive Number 3000.09."

273 **"naked soldier":** Michael Walzer, *Just and Unjust Wars: A Moral Argument With Historical Illustrations,* 4th ed. (New York: Basic Books, 1977), 138–142.

273 **"It is not against the rules of war":** Ibid, 142.

274 **"War is cruelty":** "William Tecumseh Sherman," Biography.com, https://www.biography.com/people/william-tecumseh-sherman-9482051.

274 **sergeant chastised the soldiers:** Walzer, *Just and Unjust Wars,* 141.

274 **"He got him, but":** Ibid, 140.

274 **They are the exception:** It is also worth noting that this concern about the role of mercy only applies to antipersonnel autonomous weapons. All of the examples of these moments of mercy are ones where a single enemy individual is targeted. They do not appear to arise in situations where soldiers are targeting objects, even those that have people in them such as ships or tanks. Anti-vehicle or anti-material weapons, therefore, would not run afoul of this concern.

275 **"posturing":** Dave Grossman, *On Killing: The Psychological Cost of Learning to Kill in War and Society* (Boston: Little, Brown and Company, 1996), 3–4.

275 **evidence from a variety of wars:** Ibid, 9–28.

275 **innate biological resistance:** Ibid, 177–178.

275 **one animal submits first:** Ibid, 5–6.

276 **drone crews:** Kelly Faircloth, "Everyone Names Their Roomba. What Would You Name Yours?," http://jezebel.com/everyone-names-their-roomba-what-would-you-name-yours. James Dao, "Drone Pilots Are Found to Get Stress Disorders Much as Those in Combat Do," *New York Times*, February 22, 2013, https://www.nytimes.com/2013/02/23/us/drone-pilots-found-to-get-stress-disorders-much-as-those-in-combat-do.html. Rebecca Hawkes, "Post-Traumatic Stress Disorder Is Higher in Drone Operators," *The Telegraph*, May 30, 2015, http://www.telegraph.co.uk/culture/hay-festival/11639746/Post-traumatic-stress-disorder-is-higher-in-drone-operators.html. "Can Drone Pilots Be Diagnosed With Post-Traumatic Stress Disorder?," *NPR*.org, accessed June 16, 2017, http://www.npr.org/2015/06/06/412525635/can-drone-pilots-be-diagnosed-with-post-traumatic-stress-disorder. Ed Pilkington, "Life as a Drone Operator: 'Ever Step on Ants and Never Give It Another Thought?,'" *The Guardian*, November 19, 2015, https://www.theguardian.com/world/2015/nov/18/life-as-a-drone-pilot-creech-air-force-base-nevada.

276 **firing rates:** Grossman, *On Killing*, 153.

277 **"if he can get others to share":** Another psychological factor that contributed to higher firing rates among machine gun crews, Grossman argues, was mutual accountability for their actions among the teammates. If one of them wasn't performing his job properly, the others could immediately tell. Ibid, 149–154.

277 **"We see some really dangerous behaviors":** Mary "Missy" Cummings, interview, June 1, 2016.

277 **"[P]hysical and emotional distancing":** M. L. Cummings, "Creating Moral Buffers in Weapon Control Interface Design," *IEEE Technology and Society Magazine* (Fall 2004), 29–30.

278 **"[It] is more palatable":** Ibid, 31.

278 **name their Roomba:** Celeste Biever, "My Roomba's Name Is Roswell," *Slate*, March 23, 2014, http://www.slate.com/articles/health_and_science/new_scientist/2014/03/roomba_vacuum_cleaners_have_names_irobot_ceo_on_people_s_ties_to_robots.html.

278 **"It is possible that without consciously":** Cummings, "Creating Moral Buffers in Weapon Control Interface Design," 31.

278 **"could permit people to perceive":** Ibid, 32.

278 **"cheerful, almost funny graphic":** Ibid, 32–33.

278 **Cummings criticized the Army's decision:** Cummings uses the terms "management by exception" to refer to a human-supervised autonomous control mode and "management by consent" to refer to a semi-autonomous control mode. Ibid, 33.

278 **"[E]nabling a system to essentially fire at will":** Ibid, 33.

278 **take a positive action before the weapon fires:** Ibid, 33.

279 **changed its marksmanship training:** Grossman, *On Killing*, 35.

279 **U.S. aerial firebombing killed:** Robert S. McNamara in *Fog of War: Eleven Lessons From the Life of Robert S. McNamara,* documentary, directed by Errol Morris (2003; Sony Pictures Classics).

279 **"were behaving as war criminals":** Ibid.

279 **"the television coverage was starting":** Colin L. Powell with Joseph E. Persico, *My American Journey* (New York: Ballantine Books, 2003).

279 **The dehumanization that enables killing:** Grossman, *On Killing*, 156–164.

279 **war crimes:** Ibid, 210–211.

280 **mental health surveys of deployed U.S. troops:** Office of the Surgeon, Multi-National Force Iraq and Office of the Surgeon General, United States Army Medical Command, "Mental Health Advisory Team (MHAT) IV, Operation Iraqi Freedom 05-07, Final Report," November 17, 2006, http://armymedicine.mil/Documents/MHAT-IV-Report-17NOV06-Full-Report.pdf, 34-42. Office of the Surgeon, Multi-National Force Iraq and Office of the Surgeon General, United States Army Medical Command, "Mental Health Advisory Team (MHAT) V, Operation Iraqi Freedom 06-08, Final Report," February 14, 2008, http://armymedicine.mil/Documents/Redacted1-MHATV-OIF-4-FEB-2008Report.pdf, 30–32.

280 **"What if [robotics] actually works?":** Ron Arkin, interview, June 8, 2016.

281 **"I don't care about the robots":** Ibid.

281 **"ethical governor":** Ronald C. Arkin, "Governing Lethal Behavior: Embedding Ethics in a Hybrid Deliberative/Reactive Robot Architecture," Technical Report GIT-GVU-07-11, http://www.cc.gatech.edu/ai/robot-lab/online-publications/formalizationv35.pdf.

281 **"moral imperative to use this":** Ron Arkin, interview, June 8, 2016.

281 **A typical bomb had only:** "Accuracy and Employment of Air-Dropped Guided Munitions by the United States," Center for a New American Security, https://s3.amazonaws.com/files.cnas.org/images/Accuracy-and-employment-air-dropped-guided-munitions.jpg.

281 **More than 9,000 bombs:** Richard P. Hallion, "Precision Guided Munitions and the New Era of Warfare," APSC Paper Number 53, Air Power Studies Centre, http://fas.org/man/dod-101/sys/smart/docs/paper53.htm.

281 **accurate to within five feet:** "Accuracy and Employment of Air-Dropped Guided Munitions by the United States."

282 **U.S. drone strikes:** "Drone Wars: The Full Data," *The Bureau of Investigative Journalism*, accessed June 16, 2017, https://www.thebureauinvestigates.com/stories/2017-01-01/drone-wars-the-full-data.

282 **"the use of indiscriminate rockets":** "Ukraine: Unguided Rockets Killing Civilians," *Human Rights Watch*, July 24, 2014, https://www.hrw.org/news/2014/07/24/ukraine-unguided-rockets-killing-civilians.

282 **"some of it is quite dishonorable":**, Ron Arkin, interview June 8, 2016.

282 **"utterly and wholly unacceptable":** Ronald Arkin, "The Case for Banning Killer Robots: Counterpoint," accessed June 16, 2017, https://cacm.acm.org/magazines/2015/12/194632-the-case-for-banning-killer-robots/abstract.

283 **"software safety":** Ron Arkin, interview, June 8, 2016.

283 **"back in some general's office":** Ibid.

283 **"There is no doubt in my mind":** Jody Williams, interview, October 27, 2016.

283 **"Should we create caged tigers":** Ron Arkin, interview, June 8, 2016.

284 **"Where does the danger lurk?":** Ibid.

284 **"we need to do the research":** Ibid.

284 **"could possibly support":** Ibid.

285 **"utilitarian, consequentialist":** Ibid.

285 **"fundamentally inhuman":** Jody Williams, interview, October 27, 2016.

285 **"fundamental question of whether it's appropriate":** Peter Asaro, interview, December 19, 2016.

286 **"What decisions require uniquely human judgment?":** Kenneth Anderson, email to author, January 4, 2016.

287 **if a decision is made to take a human life:** Duncan Purves, Ryan Jenkins & Brad-

ley J. Strawser, "Autonomous Machines, Moral Judgment, and Acting for the Right Reasons" *Ethical Theory and Moral Practice* 18, no. 4 (2015): 851–872.

287 **"the most fundamental and salient moral question"**: Peter Asaro, interview, December 19, 2016.

287 **Heyns called on states to declare**: Christof Heyns, "Report of the Special Rapporteur on extrajudicial, summary or arbitrary executions," United Nations Human Rights Council, April 9, 2013, http://www.ohchr.org/Documents/HRBodies/HRCouncil/RegularSession/Session23/A-HRC-23-47_en.pdf.

287 **"arbitrary for a decision to be taken"**: Christof Heyns, interview, May 18, 2016.

288 **"fundamental violation"**: Peter Asaro, interview, December 19, 2016.

288 **the right to die a dignified death in war**: Of course, dying honorably has often been important in warrior culture, but that is not the same as extending the enemy the opportunity to die an honorable death.

288 **"war without reflection is mechanical slaughter"**: Nick Cumming-Bruce, "U.N. Expert Calls for Halt on Military Robots," *New York Times*, May 30, 2013, sec. Europe, https://www.nytimes.com/2013/05/31/world/europe/united-nations-armed-robots.html.

290 **"If you eliminate the moral burden"**: Peter Asaro, interview, December 19, 2016.

290 **"moral injury"**: David Wood, "Moral Injury," *The Huffington Post*, accessed June 17, 2017, http://projects.huffingtonpost.com/projects/moral-injury. Shira Maguen and Brett Litz, "Moral Injury in the Context of War," PTSD: National Center for PTSD, accessed June 17, 2017, https://www.ptsd.va.gov/professional/co-occurring/moral_injury_at_war.asp. "What Is Moral Injury," The Moral Injury Project, accessed June 17, 2017, http://moralinjuryproject.syr.edu/about-moral-injury/.

290 **the most traumatic thing a soldier can experience**: Grossman, *On Killing*, 87–93, 156–158.

293 **"One of the places that we spend"**: Paul Selva, "Innovation in the Department of Defense with General Paul Selva," Center for Strategic and International Studies, https://www.csis.org/events/innovation-defense-department-general-paul-selva. Remarks on autonomous weapons begin around 39:00.

293 **"Because we take our values to war"**: Paul Selva, testimony before the Senate Armed Services Committee, July 18, 2017, https://www.armed-services.senate.gov/hearings/17-07-18-nomination_--selva. Comments on autonomous weapons begin around 1:11:10.

294 **"War is about attempting to increase"**: Jody Williams, interview, October 27, 2016.

294 **"crosses a moral and ethical Rubicon"**: Ibid.

294 **"You know the difference between a good robot and a bad robot"**: Ibid.

295 **good Terminators**: Ron Arkin, interview, June 8, 2016.

295 **"beyond-IHL principle of human dignity"**: Ken Anderson, interview, January 6, 2016.

295 **"then we must ask ourselves whether"**: Christof Heyns, interview, May 18, 2016.

295 **"as long as we don't lose our soul"**: Ron Arkin, interview, June 8, 2016.

296 **"Killing Japanese didn't bother me very much"**: "General Curtis E. LeMay, (1906–1990)," PBS.org, http://www.pbs.org/wgbh//amex/bomb/peopleevents/pandeAMEX61.html.

296 **"I am tired and sick of war"**: "William Tecumseh Sherman."

18   Playing with Fire: Autonomous Weapons and Stability

298 **further incentivized the Soviet Union:** Michael S. Gerson, "The Origins of Strategic Stability: The United States and the Threat of Surprise Attack," in Elbridge A. Colby and Michael S. Gerson, eds., *Strategic Stability: Contending Interpretations* (Carlisle, PA: Strategic Studies Institute and U.S. Army War College Press, 2013), 3–35.

298 **"we have to worry about his striking us":** T. C. Schelling, *Surprise Attack and Disarmament* (Santa Monica, CA: RAND, December 10, 1958).

298 **"when neither in striking first":** Ibid.

299 **"In a stable situation":** Elbridge Colby, "Defining Strategic Stability: Reconciling Stability and Deterrence," in Colby and Gerson, *Strategic Stability*, 57.

299 **"War termination":** Thomas C. Schelling, *Arms and Influence* (New Haven and London: Yale University Press, 1966), 203–208. Fred Ikle, *Every War Must End* (New York: Columbia Classics, 2005).

299 **offense-defense balance:** Charles L. Glaser and Chaim Kaufmann, "What Is the Offense-Defense Balance and Can We Measure It? (Offense, Defense, and International Politics)," *International Security* 22, no. 4 (Spring 1998).

301 **Outer Space Treaty:** Treaty on Principles Governing the Activities of States in the Exploration and Use of Outer Space, including the Moon and Other Celestial Bodies, 1967, http://www.unoosa.org/oosa/en/ourwork/spacelaw/treaties/outerspacetreaty.html.

301 **Seabed Treaty:** Treaty on the Prohibition of the Emplacement of Nuclear Weapons and Other Weapons of Mass Destruction on the Seabed and the Ocean Floor and in the Subsoil Thereof, 1971, https://www.state.gov/t/isn/5187.htm.

301 **Environmental Modification Convention:** Convention on the Prohibition of Military or Any Other Hostile Use of Environmental Modification Techniques, 1977, https://www.state.gov/t/isn/4783.htm.

301 **Anti-Ballistic Missile (ABM) Treaty:** Treaty Between The United States of America and The Union of Soviet Socialist Republics on The Limitation of Anti-Ballistic Missile Systems (ABM Treaty), 1972, https://www.state.gov/t/avc/trty/101888.htm.

301 **Intermediate-Range Nuclear Forces (INF) Treaty:** U.S. Department of State, "Treaty Between the United States of American and the Union of Soviet Socialist Republics on the Elimination of their Intermediate-Range and Shorter-Range Missiles," accessed June 17, 2017, https://www.state.gov/www/global/arms/treaties/inf1.html#treaty.

301 **neutron bombs:** In practice, though, any adjustable dial-a-yield nuclear weapon could function as a neutron bomb.

301 **attacker could use the conquered territory:** This is not to say that neutron bombs did not have, in theory, legitimate uses. The United States viewed neutron bombs as valuable because they could be used to defeat Soviet armored formations on allied territory without leaving residual radiation.

302 **U.S. plans to deploy neutron bombs to Europe:** Mark Strauss, "Though It Seems Crazy Now, the Neutron Bomb Was Intended to Be Humane," *io9*, September 19, 2014, http://io9.gizmodo.com/though-it-seems-crazy-now-the-neutron-bomb-was-intende-1636604514.

302 **first-mover advantage in naval warfare:** Wayne Hughes, *Fleet Tactics and Coastal Combat,* 2nd ed. (Annapolis, MD: Naval Institute Press, 1999).

302 **"Artificial Intelligence, War, and Crisis Stability":** Michael C. Horowitz, "Arti-

ficial Intelligence, War, and Crisis Stability," November 29, 2016. [unpublished manuscript, as of June 2017].

303 **robot swarms will lead to:** Jean-Marc Rickli, "Some Considerations of the Impact of LAWS on International Security: Strategic Stability, Non-State Actors and Future Prospects," paper submitted to Meeting of Experts on Lethal Autonomous Weapons Systems, United Nations Convention on Certain Conventional Weapons (CCW), April 16, 2015, http://www.unog.ch/80256EDD006B8954/(httpAssets)/B6E6B97 4512402BEC1257E2E0036AAF1/$file/2015_LAWS_MX_Rickli_Corr.pdf.

303 **lower the threshold for the use of force:** Peter M. Asaro, "How Just Could a Robot War Be?" http://peterasaro.org/writing/Asaro%20Just%20Robot%20War.pdf.

303 **"When is it that you would deploy these systems":** Michael Horowitz, interview, December 7, 2016.

305 **"The premium on haste":** Schelling, *Arms and Influence,* 227.

305 **"when speed is critical":** Ibid, 227.

305 **control escalation:** See also Jürgen Altmann and Frank Sauer, "Autonomous Weapon Systems and Strategic Stability," *Survival* 59:5 (2017), 117–42.

305 **"restraining devices for weapons":** Ibid, 231.

306 **war among nuclear powers:** Elbridge Colby, "America Must Prepare for Limited War," *The National Interest,* October 21, 2015, http://nationalinterest.org/feature/america-must-prepare-limited-war-14104.

306 **"[S]tates [who employ autonomous weapons]":** Michael Carl Haas, "Autonomous Weapon Systems: The Military's Smartest Toys?" *The National Interest,* November 20, 2014, http://nationalinterest.org/feature/autonomous-weapon-systems-the-militarys-smartest-toys-11708.

307 **test launch of an Atlas ICBM:** Sagan, *The Limits of Safety,* 78–80.

307 **U-2 flying over the Arctic Circle:** Martin J. Sherwin, "The Cuban Missile Crisis at 50," *Prologue Magazine* 44, no. 2 (Fall 2012), https://www.archives.gov/publications/prologue/2012/fall/cuban-missiles.html.

307 **General Thomas Power:** Powers' motivations for sending this message have been debated by historians. See Sagan, *The Limits of Safety,* 68–69.

308 **"commander's intent":** Headquarters, Department of the Army, Field Manual 100-5 (June 1993), 6-6.

308 **"If . . . the enemy commander has":** Lawrence G. Shattuck, "Communicating Intent and Imparting Presence," *Military Review* (March–April 2000), http://www.au.af.mil/au/awc/awcgate/milreview/shattuck.pdf, 66.

308 **"relocating important assets":** Haas, "Autonomous Weapon Systems."

309 **"strategic corporal":** Charles C. Krulak, ""The Strategic Corporal: Leadership in the Three Block War," *Marines Magazine* (January 1999).

309 **"affirmative human decision":** "No agency of the Federal Government may plan for, fund, or otherwise support the development of command control systems for strategic defense in the boost or post-boost phase against ballistic missile threats that would permit such strategic defenses to initiate the directing of damaging or lethal fire except by affirmative human decision at an appropriate level of authority." 10 U.S.C. 2431 Sec. 224.

310 **"you still have the problem that that's":** David Danks, interview, January 13, 2017.

310 **"[B]efore we sent the U-2 out":** Robert McNamara, Interview included as special feature on *Dr. Strangelove or: How I Learned to Stop Worrying and Love the Bomb* (DVD). Columbia Tristar Home Entertainment, (2004) [1964].

311 **"We're going to blast them now!":** William Burr and Thomas S. Blanton, eds., "The Submarines of October," The National Security Briefing Book, No. 75, October 31, 2002, http://nsarchive.gwu.edu/NSAEBB/NSAEBB75/. "The Cuban Missile Crisis, 1962: Press Release, 11 October 2002, 5:00 PM," accessed June 17, 2017, http://nsarchive.gwu.edu/nsa/cuba_mis_cri/press3.htm.

311 **game of chicken:** Schelling, *Arms and Influence*, 116–125. Herman Kahn, *On Escalation: Metaphors and Scenarios* (New Brunswick and London: Transaction Publishers, 2010), 10-11.

311 **"takes the steering wheel":** Kahn, *On Escalation*, 11.

312 **"how would the Kennedy Administration":** Horowitz, "Artificial Intelligence, War, and Crisis Stability," November.

312 **"because of the automated and irrevocable":** *Dr. Strangelove or: How I Learned to Stop Worrying and Love the Bomb*, directed by Stanley Kubrick (1964).

313 **"Dead Hand":** Nicholas Thompson, "Inside the Apocalyptic Soviet Doomsday Machine," *WIRED*, September 21, 2009, https://www.wired.com/2009/09/mf-deadhand/. Vitalii Leonidovich Kataev, interviewed by Ellis Mishulovich, May 1993, http://nsarchive.gwu.edu/nukevault/ebb285/vol%20II%20Kataev.PDF. Varfolomei Vlaimirovich Korobushin, interviewed by John G. Hines, December 10, 1992, http://nsarchive.gwu.edu/nukevault/ebb285/vol%20II%20Korobushin.PDF.

313 **Accounts of Perimeter's functionality differ:** Some accounts by former Soviet officials state that the Dead Hand was investigated and possibly even developed, but never deployed operationally. Andrian A. Danilevich, interview by John G. Hines, March 5, 1990, http://nsarchive.gwu.edu/nukevault/ebb285/vol%20iI%20 Danilevich.pdf, 62-63; and Viktor M. Surikov, interview by John G. Hines, September 11, 1993, http://nsarchive.gwu.edu/nukevault/ebb285/vol%20II%20 Surikov.PDF, 134-135. It is unclear, though, whether this refers in reference or not to a fully automatic system. Multiple sources confirm the system was active, although the degree of automation is ambiguous in their accounts: Kataev, 100–101; and Korobushin, 107.

313 **remain inactive during peacetime:** Korobushin, 107; Thompson, "Inside the Apocalyptic Soviet Doomsday Machine."

313 **network of light, radiation, seismic, and pressure:** Ibid.

313 **leadership would be cut out of the loop:** Ibid.

313 **rockets that would fly over Soviet territory:** Kataev. Korobushin.

314 **Perimeter is still operational:** Thompson, "Inside the Apocalyptic Soviet Doomsday Machine."

314 **"stability-instability paradox":** Michael Krepon, "The Stability-Instability Paradox, Misperception, and Escalation Control in South Asia," The Stimson Center, 2003, https://www.stimson.org/sites/default/files/file-attachments/stability-instability-paradox-south-asia.pdf. B.H. Liddell Hart, *Deterrent or Defence* (London: Stevens and Sons, 1960), 23.

315 **"madman theory":** Harry R. Haldeman and Joseph Dimona, *The Ends of Power* (New York: Times Books, 1978), 122. This is not a new idea. It dates back at least to Machiavelli. Niccolo Machiavelli, *Discourses on Livy*, Book III, Chapter 2.

315 **"the threat that leaves something to chance":** Thomas C. Schelling, *The Strategy of Conflict* (Cambridge, MA: Harvard University Press, 1960).

316 **"There's a real problem here":** David Danks, interview, January 13, 2017.

316 **"Autonomous weapon systems are very new":** Ibid.

316 **"it's just completely unreasonable"**: Ibid.

317 **"Mr. President, we and you ought not"**: Department of State Telegram Transmitting Letter From Chairman Khrushchev to President Kennedy, October 26, 1962, http://microsites.jfklibrary.org/cmc/oct26/doc4.html.

317 **"there are scenarios in which"**: Haas, "Autonomous Weapon Systems."

19   Centaur Warfighters: Humans + Machines

321 **Gary Kasparov:** Mike Cassidy, "Centaur Chess Brings out the Best in Humans and Machines," BloomReach, December 14, 2014, http://bloomreach.com/2014/12/centaur-chess-brings-best-humans-machines/.

321 **centaur chess:** Tyler Cowen, "What are Humans Still Good for? The Turning Point in Freestyle Chess may be Approaching," *Marginal Revolution*, November 5, 2013, http://marginalrevolution.com/marginalrevolution/2013/11/what-are-humans-still-good-for-the-turning-point-in-freestyle-chess-may-be-approaching.html.

322 **"On 17 April 1999"**: Mike Pietrucha, "Why the Next Fighter will be Manned, and the One After That," *War on the Rocks,* August 5, 2015, http://warontherocks.com/2015/08/why-the-next-fighter-will-be-manned-and-the-one-after-that/.

323 **Commercial airliners use automation:** Mary Cummings and Alexander Stimpson, "Full Auto Pilot: Is it Really Necessary to Have a Human in the Cockpit?," *Japan Today*, May 20, 2015, http://www.japantoday.com/category/opinions/view/full-auto-pilot-is-it-really-necessary-to-have-a-human-in-the-cockpit.

324 **"Do Not Engage Sector"**: Mike Van Rassen, "Counter-Rocket, Artillery, Mortar (C-RAM)," Program Executive Office Missiles and Space, accessed June 16, 2017, Slide 28, http://www.msl.army.mil/Documents/Briefings/C-RAM/C-RAM%20Program%20Overview.pdf.

324 **"The human operators do not aim"**: Sam Wallace, "The Proposed Ban on Offensive Autonomous Weapons is Unrealistic and Dangerous," Kurzweilai, August 5, 2015, http://www.kurzweilai.net/the-proposed-ban-on-offensive-autonomous-weapons-is-unrealistic-and-dangerous.

325 **"unwarranted and uncritical trust"**: Hawley, "Not by Widgets Alone."

325 **the human does not add any value:** Given recent advances in machine learning, it is possible we are at this point now. In December 2017, the AI research company DeepMind unveiled AlphaZero, a single algorithm that had achieved superhuman play in chess, *go*, and the Japanese strategy game shogi. Within a mere four hours of self-play and with no training data, AlphaZero eclipsed the previous top chess program. The method behind AlphaZero, deep reinforcement learning, appears to be so powerful that it is unlikely that humans can add any value as members of a "centaur" human-machine team for these games. Tyler Cowen, "The Age of the Centaur Is *Over* Skynet Goes Live," MarginalRevolution.com, December 7, 2017, http://marginalrevolution.com/marginalrevolution/2017/12/the-age-of-the-centaur-is-over.html. David Silver et al., "Mastering Chess and Shogi by Self-Play with a General Reinforcement Learning Algorithm," December 5, 2017, https://arxiv.org/pdf/1712.01815.pdf.

325 **as computers advance:** Cowen, "What Are Humans Still Good For?"

327 **jam-resistant communications:** Sayler, "Talk Stealthy to Me." Paul Scharre, "Yes, Unmanned Aircraft Are The Future," *War on the Rocks,* August 11, 2015, https://warontherocks.com/2015/08/yes-unmanned-combat-aircraft-are-the-future/. Amy Butler, "5th-To-4th Gen Fighter Comms Competition Eyed In Fiscal 2015,"

Aviation Week Network, June 18, 2014, http://aviationweek.com/defense/5th-4th-gen-fighter-comms-competition-eyed-fiscal-2015.

330 **"What application are we trying"**: Peter Galluch, interview, July 15, 2016.

330 **"If they don't exist, there is no"**: Jody Williams, interview, October 27, 2016.

330 **"mutual restraint"**: Evan Ackerman, "We Should Not Ban 'Killer Robots,' and Here's Why," *IEEE Spectrum: Technology, Engineering, and Science News*, July 29, 2015, http://spectrum.ieee.org/automaton/robotics/artificial-intelligence/we-should-not-ban-killer-robots.

20 The Pope and the Crossbow: The Mixed History of Arms Control

331 **"The key question for humanity today"**: "Autonomous Weapons: An Open Letter From AI & Robotics Researchers."

332 **Whether or not a ban succeeds**: For analysis of why some weapons bans work and some don't, see: Rebecca Crootof, http://isp.yale.edu/sites/default/files/publications/killer_robots_are_here_final_version.pdf; Sean Watts, "Autonomous Weapons: Regulation Tolerant or Regulation Resistant?" *Temple International & Comparative Law Journal* 30 (1), Spring 2016, 177–187; and Rebecca Crootof, https://www.lawfareblog.com/why-prohibition-permanently-blinding-lasers-poor-precedent-ban-autonomous-weapon-systems.

341 **"the most powerful limitations"**: Schelling, *Arms and Influence*, 164.

341 **"'Some gas' raises complicated questions"**: Thomas C. Schelling, *The Strategy of Conflict* (Cambridge, MA: Harvard University, 1980), 75.

341 **Escalation from one step to another**: Schelling makes this point in Arms and Influence in a critique of the McNamara "no cities" doctrine. Schelling, *Arms and Influence,* 165.

342 **"If they declare that they will attack"**: "Hitlers Bombenterror: 'Wir Werden Sie Ausradieren,'" *Spiegel Online*, accessed April 1, 2003, http://www.spiegel.de/spiegelspecial/a-290080.html.

342 **Complete bans on weapons**: This seems to suggest that if lasers were used in future wars for non-blinding purposes and ended up causing incidental blinding, then they would quickly evolve into use for intentional blinding.

342 **"antipersonnel land mine"**: Human Rights Watch, "Yemen: Houthi Landmines Claim Civilian Victims," September 8, 2016, https://www.hrw.org/news/2016/09/08/yemen-houthi-landmines-claim-civilian-victims.

343 **SMArt 155 artillery shells**: "Fitzgibbon Wants to Keep SMArt Cluster Shells," Text, *ABC News*, (May 29, 2008), http://www.abc.net.au/news/2008-05-29/fitzgibbon-wants-to-keep-smart-cluster-shells/2452894.

343 **poison gas attack at Ypres**: Jonathan B. Tucker, *War of Nerves: Chemical Warfare from World War I to Al-Qaeda* (New York: Pantheon Books, 2006).

343 **"or by other new methods"**: Declaration (IV,1), to Prohibit, for the Term of Five Years, the Launching of Projectiles and Explosives from Balloons, and Other Methods of Similar Nature. The Hague, July 29, 1899, https://ihl-databases.icrc.org/applic/ihl/ihl.nsf/385ec082b509e76c41256739003e636d/53024c9c9b216ff2c125641e0035be1a?OpenDocument.

343 **"attack or bombardment"**: Regulations: Article 25, Convention (IV) respecting the Laws and Customs of War on Land and its annex: Regulations concerning the Laws and Customs of War on Land. The Hague, October 18, 1907, https://ihl-databases

.icrc.org/applic/ihl/ihl.nsf/Article.xsp?action=openDocument&documentId
=D1C251B17210CE8DC12563CD0051678F.

343 **"the bomber will always get through":** "The bomber will always get through," Wikipedia, https://en.wikipedia.org/wiki/The_bomber_will_always_get_through.

344 **Nuclear Non-Proliferation Treaty:** "Treaty on the Non-Proliferation of Nuclear Weapons (NPT)," United Nations Office for Disarmament Affairs, https://www.un .org/disarmament/wmd/nuclear/npt/text, accessed June 19, 2017.

344 **Chemical Weapons Convention:** "Chemical Weapons Convention," Organisa- tion for the Prohibition of Chemical Weapons, https://www.opcw.org/chemical -weapons-convention/, accessed June 19, 2017.

344 **INF Treaty:** "Treaty Between the United States of American and the Union of Soviet Socialist Republics on the Elimination of their Intermediate-Range and Shorter-Range Missiles."

344 **START:** "Treaty Between the United States of American and the Union of Soviet Socialist Republics on the Reduction and Limitation of Strategic Offensive Arms," U.S. Department of State, https://www.state.gov/www/global/arms/starthtm/ start/start1.html, accessed June 19, 2017.

344 **New START:** "New Start," U.S. Department of State, https://www.state.gov/t/avc/ newstart/, accessed June 19, 2017.

344 **Outer Space Treaty:** Treaty on Principles Governing the Activities of States in the Exploration and Use of Outer Space, including the Moon and Other Celestial Bodies.

344 **expanding bullets:** Declaration (IV,3) concerning Expanding Bullets, The Hague, July 29, 1899, https://ihl-databases.icrc.org/applic/ihl/ihl.nsf/Article.xsp?action =openDocument&documentId=F5FF4D9CA7E41925C12563CD0051616B.

344 **Geneva Gas Protocol:** Protocol for the Prohibition of the Use of Asphyxiating, Poi- sonous or Other Gases, and of Bacteriological Methods of Warfare, Geneva, June 17, 1925, https://ihl-databases.icrc.org/ihl/INTRO/280?OpenDocument.

344 **CCW:** "The Convention on Certain Conventional Weapons," United Nations Office for Disarmament Affairs, https://www.un.org/disarmament/geneva/ccw/, accessed June 19, 2017.

344 **SORT:** "Treaty Between the United States of America and the Russian Federation On Strategic Offensive Reductions (The Moscow Treaty)," U.S. Department of State, https://www.state.gov/t/isn/10527.htm, accessed June 19, 2017.

344 **weapons of mass destruction (WMD) in orbit:** Treaty on Principles Governing the Activities of States in the Exploration and Use of Outer Space, including the Moon and Other Celestial Bodies.

344 **Environmental Modification Convention:** Convention on the Prohibition of Mili- tary or Any Other Hostile Use of Environmental Modification Techniques.

344 **Biological Weapons Convention:** "The Biological Weapons Convention," United Nations Office for Disarmament Affairs, https://www.un.org/disarmament/wmd/ bio/, accessed June 19, 2017.

344 **secret biological weapons program:** Tim Weiner, "Soviet Defector Warns of Bio- logical Weapons," *New York Times*, February 24, 1998. Milton Leitenberg, Raymond A. Zilinskas, and Jens H. Kuhn, *The Soviet Biological Weapons Program: A History* (Cambridge: Harvard University Press, 2012). Ken Alibek, *Biohazard: The Chilling True Story of the Largest Covert Biological Weapons Program in the World* (Delta,

2000). Raymond A. Zilinskas, "The Soviet Biological Weapons Program and Its Legacy in Today's Russia," CSWMD Occasional Paper 11, July 18, 2016.

345 **Other weapons can be:** The lack of a verification regime has been a long-standing concern regarding the Biological Weapons Convention. "Biological Weapons Convention (BWC) Compliance Protocol," NTI, August 1, 2001, http://www.nti.org/analysis/articles/biological-weapons-convention-bwc/. "The Biological Weapons Convention: Proceeding without a Verification Protocol," *Bulletin of the Atomic Scientists*, May 9, 2011, http://thebulletin.org/biological-weapons-convention-proceeding-without-verification-protocol.

21 Are Autonomous Weapons Inevitable? The Search for Lethal Laws of Robotics

346 **Convention on Certain Conventional Weapons (CCW):** "2014 Meeting of Experts on LAWS," The United Nations Office at Geneva, http://www.unog.ch/__80256ee600585943.nsf/(httpPages)/a038dea1da906f9dc1257dd90042e261?OpenDocument&ExpandSection=1#_Section1. "2015 Meeting of Experts on LAWS," The United Nations Office at Geneva, http://www.unog.ch/80256EE600585943/(httpPages)/6CE049BE22EC75A2C1257C8D00513E26?OpenDocument; "2016 Meeting of Experts on LAWS," The United Nations Office at Geneva, http://www.unog.ch/80256EE600585943/(httpPages)/37D51189AC4FB6E1C1257F4D004CAFB2?OpenDocument.

347 **"appropriate human involvement":** CCW, "Report of the 2016 Informal Meeting of Experts on Lethal Autonomous Weapon Systems (LAWS), June 10, 2016.

348 **"How near to a city is":** Schelling, *Arms and Influence*, 165.

349 **"partition":** Article 36, "Autonomous weapons—the risks of a management by 'partition,'" October 10, 2012, http://www.article36.org/processes-and-policy/protection-of-civilians/autonomous-weapons-the-risks-of-a-management-by-partition/.

349 **Campaign to Stop Killer Robots has called:** "A comprehensive, pre-emptive prohibition on the development, production and use of fully autonomous weapons." The Campaign to Stop Killer Robots, "The Solution," http://www.stopkillerrobots.org/the-solution/.

349 **technology is too diffuse:** Ackerman, "We Should Not Ban 'Killer Robots,' and Here's Why."

349 **"not a wise campaign strategy":** Steve Goose, interview, October 26, 2016.

350 **care more about human rights:** Ian Vasquez and Tanja Porcnik, "The Human Freedom Index 2016," Cato Institute, the Fraser Institute, and the Friedrich Naumann Foundation for Freedom, 2016, https://object.cato.org/sites/cato.org/files/human-freedom-index-files/human-freedom-index-2016.pdf.

351 **"You know you're not":** Steve Goose, interview, October 26, 2016.

351 **A few experts have presented:** For example, Rickli, "Some Considerations of the Impact of LAWS on International Security: Strategic Stability, Non-State Actors and Future Prospects."

351 **"it's not really a significant feature":** John Borrie, interview, April 12, 2016.

352 **"offensive autonomous weapons beyond meaningful":** "Autonomous Weapons: An Open Letter From AI & Robotics Researchers."

352 **"will move toward some type":** Bob Work, interview, June 22, 2016.

355 **no antipersonnel equivalents:** Precision-guided weapons are evolving down to the level of infantry combat, including some laser-guided munitions such as the DARPA XACTO and Raytheon Spike missile. Because these are laser-guided, they are still remotely controlled by a person.

355 **"let machines target machines":** Canning, "A Concept of Operations for Armed Autonomous Systems."

356 **"stopping an arms race":** Stuart Russell Walsh Max Tegmark and Toby, "Why We Really Should Ban Autonomous Weapons: A Response," *IEEE Spectrum: Technology, Engineering, and Science News*, August 3, 2015, http://spectrum.ieee.org/automaton/robotics/artificial-intelligence/why-we-really-should-ban-autonomous-weapons.

357 **focus on the unchanging element in war:** The ICRC, for example, has called on states to "focus on the role of the human in the targeting process." International Committee of the Red Cross, "Views of the International Committee of the Red Cross (ICRC) on autonomous weapon system," paper submitted to the Convention on Certain Conventional Weapons (CCW) Meeting of Experts on Lethal Autonomous Weapons Systems (LAWS), April 11, 2016, 5, available for download at https://www.icrc.org/en/document/views-icrc-autonomous-weapon-system.

358 **Phrases like . . . "appropriate human involvement":** Heather M. Roff and Richard Moyes, "Meaningful Human Control, Artificial Intelligence and Autonomous Weapons," Briefing paper prepared for the Informal Meeting of Experts on Lethal Autonomous Weapons Systems, UN Convention on Certain Conventional Weapons, April 2016, http://www.article36.org/wp-content/uploads/2016/04/MHC-AI-and-AWS-FINAL.pdf. Human Rights Watch, "Killer Robots and the Concept of Meaningful Human Control," April 11, 2016, https://www.hrw.org/news/2016/04/11/killer-robots-and-concept-meaningful-human-control. UN Institute for Disarmament Research, "The Weaponization of Increasingly Autonomous Technologies: Considering how Meaningful Human Control might move the discussion forward," 2014, http://www.unidir.org/files/publications/pdfs/considering-how-meaningful-human-control-might-move-the-discussion-forward-en-615.pdf. Michael Horowitz and Paul Scharre, "Meaningful Human Control in Weapon Systems: A Primer," Center for a New American Security, Washington DC, March 16, 2015, https://www.cnas.org/publications/reports/meaningful-human-control-in-weapon-systems-a-primer. CCW, "Report of the 2016 Informal Meeting of Experts on Lethal Autonomous Weapon Systems (LAWS).

358 **"The law of war rules":** Department of Defense, "Department of Defense Law of War Manual," 330.

Conclusion: No Fate but What We Make

360 **Sarah and John are forever trapped:** Kudos to Darren Franich for a mind-melting attempt to map the *Terminator* timelines: " 'Terminator Genisys': The Franchise Timeline, Explained," *EW.com*, June 30, 2015, http://ew.com/article/2015/06/30/terminator-genisys-franchise-timeline-explained/.

Afterword: How Robotic Weapons Are Transforming the Battlefield Today

364 **The Russians managed to take out all thirteen drones:** David Reid, "A swarm of armed drones attacked a Russian military base in Syria," CNBC.com, January

22, 2018, https://www.cnbc.com/2018/01/11/swarm-of-armed-diy-drones-attacks-russian-military-base-in-syria.html.

364 **Operators frequently lost communications with the Uran-9:** "Combat tests in Syria brought to light deficiencies of Russian unmanned mini-tank," Defence Blog, June 18, 2018, http://defence-blog.com/army/combat-tests-syria-brought-light-deficiencies-russian-unmanned-mini-tank.html.

364 **attackers used DJI M600 drones:** Some accounts suggest there were three drones used in the attack. Eliott C. McLaughlin, "Venezuela says it has ID'd mastermind, accomplices in apparent Maduro assassination try," CNN.com, August 7, 2018, https://www.cnn.com/2018/08/06/americas/venezuela-maduro-apparent-assas-sination-attempt/index.html.

365 **"If they can see a kill, they take it":** James Vincent, "OpenAI's Dota 2 Defeat is a Win for Artificial Intelligence," August 28, 2018, https://www.theverge.com/2018/8/28/17787610/openai-dota-2-bots-ai-lost-international-reinforcement-learning.

365 **"100 human lifetimes of experience every single day":** Ibid.

365 **the sincerity of the Chinese offer:** "Position Paper, Submitted by China," CCW, April 11 2018, https://www.unog.ch/80256EDD006B8954/(httpAssets)/E42AE83BDB3525D0C125826C0040B262/$file/CCW_GGE.1_2018_WP.7.pdf; Elsa Kania, "China's Strategic Ambiguity and Shifting Approach to Lethal Autonomous Weapons Systems," Lawfare, April 17, 2018, https://www.lawfareblog.com/chinas-strategic-ambiguity-and-shifting-approach-lethal-autonomous-weapons-systems; Campaign to Stop Killer Robots, Twitter, August 27, 2018, 6:59AM, https://twitter.com/BanKillerRobots/status/1034077916395720704.

365 **a politically-binding resolution:** "Statement by France and Germany," Meeting of the Group of Governmental Experts on Lethal Autonomous Weapons Systems, April 9-13, 2018, https://www.unog.ch/80256EDD006B8954/(httpAssets)/895931D082ECE219C12582720056F12F/$file/2018_LAWSGeneralExchange_Germany-France.pdf; "Joint statement by the delegations of France and Germany on agenda item 'Possible options for addressing the humanitarian and international security challenges,'" GGE on LAWS, August 29, 2018, http://reachingcriticalwill.org/images/documents/Disarmament-fora/ccw/2018/gge/statements/29August_France_Germany.pdf.

366 **"accountability cannot be transferred to machines":** "Report of the 2018 Group of Governmental Experts on Lethal Autonomous Weapons Systems," August 31, 2018, https://www.unog.ch/80256EDD006B8954/(httpAssets)/20092911F6495FA7C125830E003F9A5B/$file/2018_GGE+LAWS_Final+Report.pdf.

366 **Autonomous weapons are one slice:** For some examples of security-related applications of artificial intelligence, see Miles Brundage et al., "The Malicious Use of Artificial Intelligence: Forecasting, Prevention, and Mitigation," February 2018, https://maliciousaireport.com/; and Michael C. Horowitz et al., "Artificial Intelligence and International Security," Center for a New American Security, July 10, 2018, https://www.cnas.org/publications/reports/artificial-intelligence-and-international-security.

366 **Russia is using bots to spread disinformation:** Samuel C. Woolley and Philip N. Howard, "Computational Propaganda Worldwide: Executive Summary," Working Paper No. 2017.11, Computational Propaganda Research Project, University of Oxford, http://comprop.oii.ox.ac.uk/wp-content/uploads/sites/89/2017/06/

Casestudies-ExecutiveSummary.pdf. For real-time tracking of Russian influence operations, see the Hamilton 68 dashboard, Alliance for Security Democracy, German Marshall Fund, https://dashboard.securingdemocracy.org/.

366 **control its citizens:** China's Social Credit System is often over-hyped. For a more sober assessment, see Rogier Creemers, "China's Social Credit System: An Evolving Practice of Control," May 22, 2018, https://papers.ssrn.com/sol3/papers.cfm?abstract_id=3175792. For a quick overview of the Social Credit System, see Mareike Ohlberg et al., "Central Planning, Local Experiments: The complex implementation of China's Social Credit System," Mercator Institute for China Studies, December 12, 2017, https://www.merics.org/en/microsite/china-monitor/central-planning-local-experiments.

366 **The Pentagon is using artificial intelligence:** Daisuke Wakabayashi and Scott Shane, "Google Will Not Renew Pentagon Contract That Upset Employees," New York Times, June 1, 2018, https://www.nytimes.com/2018/06/01/technology/google-pentagon-project-maven.html.

366 **using opaque algorithms designed to maximize profit:** As one example, see Zeynep Tufecki, "YouTube, the Great Radicalizer," New York Times, March 10, 2018, https://www.nytimes.com/2018/03/10/opinion/sunday/youtube-politics-radical.html.

366 **shape the very course of our future:** Tim Dutton, "An Overview of National AI Strategies," Medium.com, June 28, 2018, https://medium.com/politics-ai/an-overview-of-national-ai-strategies-2a70ec6edfd.

366 **aggressively courting top experts from Silicon Valley:** "New Generation Artificial Intelligence Development Plan," translated by Graham Webster, Rogier Creemers, Paul Triolo, and Elsa Kania, August 1, 2017, https://www.newamerica.org/cybersecurity-initiative/blog/chinas-plan-lead-ai-purpose-prospects-and-problems/. For an in-depth analysis of Chinese military developments in artificial intelligence, see Elsa Kania, "Battlefield Singularity," Center for a New American Security, November 28, 2017, https://www.cnas.org/publications/reports/battlefield-singularity-artificial-intelligence-military-revolution-and-chinas-future-military-power. Meng Jing, "Chinese firms fight to lure top artificial intelligence talent from Silicon Valley," South China Morning Post, April 2, 2017, http://www.scmp.com/tech/china-tech/article/2084171/chinese-firms-fight-luretop-artificial-intelligence-talent-silicon.

366 **cracking down on foreign investment:** The White House Office of Science and Technology Policy, "2018 White House Summit on Artificial Intelligence for American Industry," May 10, 2018, https://www.whitehouse.gov/wp-content/uploads/2018/05/Summary-Report-of-White-House-AI-Summit.pdf; Aaron Boyd, "Here's What the White House's AI Committee Will Focus On," NextGov, June 28, 2018, https://www.nextgov.com/emerging-tech/2018/06/heres-what-white-houses-ai-committee-will-focus/149382/; and Section 1051, National Security Commission on Artificial Intelligence, John S. McCain National Defense Authorization Act for Fiscal Year 2019, https://congress.gov/bill/115th-congress/house-bill/5515/text?r=2.

366 **require bots to disclose that they're not human:** California State Legislature, "SB-1001, Bots: disclosure," https://leginfo.legislature.ca.gov/faces/billCompareClient.xhtml?bill_id=201720180SB1001.

367 **Europe's new data privacy laws:** EU General Data Protection Regulation (GDPR), https://eugdpr.org/.

367 **disclosure of hacks or disinformation campaigns:** Alex Hern, "Google plans to 'de-rank' Russia Today and Sputnik to combat misinformation," *The Guardian*, November 21, 2017, https://www.theguardian.com/technology/2017/nov/21/google-de-rank-russia-today-sputnik-combat-misinformation-alphabet-chief-executive-eric-schmidt; Julia Carrie Wong, "Twitter announces global change to algorithm in effort to tackle harassment," *The Guardian*, May 15, 2018, https://amp.theguardian.com/technology/2018/may/15/twitter-ranking-algorithm-change-trolling-harassment-abuse; and Adam Mosseri, "Showing More Informative Links in News Feed," Facebook newsroom, June 30, 2017, https://newsroom.fb.com/news/2017/06/news-feed-fyi-showing-more-informative-links-in-news-feed/.

367 **Microsoft's call for regulation of facial recognition technology:** Brad Smith, "Facial recognition technology: The need for public regulation and corporate responsibility," Microsoft on the issues, July 13, 2018, https://blogs.microsoft.com/on-the-issues/2018/07/13/facial-recognition-technology-the-need-for-public-regulation-and-corporate-responsibility/.

367 **ruler of the world:** "'Whoever leads in AI will rule the world': Putin to Russian children on Knowledge Day," RT, September 1, 2017, https://www.rt.com/news/401731-ai-rule-world-putin/.

Acknowledgments

A book is a strange thing. It takes scores of people to make a book, but only one person's name goes on the cover. This book began ten years ago, in a conversation with my good friend, Gene Tien. Along the way, many people have helped shape the ideas within. This book would not have been possible without their help.

I want to thank the executive team at the Center for a New American Security (CNAS): Michèle Flournoy, Richard Fontaine, Shawn Brimley, David Romley, and former CNAS CEO Bob Work. They have been incredibly supportive in the development of this book. My CNAS colleague Robert Kaplan has been an amazing mentor along the way, and I am very grateful for his advice and guidance.

Elements of this book draw upon my work for CNAS's Ethical Autonomy Project, which was made possible by the John D. and Catherine T. MacArthur Foundation. I want to thank Jeff Ubois of MacArthur for his interest and support in our work. I owe a special thanks to Michael Horowitz, who led the Ethical Autonomy Project with me and has been a frequent coauthor on many reports. Kelley Sayler, Alex Velez-Green, Adam Saxton, and Matt Seeley also provided invaluable help in our research and publications. Melody Cook designed many of the graphics used in our project, which have been repurposed in this book with the permission of CNAS.

Many of the ideas in this book began years ago when I worked in the Pentagon. The patience and foresight of my leadership team—Leslie Hunter, Jen Zakriski, Todd Harvey, David Ochmanek, Kathleen Hicks, and James N. Miller—gave me the freedom to proactively engage on this issue, a rare occurrence in a bureaucracy. John Hawley and Bobby Junker, both top-notch scientists and amazing human beings, helped shape my thinking on the nature of autonomy and human control. Andrew May, Andy Marshall, and Bob Bateman were some of the earliest to recognize the importance of these issues and support work in this area.

I would like to thank the many individuals who agreed to be interviewed for this book: Ken Anderson, Peter Asaro, Ron Arkin, Stuart Armstrong, Paul Bello, John Borrie, Selmer Bringsjord, Brian Bruggeman, Ray Buettner, John Canning, Micah Clark, Jeff Clune, Kelly Cohen, Mary "Missy" Cummings, David Danks, Duane

Davis, Neil Davison, Charles Dela Cuesta, Tom Dietterich, Bonnie Docherty, Char-
lie Dunlap, Peter Galluch, Steve Goose, John Hawley, Michael Horowitz, Christof
Heyns, Frank Kendall, William Kennedy, Tom Malinowski, Bryan McGrath, Mike
Meier, Heather Roff, Stuart Russell, Larry Schuette, Bradford Tousley, Brandon
Tseng, Kerstin Vignard, Michael Walker, Mary Wareham, Steve Welby, Bob Work,
and Jody Williams. Unfortunately, I was not able to include all of their interviews
for reasons of space, but their ideas and insights helped shape the book in ways big
and small. My thoughts on autonomous weapons have also been shaped over the
years by many fellow travelers on this topic: Karl Chang, Rebecca Crootof, David
Koplow, Kathleen Lawand, Patrick Lin, Matt McCormack, Brian Hall, Tim Hwang,
David Simon, Shawn Steene, Noel Sharkey, Ryan Tewell, Alex Wagner, and Matt
Waxman, among many others. I am grateful to Elbridge Colby and Shawn Steene
for their feedback on drafts of the book. Maura McCarthy's sharp eye as an editor
helped ensure that this book's proposal landed a publisher. My colleagues Neal
Urwitz, JaRel Clay, and Jasmine Butler have been incredibly helpful in promotion
and outreach. Thanks to Jennifer-Leigh Oprihory for suggesting the title.

The U.S. Department of Defense was very open and supportive throughout the
development of this book. I would especially like to thank Jared Adams at the
Defense Advanced Research Projects Agency (DARPA), and Kimberly Lansdale at
the Aegis Training and Readiness Center, for their support and assistance in facili-
tating visits and interviews.

I'd like to thank my agent, Jim Hornfischer at Hornfischer Literary Management,
and my editor, Tom Mayer at W. W. Norton, for taking a chance on me and this book.
I owe a special thanks to Emma Hitchcock, Sarah Bolling, Kyle Radler, and the rest
of the team at Norton for bringing this book to fruition.

My brother, Steve, has been incredibly patient over the years listening to me prat-
tle on about autonomous weapons, and provided valuable feedback on early chap-
ters. I owe everything to my parents, Janice and David. Nothing I've done would
have been possible without their love and support.

Most of all, I am forever grateful to my wife Heather, who now knows more about
autonomous weapons than she ever cared to. Her patience, love, and support made
this possible.

Abbreviations

AAA	antiaircraft artillery
ABM	Anti-Ballistic Missile
ACTUV	Anti-submarine warfare Continuous Trail Unmanned Vessel
AGI	artificial general intelligence
AGM	air-to-ground missile
AI	artificial intelligence
AMRAAM	Advanced Medium-Range Air-to-Air Missile
ARPA	Advanced Research Projects Agency
ASI	artificial superintelligence
ASW	anti-submarine warfare
ATR	automatic target recognition
BDA	battle damage assessment
BWC	Biological Weapons Convention
CCW	Convention on Certain Conventional Weapons
C&D	Command and Decision
CIC	combat information center
CIWS	Close-In Weapon System
CODE	Collaborative Operations in Denied Environments
DARPA	Defense Advanced Research Projects Agency
DDoS	distributed denial of service
DIY	do-it-yourself
DMZ	demilitarized zone
DoD	Department of Defense
FAA	Federal Aviation Administration
FIAC	fast inshore attack craft
FIS	Fire Inhibit Switch
FLA	Fast Lightweight Autonomy
GGE	Group of Governmental Experts
GPS	global positioning system
ICRAC	International Committee for Robot Arms Control
ICRC	International Committee of the Red Cross

IEEE	Institute of Electrical and Electronics Engineers
IFF	identification friend or foe
IHL	international humanitarian law
IMU	inertial measurement unit
INF	Intermediate-Range Nuclear Forces
IoT	Internet of Things
J-UCAS	Joint Unmanned Combat Air Systems
LIDAR	light detection and ranging
LOCAAS	Low Cost Autonomous Attack System
LRASM	Long-Range Anti-Ship Missile
MAD	mutual assured destruction
MARS	Mobile Autonomous Robotic System
MMW	millimeter-wave
NASA	National Aeronautics and Space Administration
NGO	nongovernmental organization
NORAD	North American Aerospace Defense Command
ONR	Office of Naval Research
OODA	observe, orient, decide, act
OPM	Office of Personnel Management
PGM	precision-guided munition
PLC	programmable logic controllers
RAS	IEEE Robotics and Automation Society
R&D	research and development
ROE	rules of engagement
SAG	surface action group
SAR	synthetic aperture radar
SAW	Squad Automatic Weapon
SEC	Securities and Exchange Commission
SFW	Sensor Fuzed Weapon
SORT	Strategic Offensive Reductions Treaty
START	Strategic Arms Reduction Treaty
SUBSAFE	Submarine Safety
TASM	Tomahawk Anti-Ship Missile
TBM	tactical ballistic missile
TJ	Thomas Jefferson High School
TLAM	Tomahawk Land Attack Missile
TRACE	Target Recognition and Adaption in Contested Environments
TTO	Tactical Technology Office
TTP	tactics, techniques, and procedures
UAV	uninhabited aerial vehicle
UCAV	uninhabited combat aerial vehicle
UK	United Kingdom
UN	United Nations
UNIDIR	UN Institute for Disarmament Research
U.S.	United States
WMD	weapons of mass destruction

Illustration Credits

(All photographs courtesy of Paul Scharre unless otherwise indicated.)

Text images:

page 48	Center for a New American Security
page 49	U.S. Navy
page 66	© Lockheed Martin
page 67	© Lockheed Martin
page 103	Center for a New American Security
page 140	Center for a New American Security
page 181	Anh Nguyen, Jason Yosinski, Jeff Clune
page 183	Christian Szegedy, Wojciech Zaremba, Ilya Sutskever, Joan Bruna, Dumitru Erhan, Ian Goodfellow, Rob Fergus
page 185	Anh Nguyen, Jason Yosinski, Jeff Clune
page 327	U.S. Air Force

Insert images:

1.	U.S. Marine Corps Historical Division Archives
2.	John Warwick Brooke / Imperial War Museum collection
3.	Mass Communication Specialist 3rd Class Eric Coffer / U.S. Navy
4.	U.S. Navy
5.	Mass Communication Specialist Seaman Anthony N. Hilkowski / U.S. Navy
6.	Israel Aerospace Industries
7.	U.S. Navy
8.	Glenn Fawett / Department of Defense
9.	Senior Airman Christian Clausen / U.S. Air Force
10.	NASA
11.	Alan Radecki / U.S. Navy courtesy of Northrop Grumman
12.	Elizabeth A. Wolter / U.S. Navy
13.	Israel Aerospace Industries
14.	John F. Williams / U.S. Navy

15.	John F. Williams / U.S. Navy
16.	John F. Williams / U.S. Navy
17.	DARPA
18.	DARPA
19.	DARPA
20.	Specialist Lauren Harrah / U.S. Army
21.	DARPA
22.	Daryl Roy / Aegis Training and Readiness Center
23.	Mass Communication Specialist 2nd Class Michael Ehrlich / U.S. Navy
24.	Mass Communication Specialist 2nd Class Michael Ehrlich / U.S. Navy
25.	John F. Williams / U.S. Navy
26.	Ben Santos / U.S. Forces Afghanistan
27.	Campaign to Stop Killer Robots
28.	DJI Technology, Inc.
29.	Courtesy Paul Scharre
30.	Staff Sergeant Sean Harp / Department of Defense
31.	Mass Communication Specialist 3rd Class Sean Weir / U.S. Navy

Index

Page numbers followed by *f* indicate figures and illustrations; page numbers followed by *n* indicate notes; page numbers followed by *m* indicate maps; page numbers followed by *t* indicate tables.